Introduction to Atomic and Molecular Spectroscopy

Introduction to
Atomic and Molecular Spectroscopy

Vimal Kumar Jain

Alpha Science International Ltd.
Oxford, U.K.

Vimal Kumar Jain
Department of Physics
M.D. University
Rohtak, India

Copyright © 2007
Reprint 2015

ALPHA SCIENCE INTERNATIONAL LTD.
7200 The Quorum, Oxford Business Park North
Garsington Road, Oxford OX4 2JZ, U.K.

www.alphasci.com

ISBN 978-1-84265-357-9

Printed in India

To my parents

Harbans Lal and Trishla Wanti Jain

Preface

'*Introduction to Atomic and Molecular Spectroscopy*' has been prepared to serve as a text for graduate, postgraduate and research students of physics and chemistry. This book has been written with a view to cover the syllabii of compulsory and optional papers on Atomic and Molecular Physics for B.Sc. and M.Sc. students. The topics covered in the book will also be of interest to the beginners as all the topics are covered right from the elementary level and readers need not consult different books for different topics. It also serves to provide an introductory background to research students of physics, chemistry, biology and mineralogy who wish to use electron spin resonance, nuclear magnetic resonance and Mössbauer spectroscopy techniques as a part of their studies.

This text book has grown out of many years of course notes that I developed for M.Sc. and M.Phil. students covering many fundamental aspects of atomic and molecular spectra, X-Ray spectroscopy, electron spin resonance, nuclear magnetic resonance, Mössbauer spectroscopy and laser physics. An attempt has been made to emphasize the physical basis of the subject without undue neglect of the mathematical aspects. An elementary background of quantum mechanics is, however, required for understanding some of the sections of this book. A large number of solved and unsolved problems based on the articles are given at the end of each chapter to supplement the text. The vectors are denoted with bold face in the text.

In developing such a wide range of topics, I have consulted a large number of books by various authors to whom I am grateful. A list of some of these is given at the end of the book. My daughters Manu and Ira have assisted me at the manuscript stage and without their comments and corrections there would be more errors than those that may still be present. The person whom I must acknowledge is my wife Minakshi who patiently supported and encouraged when at times I felt like giving up.

Finally, I must thank Mr. N.K.Mehra, Managing Director, Narosa Publishing House for readily agreeing to undertake this project.

Vimal Kumar Jain

Contents

1

Introduction

Spectroscopy is the study of the interaction of the photons of the electromagnetic radiation and matter. Spectroscopy may be said to date from Isaac Newton's observation of the spectrum, which is obtained when white light is passed through a prism. White light after passing through the prism separate or disperse into its component colours in a definite pattern. The spectrum in the domain of electro-magnetic radiation is defined as a series of radiant energy arranged in order of wavelength or of frequency. Interpretation of spectra provides fundamental information on energy levels, transition probabilities, for atoms and molecules are of direct importance in astrophysics, plasma and laser physics. The spectroscopic information can be used in various kinds of analysis for example, optical absorption or emission spectroscopy is used for qualitatively and quantitative chemical analysis.

Table 1.1 The Electromagnetic Spectrum Regions

Region	Wavelength range	Frequency range (Hz)	Wave number range (cm^{-1})	Types of spectrum
gamma-ray	<10pm	3×10^{19}	$>10^9$	Mössbauer
x-ray	10pm-10nm	3×10^{19}-3×10^{16}	10^9-10^6	Electronic (core orbitals)
Vacuum ultraviolet	10-200 nm	3×10^{19} 1.5×10^{15}	10^6-5×10^4	Electronic spectra (valence orbitals)
Ultraviolet	200-400nm	1.5×10^{15}-7.5×10^{14}	5×10^4-2.5×10^4	
Visible	400-800 nm	7.5×10^{14}-3.75×10^{14}	2.5×10^4-1.25×10^4	
Near infrared	0.8-2.5 μm	3.75×10^{14}-1.2×10^{14}	1.25×10^4-4×10^3	
Infrared	2.5-50 μm	1.2×10^{14}-6×10^{12}	4×10^3-2×10^2	Vibrational spectroscopy
Far Infrared	50-100 μm	6×10^{12}-3×10^{12}	200-100	Rotational spectroscopy
Microwave	0.1-30 cm	3×10^{11}-10^9	10-0.033	Rotation, Electron spin resonance
Radio frequency	0.1-1000 m	3×10^9-3×10^5	0.1-10^{-5}	Nuclear magnetic resonance, Nuclear quadrupole resonance

The choice of spectroscopic method is primarily determined by energy range of the phenomenon to be studied. In Table 1.1, the spectral ranges that are of interest to atomic and molecular spectroscopy are shown. The energy ranges for different types of structure and transitions are also mentioned.

The energy interval ΔE is given by

$$\Delta E = h\nu \tag{1.1}$$

where h is Planck's constant, c is velocity of light , ν is the frequency. The relation between frequency and wavelength λ is

$$\lambda = \frac{c}{\nu} \tag{1.2}$$

$$\frac{1}{\lambda} = \frac{\nu}{c} \tag{1.3}$$

$$\nu = \frac{c}{\lambda} \tag{1.4}$$

Using Eqs. (1.1) to (1.4), the energy interval ΔE can be expressed in eV, nm(10^{-9} m),cm^{-1} (wavenumber) or Hz (frequency). 1cm^{-1} sometimes called 1 Kayser. In Table 1.2 conversion factors between different units are given.

Table 1.2 Conversion factors between different energy units

Unit	Joule	eV	cm^{-1}	Hz
1Joule	1	6.24146×10^{18}	5.003417×10^{22}	1.50916×10^{33}
1eV	1.60219×10^{-19}	1	8.60545×10^{3}	2.41796×10^{14}
1cm^{-1}	1.98648×10^{-23}	1.239853×10^{-4}	1	2.997925×10^{10}
1Hz	6.6262×10^{-19}	4.13571×10^{-15}	3.33564×10^{-11}	1

The choice of unit depends on the energy region and the traditional factors and is given in Table 1.3.

Table 1.3 Units used in various energy regions

Electromagnetic region	Unit
gamma-ray region	MeV
X-ray region	KeV
Visible and UV region	nm, Å (in solid state physics eV)
Infrared region	cm^{-1}, μm
Radio frequency region	cm^{-1}, MHz

Examples

1. Convert 10 Joules in to Hz

    From $E = h\nu$, we have

$$\nu = \frac{E}{h} = \frac{1 \text{Joule}}{6.6262 \times 10^{-34} \text{Js}} = 1.5092 \times 10^{34} \text{Hz}$$

2. Convert 10GHz into wavelength and wave number.

From Eq. (1.3)

$$\lambda = \frac{c}{v} = \frac{2.997925 \times 10^8 \, \text{ms}^{-1}}{10 \times 10^9 \, \text{Hz}} = 2.997925 \, \text{cm}$$

$$\bar{v} = \frac{1}{\lambda} = \frac{1}{2.997925 \, \text{cm}} = 0.33356 \, \text{cm}^{-1}$$

3. The wavelength of radiation absorbed during a particular spectroscopic transition is observed to be 121.567 nm. Express this in frequency (Hz) and in wavenumber (cm^{-1}). Calculate the energy change during the transition in eV. State in which region of the electromagnetic spectrum you would expect this to appear, and what sort of transition this corresponds to.

From Eq. (1.3)

$$v = \frac{c}{\lambda} = \frac{2.997925 \times 10^8 \, \text{ms}^{-1}}{121.567 \times 10^{-9} \, \text{m}} = 2.46607 \times 10^{15} \, \text{Hz}$$

From Eq. (1.3)

$$\bar{v} = \frac{1}{\lambda} = \frac{1}{121.567 \times 10^{-9} \, \text{m}} = 8.226 \times 10^4 \, \text{cm}^{-1}$$

From Eq. (1.1) we have

$$\Delta E = 6.6262 \times 10^{-34} \, \text{Js} \times 2.46607 \times 10^{15} \, \text{Hz} = 16.34067 \times 10^{-19} \, \text{J} = \frac{16.34067 \times 10^{-19} \, \text{eV}}{1.60219 \times 10^{-19}}$$

$$= 10.19896 \, \text{eV}$$

The corresponding EM region is UV and the transition corresponding to valence electrons.

Problems

1.1 Convert 1 eV into cm^{-1}, nm and Hz.

1.2 Show that $k_B T$ at 300 K is =0.025 eV.

1.3 The energy absorbed during a particular spectroscopic transition is observed be 1.889 eV. Express this in frequency, wavenumber and in wavelength. State in which region of the electromagnetic spectrum you would expect this to appear, and what sort of transition this corresponds to.

2

One Electron Atoms

In this chapter we begin our study of atoms by treating the simplest case, the one-electron atom. The one electron atom is the simplest bound system that occurs in nature. The system consists of a positively charged nucleus and a negatively charged electron moving under the influence of their natural Coulomb attraction and bounded together by that attraction. The simplest one electron atom is hydrogen atom, which consists of a proton and an electron. Other similar one – electron systems called hydrogenic atoms include the isotopes of hydrogen (deuterium, tritium) and the hydrogenic ions (He^+ , Li^{++} , Be^{+++} , B^{++++} , C^{+++++} , N^{++++++}) . In spectroscopy the spectrum of a given stage of ionisation is indicated by Roman Numerals after the element symbol, e.g. , He I represents neutral helium atom , He II represents single charged helium ion , He III represents doubly charged ion etc.

2.1 Bohr Model of Hydrogen Atom

Neils Bohr developed a quantitative atomic model for hydrogen atom which satisfactorily explained the observed spectrum .The model incorporated the nuclear model of the atom proposed by Rutherford, as well as the concept of light photon developed by Einstien to explain the photoelectric effect. The Bohr theory is based on the following postulates:

1. An electron in an atom moves in circular orbits about the nucleus under the action of a coulomb field of force.
2. Of the infinite number of orbits of an electron about an atomic nucleus which would be possible in classical mechanics, only those for which the orbital angular momentum L is equal to an integer times $\hbar$ (Planck's constant h divided by 2π) are allowed .
3. Electron in such an allowed circular orbits is in a stable state and would not radiate despite constantly accelerating.
4. Electromagnetic radiations is emitted or absorbed by a transition of the electron from one orbit of total energy E_i to another object of total energy E_f by a quantum jump. The frequency ν of radiation is equal to E_i - E_f divided by Planck's constant i.e.

$$\nu = \frac{(E_i - E_f)}{h} \tag{2.1}$$

Consider a hydrogenic atom consisting of a nucleus of charge +Ze and mass M and a single electron of charge -e and mass m_e . It is assumed that electron revolves around the nucleus in circular orbit of radius r (Fig.2.1). Since mass of the electron is negligible in comparison with the mass of the nucleus; therefore, in the first approximation it is assumed that nucleus is at rest.

The acceleration of the electron in circular orbit is centripetal acceleration a_c given by

$$a_c = \frac{v^2}{r} \tag{2.2}$$

where v is the speed of the electron and r is the radius of its circular orbit. The force F producing the acceleration of the electron is the Coulomb force exerted by the nuclear charge Ze on the charge -e on the electron. The magnitude of the force is:

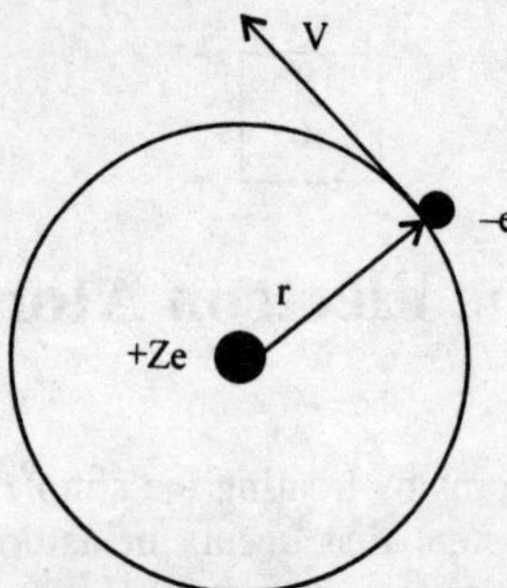

Fig. 2.1 Bohr's atomic model

$$F = \frac{Ze^2}{4\pi\varepsilon_0 r^2} \qquad (2.3)$$

According to Newton's second law:

$$F = m_e\mathbf{a} \qquad (2.4)$$

Since force **F** and acceleration **a** are in the same direction, we take their magnitudes to obtain

$$F = \frac{Ze^2}{4\pi\varepsilon_0 r^2} = m_e\frac{v^2}{r} \qquad (2.5)$$

From Eq. (2.5)

$$v^2 = \frac{Ze^2}{4\pi\varepsilon_0 m_e r} \qquad (2.6)$$

According to Bohr's second postulate:

$$L = \frac{nh}{2\pi} \qquad (2.7)$$

where n is a positive integer known as quantum number. The orbital angular momentum is:

$$L = \mathbf{r}\mathbf{x}\mathbf{p} \qquad (2.8)$$

In circular motion, the position vector r of a particle is perpendicular to its momentum p, therefore Eq. (2.8) is

$$L = rp = m_e rv \qquad (2.9)$$

Eq. (2.7) becomes:

$$m_e rv = \frac{nh}{2\pi}$$

Solving for v:

$$v = \frac{nh}{2\pi\, m_e r}$$

(2.10)

Using Eqs. (2.10) and (2.6)

$$r = \frac{(4\pi\, \varepsilon_0)n^2 h^2}{4\pi^2 m_e Z e^2} = \frac{a_0 n^2}{Z}$$

(2.11)

For hydrogen atom Z =1 and radius of n =1 orbit

$$a_0 = \frac{(4\pi\, \varepsilon_0)\ h^2}{4\pi^2 m_e e^2}$$

(2.12)

substituting the values of constants in Eq. (2.12)

$$a_0 = 5.29 \times 10^{-11}\, m$$

(2.13)

which is radius of orbit of hydrogen in the ground state and is known as first Bohr radius of hydrogen.

From Eqs. (2.11) and (2.10)

$$v = \frac{2\pi\, e^2 Z}{(4\pi\, \varepsilon_0)nh}$$

(2.14)

as compared with the velocity of light for n = 1 and Z = 1

$$\frac{v}{c} = \frac{2\pi\, e^2}{(4\pi\, \varepsilon_0)hc} = \alpha = \frac{1}{137.29}$$

(2.15)

α is known as Sommerfeld fine structure constant. Frequency of revolution f of the electron in an orbit of radius r is:

f = (electron speed in orbit of radius r) / (circumference of orbit radius r)

$$f = \frac{v}{2\pi\, r}$$

(2.16)

Substituting the value of r and v from Eq. (2.11) and Eq. (2.14), respectively in Eq. (2.16)

$$f = \frac{4\pi^2 m_e e^4 Z^2}{(4\pi\, \varepsilon_0)^2 n^3 h^3}$$

(2.17)

The kinetic energy T of the electron is:

$$T = \frac{m_e v^2}{2}$$

The potential energy V of electron is

$$V = -\frac{Ze^2}{(4\pi\,\varepsilon_0)r}$$ (2.18)

The total energy of electron is

$$E = T + V = \frac{m_e v^2}{2} - \frac{Ze^2}{(4\pi\,\varepsilon_0)r}$$

Using Eq. (2.6) in the above Eq., we get

$$E = \frac{Ze^2}{2(4\pi\,\varepsilon_0)r} - \frac{Ze^2}{(4\pi\,\varepsilon_0)r} = -\frac{Ze^2}{2(4\pi\,\varepsilon_0)r}$$ (2.19)

E goes to zero, as r approaches infinity i.e. total energy is zero when atom is ionised. using the value of r from Eq. (2.11) in Eq. (2.19)

$$E = -\frac{2\pi^2 m_e e^4 Z^2}{(4\pi\,\varepsilon_0)^2 n^2 h^2}$$ (2.20)

Since n is an integer, only certain values of energy are permitted. According to fourth postulate:

$$h\nu = E_i - E_f = -\frac{2\pi^2 m_e e^4 Z^2}{(4\pi\,\varepsilon_0)^2 h^3}\left(\frac{1}{n_f^2} - \frac{1}{n_i^2}\right)$$

$$\nu = \frac{2\pi^2 m_e e^4 Z^2}{(4\pi\,\varepsilon_0)^2 h^3}\left(\frac{1}{n_f^2} - \frac{1}{n_i^2}\right)$$

The wavelength λ is given by

$$\frac{1}{\lambda} = \frac{\nu}{c} = \bar{\nu} = R_\infty Z^2\left(\frac{1}{n_f^2} - \frac{1}{n_i^2}\right)$$ (2.21)

where Rydberg constant

$$R_\infty = \frac{2\pi^2 m_e e^4 Z^2}{(4\pi\,\varepsilon_0)^2 ch^3}$$ (2.22)

assuming nucleus is at rest. Using Eqs. (2.20) and (2.22) we have :

$$E = -\frac{13.6(eV)Z^2}{n^2} \tag{2.23}$$

and

$$R_\infty = 109737 \, cm^{-1} \tag{2.24}$$

Following Wolfgang Grotrian, Bohr's permitted values of energy (Eq.2.23) for hydrogen atom (Z=1) are represented graphically on an energy level diagram (Fig. 2.2).The vertical axis of the graph is an energy (eV or wave number) scale, there is no horizontal scale, but horizontal lines are drawn to show the position of energy levels.

The ground level (n = 1 for hydrogen) lies near the bottom of the diagram, and the remaining (excited) levels are ranked by energy above this level .The distance between energy level rapidly decreases. Each spectral line is represented by a vertical line joining two energy levels The length of any arrow is proportional to the frequency (or the reciprocal wavelength) for the corresponding spectral lines. The ionisation limit is defined as the zero point of the energy scale. With this energies of the discrete excited states are negative numbers, whereas the positive energies occur only when the atom is ionised.

An electron in an orbit with $n_i > 1$ and energy E_i , in making a transition to an orbit of lower energy E_f produces a photon with specific energy, equivalently with a definite frequency or wavelength. The discrete emission spectrum of hydrogen therefore corresponds to electrons cascading down to lower energy levels in hydrogen atoms. All their spectral lines can be grouped in to five series depending on the values of n_f and n_i . These series are named after their discoverer. In a series spacing and intensity of lines decrease regularly. The series limit is derived from Eq. (2.21) for every series by putting $n_i = \infty$. The spectral lines lie in various spectral regions as shown in Table 2.1.

Normally, only lines corresponding to Lyman Series appear in the absorption spectrum as atom is always initially in the ground state n − 1, so that only absorption process from   n = 1 to n > 1 can occur. However at high temperature (T ~ 10^{15} K) owing to collisions some of the atoms will initially be in the first excited state (n = 2) and absorption lines corresponding to the Balmer Series will be observed. It is observed that for every line in the absorption spectrum there is a corresponding (same wavelength) line in its emission spectrum, however the reverse is not true. If hydrogen atom is initially in excited state n > 1 then in going to its ground state it can follow different paths and as a result of this it will emit [n (n-1)] / 2 number of different wavelengths.

In the theory it is assumed that nucleus remains at rest at the centre of circular orbits. This is true only if the nucleus has an infinite mass. Thus a correction must be made to the previous results, as mass of the

Table 2.1 The Spectral lines in various spectral regions

Series	$\overline{v} = \dfrac{1}{\lambda} = R_\infty \left(\dfrac{1}{n_f^2} - \dfrac{1}{n_i^2} \right)$		Spectral Region
	n_f	n_i	
Lyman	1	2 , 3 , 4....., ∞	Far UV
Balmer	2	3 , 4 , 5,.... ,∞	Near UV and Visible
Paschen	3	4 , 5 , 6,...., ∞	Infrared
Brackett	4	5 , 6 , 7,.... ,∞	Infrared
Pfund	5	6 , 7 , 8,.... ,∞	Infrared

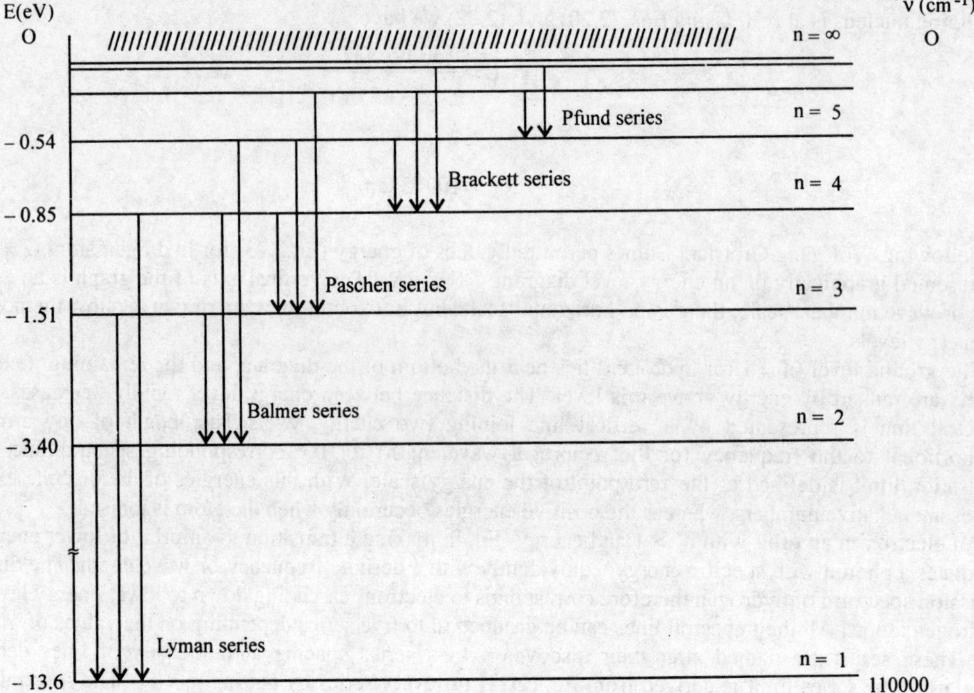

Fig. 2.2 Energy levels of the hydrogen atom according to the Bohr theory

nucleus is finite .In this case the nucleus and the electron move about their common centre of mass (CM) as shown in Figure 2.3.By definition of CM

$$Mr' = m_e r_e \qquad (2.25)$$

$$r = r' + r_e \qquad (2.26)$$

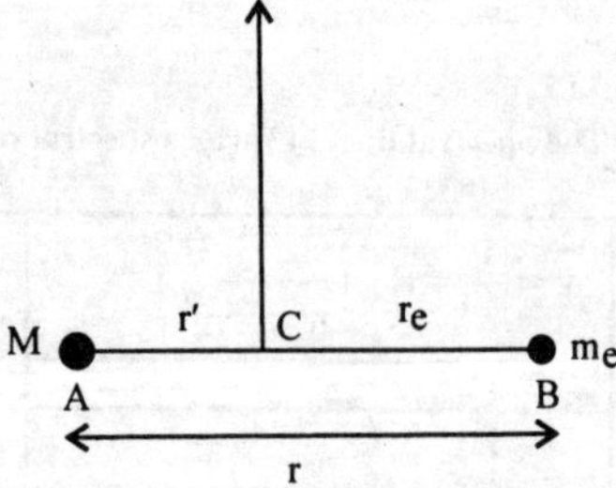

Fig. 2.3 Bohr Model with finite nuclear mass

From Eq. (2.25) and Eq. (2.26)

$$r_e = \frac{Mr}{M + m_e} \qquad (2.27)$$

$$r' = \frac{m_e r}{M + m_e}$$

(2.28)

Total angular moment of the system is:

$$L = I\omega = \omega(Mr'^2 + m_e r_e^2)$$

(2.29)

Substituting the values of r_e and r' from Eq. (2.27) and Eq. (2.28) in Eq. (2.29)

$$L = \frac{m_e M \omega r^2}{m_e + M} = \mu \omega r^2$$

(2.30)

where reduced mass $\mu = \dfrac{m_e M}{m_e + M}$.

Putting $r\omega = v$ in Eq. (2.30) :

$$L = \mu v r = n(h/2\pi)$$

(2.31)

which is same as Eq. (2.9) but with μ replacing m_e .

The centripetal force acting on the electron is:

$$m_e r_e \omega^2 = \frac{m_e M r \omega^2}{m_e + M} = \mu \omega^2 r$$

(2.32)

and equating this with Coulomb force [Eq. (2.3)]:

$$\mu r \omega^2 = \frac{\mu v^2}{r} = \frac{Ze^2}{(4\pi \varepsilon_0)r^2}$$

which is same relation as Eq. (2.5) but μ replaces m_e . Thus m_e is replaced by μ in Eq. (2.22). The Rydberg constant:

$$R_\infty = \frac{2\pi^2 m_e e^4 Z^2}{(4\pi \varepsilon_0)^2 ch^3}$$

changes to R given by

$$R = \frac{2\pi^2 m_e e^4 Z^2}{(4\pi \varepsilon_0)^2 ch^3}$$

$$R = \frac{R_\infty}{1 + \dfrac{m_e}{M}}$$

(2.33)

The Rydberg constant R is less than R_∞ by a factor of $\dfrac{1}{1+(m_e/M)}$. Since $m_e < M$, R is not very different from R_∞.Therefore, R is not exactly the same for all atoms.

2.2 Sommerfeld Model for Hydrogen Atom

The Bohr theory does not account for the fine structure of the spectra of hydrogenic atoms. The fine structure is a splitting of the spectral lines into several distinct components, which is found in all atomic spectra. Sommerfeld extended the Bohr model to include elliptical orbits. In general, the orbit of a particle under the influence of an inverse – square force is a conic section. This orbit may be in the shape of a parabola, a hyperbola, circle or an ellipse. The shape depends on the total energy of the particle. In the case of electron in a hydrogen atom, the total energy is negative, and hence the path of the electron is an ellipse. In such a case, in general there are two coordinates that vary periodically. In addition to the angle θ, which the radius vector **r** makes with horizontal axis, the length of **r** also varies periodically.
Sommerfeld postulated that:
1. The electron in an atom moves in an elliptical orbit with nucleus at one of the foci (Fig.2.4).
2. Of the infinite number of elliptical orbits which are possible, only certain discrete orbits satisfying the following quantum conditions occur:

$$\oint p_\theta \, d\theta = n_\theta \tag{2. 34}$$

$$\oint p_r dr = n_r \tag{2. 35}$$

where n_θ and n_r take integral values only and are called the azimuthal quantum number and radial quantum number respectively.

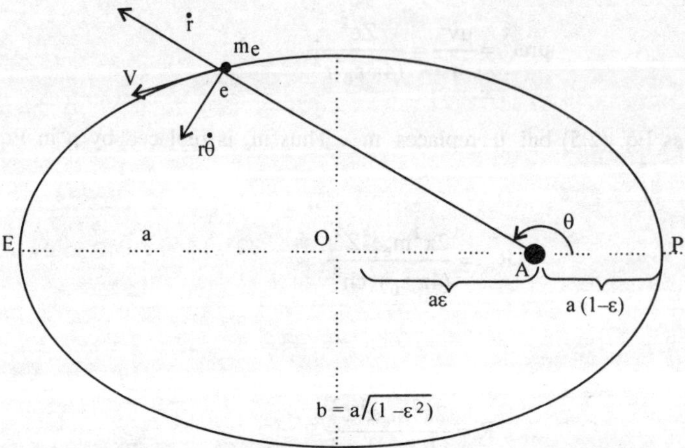

Fig. 2.4 Elleptical orbits

3. Electromagnetic radiation is emitted or absorbed when the electron undergoes a transition from an orbit of energy E_i to another orbit of energy E_f . The frequency of emitted or absorbed electro-magnetic radiation is given by :

$$v = \frac{(E_i - E_f)}{h} \tag{2.36}$$

The potential energy of the electron is given by Eq. (2.18) and kinetic energy is given by:

$$T = \frac{m_e(v_x^2 + v_y^2)}{2}$$

Using plane polar coordinates:

$$T = \frac{m_e[(dr/dt)^2 + r^2(d\theta/dt)^2]}{2} = \frac{p_r^2}{2m_e} + \frac{p_\theta^2}{2m_e r} \tag{2.37}$$

where

$$p_\theta = m_e r^2 (d\theta/dt) \tag{2.38}$$

$$p_r = m_e (dr/dt) \tag{2.39}$$

According to Kepler's second law of planetary motion, the angular momentum p_θ remains constant throughout the motion so that the integral (2.34) is

$$\int p_\theta d\theta = 2\pi p_\theta = 2\pi m_e r^2 (d\theta/dt) \tag{2.40}$$

The electron in an atom moves around the nucleus in an elliptical orbit. The general equation of an ellipse in polar coordinate

$$\frac{1}{r} = C_1 + C_2 \cos\theta \tag{2.41}$$

where C_1 and C_2 are constants to be determined for every ellipse . The eccentricity ε of an ellipse is defined as the ratio

$$\frac{OA}{a} = \varepsilon$$

where a is semimajor axis of the ellipse . Using perihelion AP distance ($\theta = 0^\circ$) and aphelion ($\theta = \pi$) distances , the constants C_1 and C_2 are :

$$C_1 = \frac{1}{a(1-\varepsilon^2)} \quad \text{and} \quad C_2 = \frac{\varepsilon}{a(1-\varepsilon^2)} \tag{2.42}$$

and Eq. (2.41) of ellipse becomes :

$$\frac{1}{r} = \frac{1+\varepsilon\cos\theta}{a(1-\varepsilon^2)} \tag{2.43}$$

Taking log of each side of this equation, and differentiating and putting $p_\theta = m_e r^2 (d\theta/dt)$ from Eq. (2.43) we obtain

$$p_r = \frac{p_\theta}{r^2}\frac{dr}{d\theta}$$

(2.44)

$$p_r dr = \frac{p_\theta}{r^2}\frac{dr}{d\theta}dr = \frac{p_\theta}{r^2}\left(\frac{dr}{d\theta}\right)^2 d\theta = \frac{p_\theta \varepsilon^2 \sin^2\theta\, d\theta}{(1+\varepsilon\cos\theta)^2}$$

(2.45)

Eq. (2.35) becomes:

$$\oint p_r dr = p_\theta \varepsilon^2 \int \frac{\sin^2\theta\, d\theta}{(1+\varepsilon\cos\theta)^2} = n_r h$$

$$= \frac{n_\theta h}{2\pi\varepsilon^2}\int \frac{\sin^2\theta\, d\theta}{(1+\varepsilon\cos\theta)^2} = n_r h$$

$$= \frac{\varepsilon^2}{2\pi}\int_0^{2\pi} \frac{\sin^2\theta\, d\theta}{(1+\varepsilon\cos\theta)^2} = \frac{n_r}{n_\theta}$$

(2.46)

Let

$$\frac{1}{1+\varepsilon\cos\theta} = x$$

(2.47)

On differentiation

$$\frac{\varepsilon\sin\theta\, d\theta}{(1+\varepsilon\cos\theta)^2} = dx$$

(2.48)

using Eqs. (2.48) and (2.46)

$$\oint p_r dr = \frac{\int_0^{2\pi}\varepsilon\sin\theta\, dx}{2\pi}$$

integrating by parts and using the integral

$$\int \frac{dx}{(a+b\cos\theta)} = \frac{2\tan^{-1}\sqrt{(a^2-b^2)}\tan(x/2)}{\sqrt{(a^2-b^2)}(a+b)}$$

The above integral becomes

$$\oint p_r dr = \frac{1}{\sqrt{1-\varepsilon^2}} - 1 = \frac{n_r}{n_\theta}$$

$$\frac{1}{\sqrt{1-\varepsilon^2}} = \frac{n_r + n_\theta}{n_\theta} = \frac{n}{n_\theta} \tag{2.49}$$

where n is total quantum number .and can take value 1 , 2 ,3 ....
For an ellipse major axis a and minor axis b are related by:

$$1-\varepsilon^2 = \frac{b^2}{a^2}$$

from Eq.(2.49)

$$1-\varepsilon^2 = \frac{b^2}{a^2} = \frac{n_\theta^2}{n^2}$$

$$\frac{b}{a} = \frac{n_\theta}{n} \tag{2.50}$$

Thus only those orbits are allowed which can satisfy the above conditions.
   Using Eqs. (2. 18), (2.37) and (2. 44) the total energy E is

$$E = T + V = \frac{p_\theta^2 \{[\frac{1}{r}\frac{dr}{d\theta}]^2 + 1\}}{2m_e r^2} - \frac{Ze^2}{(4\pi\varepsilon_0)r} \tag{2.51}$$

$$E = \frac{p_\theta^2}{m_e a^2 (1-\varepsilon^2)^2}[\frac{1-\varepsilon^2}{2} + \varepsilon\cos\theta] - \frac{Ze^2(1+\varepsilon\cos\theta)}{4\pi\varepsilon_0 r\, a(1-\varepsilon^2)} \tag{2.52}$$

   For a conservative system the total energy is constant and is independent of time and angle θ. Since the total energy is constant and cos θ varies, the coefficients of cos θ must vanish. Collecting the terms in $\cos\theta$ from Eq.(2.52) and equating to zero gives

$$a = \frac{p_\theta^2}{4\pi\varepsilon_0 m_e Ze^2(1-\varepsilon^2)} \tag{2.53}$$

from Eqs. (2.53) and (2.12)

$$a = \frac{h^2(n_r + n_\theta)^2}{(4\pi\varepsilon_0)4\pi^2 m_e e^2 Z} = \frac{a_0 n^2}{Z} \tag{2.54}$$

$$b = a\sqrt{1-\varepsilon^2} = \frac{a_0 n\, n_\theta}{Z} \tag{2.55}$$

a_0 is the radius of first Bohr circular orbit . Equation (2.52) reduces to:

$$E = -\frac{2\pi^2 m_e e^4 Z^2}{(4\pi \, \varepsilon_0)^2 n^2 h^2} \qquad (2.56)$$

the energy is exactly the same as that obtained by Bohr for circular orbit.

For any given value of the energy there are n different quantised orbits for the electron. The several orbits characterized by a common value of n are said to be degenerate. If for example n = 3, the azimuthal quantum number n_θ = 1 , 2 or 3 while the radial quantum number n_r = 0 , 1 , 2 . States with n_θ = 0 reduces to motion along a straight line with the nucleus at one end. Such states are excluded, as electron will collide with the nucleus.

The introduction of elliptical orbits does not introduce any new energy levels and hence no new transitions. Sommerfeld showed that degeneracy could be removed if relativistic corrections to the energy are taken into consideration. An electron in a circular orbit has constant velocity, but an electron in an elliptical orbit has velocity that is different at different positions, speeding up when the electron is near the nucleus and slows down when it is far away. This is due to the relativistic variation of electron mass which will be of the order of $(v \, / \, c)^2$. The actual size of the relativistic correction depends on the average velocity of the electron, which in turn depends on the eccentricity of the orbit. The elliptical path actually becomes a precessing ellipse as shown in the Fig.(2.5) .

The precession rate is different for different elliptical orbits (i.e. rate depends on n_θ) even though n may be small. Thus degeneracy is removed. Sommerfeld showed that the total energy of an electron in

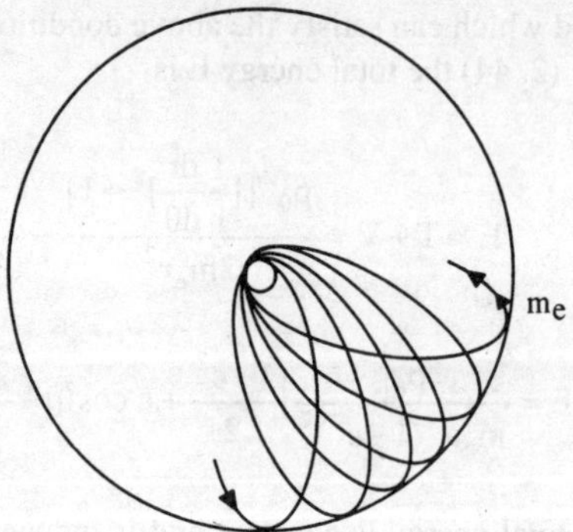

**Fig. 2.5 Schematic representation of the precession of an electron orbit
when a relativistic mass correction is applied to the electron**

an orbit characterized by n and n_θ is equal to :

$$E = -\frac{4\pi^2 \mu Z^2 e^4}{(4\pi \, \varepsilon_0)^2 2n^2 h^2}\left[1+\frac{\alpha^2 Z^2}{n}\left(\frac{1}{n_\theta}-\frac{3}{4n}\right)\right] \qquad (2.57)$$

Fig. 2.6 shows the energy level diagram of the hydrogen atom according to Sommerfeld model for n = 2 and n = 3 states. Arrows indicate transitions between the various energy states, which produce the lines of atomic spectrum. Lines corresponding to transitions represented by the solid arrows are observed in the hydrogen spectrum. The wavelengths of the lines are in very good agreement with the prediction derived from Eq. (2.57). However the lines corresponding to the transitions represented by broken arrows are not found in the spectrum. Study of the figure indicates that transition occurs only if

$$\Delta n_\theta = n_{\theta_i} - n_{\theta_f} = \pm 1 \tag{2.58}$$

This means that H_α ($n = 3 \rightarrow n = 2$ transition) line which was a singlet according to Bohr theory, is a close doublet according to the Bohr – Sommerfeld Model.

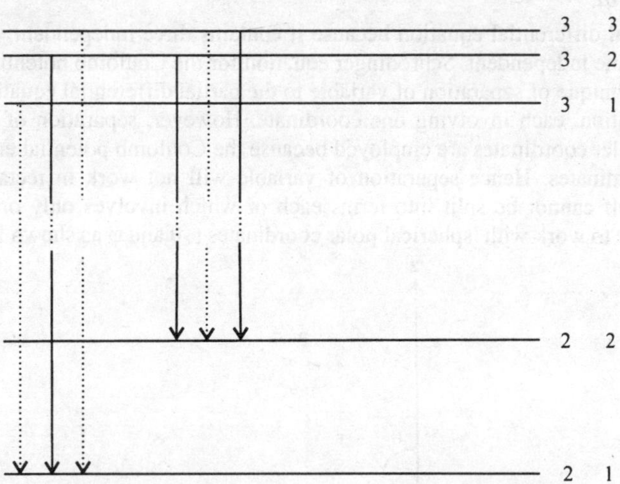

Fig. 2.6 The fine structure splitting of energy level n = 2 and n = 3 of the hydrogen atom. The splitting is greatly exaggerated.Transitions that produce observed lines of the hydrogen spectrum are indicated by solid arrows

2.3 Quantum Theory of Hydrogen Atom

The Bohr-Sommerfeld theory succeeded in many aspects of the atomic spectra of hydrogen and hydrogenic atoms. For example, it explained different series of transitions observed in hydrogen atom, the correct value of excitation and ionisation potentials etc. However, the theory is unsatisfactory in many respects. Firstly the theory could not account for the energy levels of atoms with two or more electrons .The logical reason for assuming different quantum numbers. The theory is also not able to explain the relative intensities of the lines etc. These difficulties were overcome by the development of quantum mechanics. In this section we discuss the hydrogenic atom using nonrelativistic Schrödinger wave equation.

Let us consider a hydrogenic atom containing an atomic nucleus of charge Ze (Z=1 for hydrogen atom) and electron of charge –e interacting by means of the coulomb potential

$$V(r) = -\frac{Ze^2}{4\pi \, \varepsilon_0 r} \tag{2.59}$$

where r is the electron nucleus separation distance. The non-relativistic Schrödinger wave equation in this case is

$$\frac{\partial^2 \Psi}{\partial x^2} + \frac{\partial^2 \Psi}{\partial y^2} + \frac{\partial^2 \Psi}{\partial z^2} + \frac{8\pi^2 \mu}{h^2}[E - V(r)]\Psi = 0 \tag{2.60}$$

here $\Psi = \Psi(x,y,z)$ is the wave function of the electron, μ is the reduced mass of the electron and nucleus given $\mu = (m_e M) / (m_e + M)$. It is convenient to write Eq. (2.60) as

$$\nabla^2\Psi + \frac{8\pi^2\mu}{h^2}[E - V(r)]\Psi = 0 \tag{2.61}$$

where $\nabla^2 = \dfrac{\partial^2}{\partial x^2} + \dfrac{\partial^2}{\partial y^2} + \dfrac{\partial^2}{\partial z^2}$ is called Laplacian operator or 'del squared' in rectangular coordinates..The

above equation is a partial differential equation because it contains three independent variables, the space coordinates x, y, z. The time independent Schrödinger equation for the Coulomb potential can be solved by making application of technique of separation of variable to the partial differential equation into set of three ordinary differential equation, each involving one coordinate. However, separation of variable cannot be carried out when rectangular coordinates are employed because the Coulomb potential energy is function of all three x, y and z coordinates. Hence separation of variable will not work in rectangular coordinates because the potential itself cannot be split into terms each of which involves only one such coordinate. Therefore, it is convenient to work with spherical polar coordinates r, θ and φ as shown in the Fig.2.7.

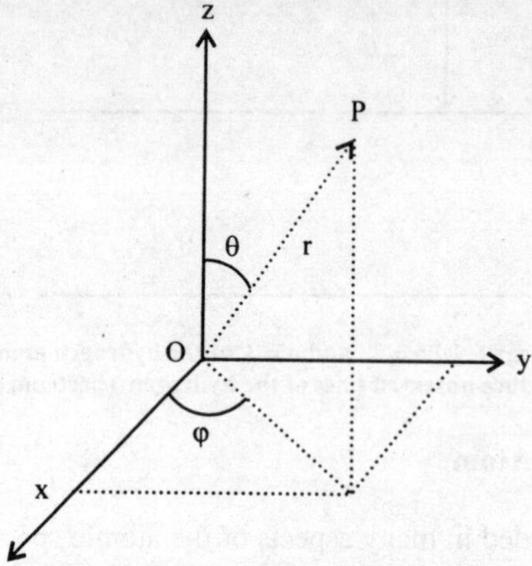

Fig. 2.7 Polar coordinates of a point P

The length of the straight line connecting the electron with origin (nucleus) is r; and θ and φ are the polar and azimuthal angles specifying the orientation of the lines. The relation between rectangular coordinates and spherical polar coordinates is

$$x = r\sin\theta\cos\varphi$$
$$y = r\sin\theta\sin\varphi$$
$$z = r\cos\theta$$

In spherical polar coordinates the Coulomb potential can be expressed as a function of a single coordinate r. It is then possible to carry out the separation of variables on the time independent Schrödinger equation. The Schrödinger Eq. (2.61) in spherical polar coordinates can be written as

$$\frac{1}{r^2}\frac{d}{dr}\left(r^2\frac{\partial\Psi}{\partial r}\right) + \frac{1}{r^2\sin\theta}\frac{d}{d\theta}\left(\sin\theta\frac{\partial\Psi}{\partial\theta}\right) + \frac{1}{r^2\sin^2\theta}\frac{\partial^2\Psi}{\partial\varphi^2} + \frac{8\pi^2\mu}{h^2}[E - V(r)]\Psi = 0 \tag{2.62}$$

where $\Psi = \Psi(r,\theta,\varphi)$ and it can be written as a product of three functions, each of which depends only on one coordinate

$$\Psi(r,\theta,\varphi) = R(r)\,\Theta(\theta)\,\Phi(\varphi) = R(r)\,Y(\theta,\varphi) \tag{2.63}$$

and

$$Y(\theta,\varphi) = \Theta(\theta)\,\Phi(\varphi) \tag{2.64}$$

Substituting the values of Ψ from Eq.(2.63) into Eq. (2.62) we obtain

$$\frac{Y}{r^2}\frac{\partial}{\partial r}(r^2\frac{\partial R}{\partial r}) + \frac{R}{r^2\sin\theta}\frac{\partial}{\partial\theta}(\sin\theta\frac{\partial Y}{\partial\theta}) + \frac{R}{r^2\sin^2\theta}\frac{\partial^2 Y}{\partial\varphi^2} + \frac{8\pi^2\mu}{h^2}[E - V(r)]RY = 0 \tag{2.65}$$

Dividing the above equation by Ψ

$$\frac{1}{R}\frac{\partial}{\partial r}(r^2\frac{\partial R}{\partial r}) + \frac{8\pi^2\mu}{h^2}[E - V(r)] = -\frac{1}{Y}[\frac{1}{\sin\theta}\frac{\partial}{\partial\theta}(\sin\theta\frac{\partial Y}{\partial\theta}) + \frac{1}{\sin^2\theta}\frac{\partial^2 Y}{\partial\varphi^2}] \tag{2.66}$$

Since the two sides of this equation depend on different variables they can equal each other only if they are equal to a constant Λ. Thus the above equation can be written as

$$\frac{1}{R}\frac{\partial}{\partial r}(r^2\frac{\partial R}{\partial r}) + \frac{8\pi^2\mu}{h^2}[E - V(r)] = -\frac{1}{Y}[\frac{1}{\sin\theta}\frac{\partial}{\partial\theta}(\sin\theta\frac{\partial Y}{\partial\theta}) + \frac{1}{\sin^2\theta}\frac{\partial^2 Y}{\partial\varphi^2}] = \Lambda \tag{2.67}$$

$$\frac{1}{r^2}\frac{\partial}{\partial r}(r^2\frac{\partial R}{\partial r}) + \{\frac{8\pi^2\mu}{h^2}[E - V(r)] - \frac{\Lambda}{r^2}\}R = 0 \tag{2.68}$$

$$\frac{1}{\sin\theta}\frac{\partial}{\partial\theta}(\sin\theta\frac{\partial Y}{\partial\theta}) + \frac{1}{\sin^2\theta}\frac{\partial^2 Y}{\partial\varphi^2} + \Lambda Y = 0 \tag{2.69}$$

Eq. (2.68) involves only radial variable while Eq. (2.69) contain angular parts. Substituting Y from Eq.(2.64) in Eq.(2.69) we have

$$\frac{\sin^2\theta}{\Theta\sin\theta}\frac{d}{d\theta}(\sin\theta\frac{d\Theta}{d\theta}) + \Lambda = -\frac{1}{\Phi}\frac{d^2\Phi}{d\varphi^2} \tag{2.70}$$

Left hand side of the above equation depends only on θ while right hand side depend only on φ. This is possible only if they are equal to some constant say m^2.
Thus

$$-\frac{1}{\Phi}\frac{d^2\Phi}{d\varphi^2} = m^2 \tag{2.71}$$

and

$$\frac{\sin^2\theta}{\Theta\sin\theta}\frac{d}{d\theta}(\sin\theta\frac{d\Theta}{d\theta}) + \Lambda = m^2 \tag{2.72}$$

$$\frac{1}{\sin\theta}\frac{d}{d\theta}(\sin\theta\frac{d\Theta}{d\theta})+(\Lambda-\frac{m^2}{\sin^2\theta})\Theta=0 \qquad (2.73)$$

The solution of the Eq. (2.71) is given by

$$\Phi(\varphi)=A\exp(\pm im\varphi)$$

$$=A\exp(im\varphi) \qquad (2.74)$$

if sign is included in m. For an arbitrary value of m the function $\Phi(\varphi)$ is not singled value ,that is, $\Phi(\varphi)$ does not have the same value for $\varphi=0$ and $\varphi=2\pi$ which correspond to the same point in space. For wave function to be single valued, the following condition must be satisfied

$$\Phi(\varphi+2\pi)=\Phi(\varphi) \qquad (2.75)$$

$$\exp[im(\varphi+2\pi)=\exp(im\varphi) $$

$$\exp(im\,2\pi)=1 \qquad (2.76)$$

$$m=0,\pm1,\pm2,\pm3,\ldots\ldots\ldots \qquad (2.77)$$

we label the function by subscript m

$$\Phi_m(\varphi)=A\exp(im\varphi)$$

The normalized function is

$$\Phi_m(\varphi)=\frac{1}{\sqrt{2\pi}}\exp(im\varphi) \qquad (2.78)$$

Putting $w=\cos\theta$ and writing $\Theta(\theta)=P(w)$ in equation (2.73) we have

$$\frac{d}{dw}[(1-w^2)\frac{dP}{dw}]+(\Lambda-\frac{m^2}{1-w^2})P=0 \qquad (2.79)$$

The domain of $P(w)$ is $-1\leq w\leq1$. Further, equation become infinite when w takes on either of its extreme value. However, we are considering the problem of electron nucleus bound system; therefore, the wave function must vanish at infinite distance. The general solution is therefore not physically acceptable. Nevertheless, if $\Lambda=l(l+1)$ where l is integer and $l\geq0$, one of the solution can be finite for all values of w. For $|m|=0$, P (w) will be Legendre polynomial, for $m\neq0$, a finite solution will be possible only if $|m|\leq l$. In the latter case, the solution will be associated Legendre polynomial $P_l^{|m|}(w)$ such that

$$P_l^m(w)=(1-w^2)^{m/2}\frac{d^m}{dw^m}P_l(w) \qquad (2.80)$$

where

$$P_l(w) = \frac{1}{2^l l!} \frac{d^l}{d^l w}(w^2 - 1)^l \tag{2.81}$$

The solution of the angular part of the Schrödinger equation of the form of $Y(\theta, \varphi)$ in Eq. (2.63) is labelled by subscript l and m and is denoted by $Y_{lm}(\theta, \varphi)$. $Y_{lm}(\theta, \varphi)$ are called spherical harmonics.   After normalization $Y_{lm}(\theta, \varphi)$ from Eqs. (2. 64), (2.78) and (2.80) is

$$Y_{lm}(\theta, \varphi) = \left[\frac{(2l+1)(l-m)!}{4\pi(l+m)!}\right]^{1/2}(-1)^m P_l{}^m(\cos\theta)\exp(im\varphi) \tag{2.82}$$

for $m > 0$ and for $m < 0$ one may use the relation

$$Y_{l,-m} = (-1)Y_l{}^*{}_{,m}$$

The values of $P_l{}^m$ up to $l = 2$ are given in Table 2.2

Table 2.2. Some $P_l{}^m(\cos\theta)$ values calculated by using Eq. (2.80).

l	m	$P_l{}^m(\cos\theta)$
0	0	1
1	1	$\sin\theta$
1	0	$\cos\theta$
2	2	$3\sin^2\theta$
2	1	$3\sin\theta\cos\theta$
2	0	$(3\cos^2\theta - 1)/2$

Substituting the value of $\Lambda = l(l+1)$ in Eq.(2.68) the radial part of the Schrödinger equation is

$$\frac{d^2R}{dr^2} + \frac{2}{r}\frac{dR}{dr} + \left\{-\frac{l(l+1)}{r^2} + \frac{8\pi^2\mu}{h^2}[E - V(r)]\right\}R = 0 \tag{2.83}$$

Putting

$$u = rR, \quad a_0 = \frac{(4\pi\varepsilon_0)h^2}{4\pi^2\mu e^2}, \quad \rho = \left(-\frac{32\pi^2\mu E}{h^2}\right)^{1/2} r, \quad \beta = \frac{2\pi Ze^2}{h}\left(-\frac{\mu}{2E}\right)^{1/2} \tag{2.84}$$

Eq. (2.83) reduces to

$$\frac{d^2u(\rho)}{d\rho^2} + \left[-\frac{l(l+1)}{\rho^2} + \frac{\beta}{\rho} - \frac{1}{4}\right]u(\rho) = 0 \tag{2.85}$$

For $\rho \to \infty$, Eq. (2.85) is

$$\frac{d^2u(\rho)}{d\rho^2} - \frac{u(\rho)}{4} = 0 \tag{2.86}$$

The solution of the above equation is

$$u(\rho) = \exp(\pm \frac{\rho}{2}) \tag{2.87}$$

the boundary condition at $\rho \to \infty$ allows only negative exponential

$$u(\rho) = \exp(-\frac{\rho}{2}) \tag{2.88}$$

when $\rho \to 0$, the radial Eq. (2.85) becomes

$$\frac{d^2 u(\rho)}{d\rho^2} - \frac{l(l+1)}{\rho^2} u(\rho) = 0 \tag{2.89}$$

The solution of Eq. (2.89) has the form

$$u(\rho) \to \rho^{l+1} \tag{2.90}$$

as function must vanish at the origin, otherwise the integrated probabilities over a sphere of vanishingly small radius would remain finite. The general solution of radial Eq. (2.84) can be written as

$$u(\rho) = \rho^{l+1} \exp(-\frac{\rho}{2}) L(\rho) \tag{2.91}$$

Substituting Eq. (2.91) in Eq. (2.84) we have

$$\rho L''(\rho) + [2(l+1) - \rho] L'(\rho) + [\beta - l - 1] L(\rho) = 0 \tag{2.92}$$

Putting

$$L(\rho) = \sum_{s=0}^{\infty} a_s \rho^s \tag{2.93}$$

in Eq. (2.92)   we obtain

$$\sum \{s[(s-1) + 2l + 2]a_s \rho^{s-1} + (\beta - l - 1 - s)a_s \rho^s\} = 0 \tag{2.94}$$

For this expression to be true the coefficient of each term must vanish which leads to

$$(s+1)(s+2l+2)a_{s+1} + (\beta - l - 1 - s)a_s = 0 \tag{2.95}$$

$$\frac{a_{s+1}}{a_s} = \frac{l+s+1-\beta}{(s+1)(s+2l+2)} \tag{2.96}$$

For large value of s, and fixed l, Eq. (2.96) is

$$\frac{a_{s+1}}{a_s} \to \frac{1}{s}$$

(2.97)

but this is the behaviour of an exponential series

$$\exp(\rho) = 1 + \rho + \frac{\rho^2}{2!} + \ldots\ldots + \frac{\rho^{s-1}}{(s-1)!} + \frac{\rho^s}{s!} + \ldots\ldots$$

(2.98)

the coefficient

$$\frac{a_{s+1}}{a_s} \to \frac{1}{s}$$

Thus the solution of Eq. (2.91) is of the form

$$u(\rho) = \rho^{l+1}\exp(-\frac{\rho}{2})\exp(\rho) = \rho^{l+1}\exp(\frac{\rho}{2})$$

(2.99)

the boundary condition $u(\rho) \to 0$ as $\rho \to \infty$ is violated. Therefore, the series has to be terminated and this is possible only if for some value of $s = s_{max}$, we have

$$s_{max} + l + 1 = \beta$$

(2.100)

s_{max} = 0, 1, 2 ,3, ...is radial quantum number. Since both s_{max} and l takes only integer value, we can write Eq. (2.100) as

$$s_{max} + l + 1 = \beta = n$$

(2.101)

Thus from Eqs. (2.84) and (2.101)

$$\beta^2 = -\frac{4\pi^2 \mu Z^2 e^4}{2E(4\pi\varepsilon_0)^2 n^2 h^2} = n^2$$

$$E = -\frac{4\pi^2 \mu Z^2 e^4}{2(4\pi\varepsilon_0)^2 n^2 h^2} = -\frac{R_\infty chZ^2}{n^2}$$

(2.102)

where R_∞ is Rydberg constant given by Eq. (2.24)

   Energy depend only on the index n and not on l, that is, for n >1 we have states with different values of l corresponding to the same value of n and hence these states are degenerate. From Eq. (2.101) $s_{max} = n - 1 - 1$ and $s_{max} \geq 0$, we obtain $l \leq n - 1$, so for a given value of n , l cannot exceed the value $l_{max} = n - 1$. This corresponds to the value l = 0,1, ...(n-1) . each of these values correspond to different values of s_{max} , therefore different wavefunctions. The energy E depends only on n whereas the wavefunction depends on n, l and m. For each value of n, l can take value 0,1,2,......(n-1) and for each value of l, there are (2l+1) possible values of m i.e. from –l, -l+1, ...,+l. The total degeneracy of the energy level is therefore given by

$$\sum_{l=0}^{n-1} (2l+1) = n^2 \qquad (2.103)$$

The energy levels are said to be n^2-fold degenerate. The degeneracy with respect to m is obvious, for m describes the projection of vector l on a coordinate axis (the z axis) in space, if this axis is not defined by some physical criterion but is arbitrarily chosen then the physical observable energy cannot depend on this arbitrary choice. We are assuming x, y, z space to be isotropic. The degeneracy with respect to l arising from the fact that for hydrogen like atom V(r) is purely Coulomb field i.e. V(r) is proportional to r^{-1} and further, that Coulomb's law is valid over the entire range of r of interest, in atoms. This is called accidental degeneracy.

The polynomial solution of Eq. (2.92) can be expressed in terms of Laguerre polynomial $L_q^p(\rho)$ that are known to satisfy the equation

$$\rho \frac{d^2 L_q^p}{d\rho^2} + (p+1-\rho)\frac{dL_q^p}{d\rho} + (q-p)L_q^p = 0 \qquad (2.104)$$

The equation has precisely this form with $2l + 1 = p$ and $q = n + l$. Therefore, its polynomial solution is the associated Laguerre polynomial apart from an arbitrary constant factor N_{nl}. The radial function R of Eq. (2.63) is labelled by two indices n and l. The radial wavefunction after normalisation is

$$R_{nl}(r) = -[(\frac{2Z}{na_0})^3 \frac{(n-l-1)!}{2n(n+l)!}]^{1/2} \exp(-\frac{\rho}{2})\rho^l L_{n+l}^{2l+1}(\rho) \qquad (2.105)$$

where $\rho = \dfrac{2Zr}{na_0}$ and

$$L_{n+l}^{2l+1}(\rho) = \sum_{k=0}^{n-l-1} (-1)^{k+2l+1} \frac{(n+l)^2 \rho^k}{(n-l-1-k)!(2l+1+k)!k!} \qquad (2.106)$$

The wavefunction $\Psi(r, \theta, \varphi)$ given by Eq. (2.63) to which we attach subscript nlm is given by

$$\Psi_{nlm}(r, \theta, \varphi) = R_{nl} Y_{lm}(\theta, \varphi) \qquad (2.107)$$

with $R_{nl}(r)$ and $Y_{lm}(\theta, \varphi)$ given from Eqs. (2.105) and (2.82), respectively. Explicit form of $R_{nl}(r)$ are given in Table 2. 3.

Plots of $R_{nl}(r)$ are shown in Fig. 2.8(a). It is observed that due to properties of the associated Laguerre polynomial, R_{nl} will changes its sign n-l-1 times. Further, the states with l = 0 are finite at the origin, whereas the states with $l \geq 1$ vanish there. Fig. 2.8(b) shows the plot of R_{nl}^2 against ρ (or $\cdot$r). Since R_{nl} is a real function $R_{nl} = R_{nl}^*$ and $R_{nl}^2 dr$ is the probability of finding the electron between r and r+dr. A plot of R_{nl}^2 against ρ (or r) represents the radial probability distribution of the electron. Fig.2.8c shows the plot of radial charge density $4\pi\, r^2\, R_{nl}^2$ against ρ (or r). This plot has n - 1 bumps. The quantity $4\pi\, r^2 R_{nl}^2$ is the probability of finding an electron within an element of volume that consists of spherical shell of thickness dr and a volume element $4\pi\, r^2 dr$. The dependence of the function $4\pi\, r^2 R_{nl}^2$ on r differs from that of R_{nl}^2 function. The R_{nl}^2 function is maximum in the neighbourhood of the nucleus. The $4\pi\, r^2 R_{nl}^2$ changes, however, from zero at r = 0 passes through the maximum and gradually decreases. This behaviour is due to dependence on the factors $4\pi\, r^2$ and R_{nl}^2. It starts from zero (at r = 0) and increases with r. For small values

Table 2.3 Radial wavefunctions R_{nl} in hydrogenic atom.

$R_{10}(r)$	=	$2(Z/a_0)^{3/2}\exp(-Zr/2a_0)$
$R_{20}(r)$	=	$2(Z/2a_0)^{3/2}(1-Zr/2a_0)\exp(-Zr/2a_0)$
$R_{21}(r)$	=	$(1/3)^{1/2}(Z/2a_0)^{3/2}(Zr/a_0)\exp(-Zr/2a_0)$
$R_{30}(r)$	=	$2(Z/3a_0)^{3/2}(1-Zr/3a_0 +2Z^2r^2/27a_0^2)\exp(-Zr/3a_0)$
$R_{31}(r)$	=	$(4\sqrt{2}/9)\,(Z/3a_0)^{3/2}(1-Zr/6a_0)\,(Zr/a_0)\exp(-Zr/3a_0)$
$R_{32}(r)$	=	$(4/27\sqrt{10})\,(Z/3a_0)^{3/2}(Zr/a_0)^2\exp(-Zr/3a_0)$

of r, $4\pi\,r^2$ value increases more rapidly than R_{nl}^2 decreases, thus leading to an increase of their product; then R_{nl}^2 starts decreasing more rapidly than the $4\pi\,r^2$ increases. As a result their product passes the maximum and shows a further gradual decrease. The maximum of the radial distribution corresponds to the atomic radius. The average value of various power of r is defined as

$$< r^k >_{nl} = \int r^k [R_{nl}(r)]^2 dr \qquad (2.108)$$

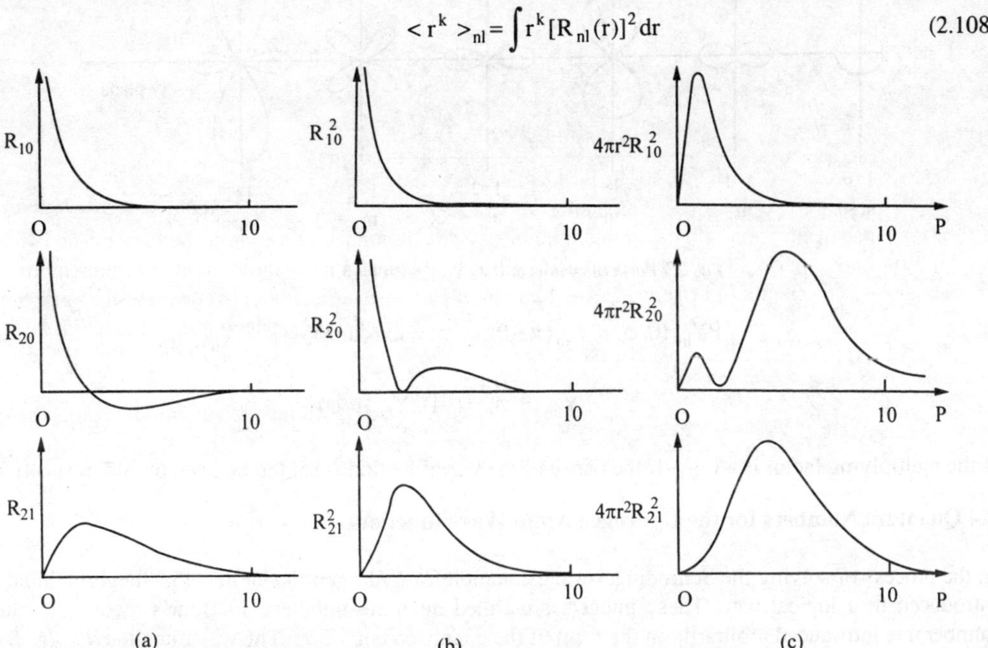

Fig. 2.8 Plots of (a) the radial wavefunction R_{nl} , (b) the radial probability distribution function R_{nl}^2 , and (c) the radial charge density function $4\pi r^2 R_{nl}^2$ against ρ

The result of several values of k are given in Table 2.4

The forms of a few $Y_{lm}(\theta, \varphi)$ are given in Table 2.5. The value $|Y_{lm}(\theta,\varphi)|^2$ is plotted as a function of angle θ (Fig.2. 9).

The value can be read off from the radial distance of the curve from the origin of the diagram.

Because of the φ dependence, such representations have rotational symmetry about the z-axis. The diagram for m and –m are identical. The function $Y_{ll}(\theta, \varphi)$ (i.e. for maximum value of m) are strongly concentrated about the xy plane.

Table 2.4   Values of $\langle r^k \rangle_{nl}$ for various k.

$\langle r \rangle_{nl}$	=	$(1/2)[3n^2 - l\,(l+1)]a_0/Z$
$\langle r^2 \rangle$	=	$(1/2)[5n^2 + 1 - 3\,l\,(l+1)]\,n^2(a_0/Z)^2$
$\langle r^{-1} \rangle_{nl}$	=	$Z/n^2 a_0$
$\langle r^{-2} \rangle_{nl}$	=	$[2\,/\,(2l+1)\,n^3]\,(Z/a_0)^2$
$\langle r^{-3} \rangle_{nl}$	=	$[1\,/\,l\,(l+1/2)\,(l+1)\,n^3]\,(Z/a_0)^3$

The wave function of an atom can also be characterised by its parity. The parity operator P transforms the position vector **r** into −**r**, or acting on a state Ψ, one has   P Ψ (**r**) = Ψ (−**r**). The vector −**r** has polar coordinates $\pi - \theta$, $\pi + \varphi$. Because of $\cos (\pi - \theta) = -\cos \theta$ we have

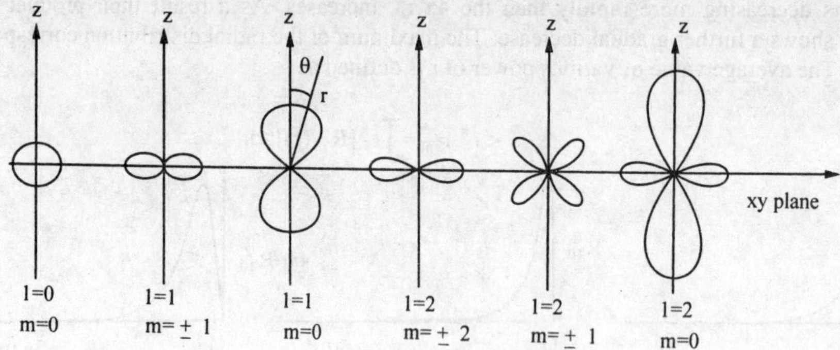

Fig. 2.9 Polar diagrams of the $|\,Y_{lm}\,|^2$ with l = 0,1,2

$$PY_{lm}\,(\theta, \varphi) = Y_{lm}\,(\pi - \theta, \pi + \varphi) = \exp(im\pi)(-1)^{l+m}\,Y_{lm}\,(\theta, \varphi) \tag{2.108}$$

$$PY_{lm}\,(\theta, \varphi) = (-1)^l\,Y_{lm}\,(\theta, \varphi) \tag{2.109}$$

If the multiplying factor is +1 or −1, the parity of the wavefunction is said to be even or odd, respectively.

2.4 Quantum Numbers for the Hydrogen Atom Wave function

In the process of solving the Schrödinger wave equation for hydrogen like atoms, the integer n, l and m are introduced in a logical way. These integers are called quantum numbers. In Bohr's theory the quantum number n is introduced arbitrarily in the form of the quantised condition. The wavefunction $\Psi_{nlm}(r, \theta, \varphi)$ are description of states of the system. They are related to n, l and m. Thus the quantum number themselves may be said to describe the state of the system. The values of these quantum numbers are:

$$n = 1, 2, 3,........$$
$$l = 0, 1, 2,......, n-1 \tag{2.110}$$
$$m = 0, \pm 1, \pm 2,.........,\pm l$$

It is now shown that the quantum number l and m are related to the magnitude L of the orbital angular momentum and m is the z component L_z. The classical orbital angular momentum of a particle is

Table 2.5 Angular wavefunction $Y_{lm}(\theta,\varphi)$

$Y_{0,0}$	$=$	$(1/4\pi)^{1/2}$
$Y_{1,0}$	$=$	$(3/4\pi)^{1/2}\cos\theta$
$Y_{1,\pm1}$	$=$	$+(3/8\pi)^{1/2}\sin\theta\exp(\pm i\varphi)$
$Y_{2,0}$	$=$	$(5/16\pi)^{1/2}(3\cos^2\theta-1)$
$Y_{2,\pm1}$	$=$	$+(15/8\pi)^{1/2}\sin\theta\cos\theta\exp(\pm i\varphi)$
$Y_{2,\pm2}$	$=$	$(15/32\pi)^{1/2}\sin^2\theta\exp(\pm 2i\varphi)$
$Y_{3,0}$	$=$	$(7/16\pi)^{1/2}(5\cos^3\theta-3\cos\theta)$
$Y_{3,\pm1}$	$=$	$-1x(\pm)(21/64\pi)^{1/2}\sin\theta(5\cos^2\theta-1)\exp(\pm i\varphi)$
$Y_{3,\pm2}$	$=$	$(105/32\pi)^{1/2}\sin^2\theta\cos\theta\exp(\pm 2i\varphi)$
$Y_{3,\pm3}$	$=$	$-1x(\pm)(35/64\pi)^{1/2}\sin^3\theta\exp(\pm 3i\varphi)$

$$L = r \times p \tag{2.111}$$

where $\mathbf{r}$ and $\mathbf{p}$ are the position and momentum vectors of the particle, respectively. The momentum $\mathbf{p}$ is represented by the operator $-i(h/2\pi)\nabla$ in quantum mechanics. Therefore, Eq. (2.111) is

$$L = -i(h/2\pi)(r\times\nabla) \tag{2.112}$$

Its cartesian components are

$$L_x = yp_z - zp_y = -i\frac{h}{2\pi}\left[y\frac{\partial}{\partial z} - z\frac{\partial}{\partial y}\right]$$

$$L_y = zp_x - xp_z = -i\frac{h}{2\pi}\left[z\frac{\partial}{\partial x} - x\frac{\partial}{\partial z}\right] \tag{2.113}$$

$$L_z = xp_y - yp_x = -i\frac{h}{2\pi}\left[x\frac{\partial}{\partial y} - y\frac{\partial}{\partial x}\right]$$

In spherical polar coordinates components of $\mathbf{L}$ are represented by

$$L_x = -i\frac{h}{2\pi}\left(\sin\theta\frac{\partial}{\partial\theta} + \cot\theta\cos\varphi\frac{\partial}{\partial\varphi}\right)$$

$$L_y = -i\frac{h}{2\pi}\left(-\cos\varphi\frac{\partial}{\partial\theta} + \cot\theta\sin\varphi\frac{\partial}{\partial\varphi}\right) \tag{2.114}$$

$$L_z = -i\frac{h}{2\pi}\frac{\partial}{\partial\varphi}$$

and

$$L^2 = L_x^2 + L_y^2 + L_z^2$$

$$L^2 = -\frac{h^2}{4\pi^2}\left[\frac{1}{\sin\theta}\frac{\partial}{\partial\theta}\left(\sin\theta\frac{\partial}{\partial\theta}\right) + \frac{1}{\sin^2\theta}\frac{\partial^2}{\partial\varphi^2}\right] \qquad (2.115)$$

The Eq. (2.69) is just the eigenvalue equation having the form [with $\Lambda = l\,(l+1)$]

$$L^2 Y_{lm} = (h/2\pi)^2 l(l+1) Y_{lm} \qquad (2.116)$$

Also

$$L_z Y_{lm}(\theta,\varphi) = m\,(h/2\pi)\,Y_{lm}(\theta,\varphi) \qquad (2.117)$$

This is again an eigenvalue equation. Thus $Y_{lm}(\theta,\varphi)$ is simultaneously the eigenfunction of L^2 and L_z operator. Therefore, L^2 and L_z commute i.e. corresponding observable can be precisely measured simultaneously. The magnitude of the orbital angular momentum is $\hbar\sqrt{l\,(l+1)}$ while that of L_z is $m\,\hbar$. Since L^2 and L_z do not operate on the radial part of the wavefunction, $\Psi_{nlm}(r,\theta,\varphi)$ itself is a simultaneous eigenfunction of L^2 and L_z. The quantum number l gives the magnitude of the angular momentum, m the orientation of the angular momentum and n gives the quantisation of energy. The quantum numbers l and m can be equal only for $l = 0$.

2.5 Atomic Orbitals

The solution to the Schrödinger equation represents the wavefunction $\Psi_{nlm}(r,\theta,\varphi)$, referred to as one electron functions called atomic orbitals. They are described by the quantum numbers n, l and m. The atomic orbitals differ from orbit in the following:

1. Orbits refer to definite path of the electron around the nucleus of an atom. They were postulated in Bohr's theory. They are characterized by the principal quantum number n. An orbital does not indicate the exact location of an electron. Its square gives the probability of finding the electron at a given point in space. It is characterized by the quantum numbers n, l and m.
2. The representation of an orbit is two-dimensional while that of an orbital is three-dimensional involving the spherical polar coordinates r, θ and φ.
3. An orbit having a principal quantum number n can accomodate $2n^2$ electrons (with spin taken into account). The size of the orbital part is determined by the radial part and their shape by the angular part of the wavefunction.

The orbitals with l quantum numbers are denoted by the letters

l =	0	1	2	3	4	5	6	7
	s	p	d	f	g	h	i	k

The principal quantum number n is placed before these letters; for example, the s orbital has n = 1, l = 0, 3d orbital n = 3, l = 2 etc. The magnetic quantum number m is placed as subscript: $2p_0$, $3d_2$ etc.

The wavefunction $\Psi_{nlm}(r,\theta,\varphi)$ (Eq. 2.107) consists of angular and radial parts. The radial part of the wave function represents its dependence on the radius only. It describes the values and the distribution of electron density in atomic orbitals. The radial function depends on the quantum number n and l and does not depend on m, i.e. for all the $(2l+1)$ functions for a given value of l, there is same radial distribution of electron density. The angular part $Y_{lm}(\theta,\varphi)$ shows the dependence on θ and φ angles only (Table 2.5) . It describes the form of atomic orbitals. Since $Y_{lm}(\theta, \varphi)$ depend on the quantum number l and m only and not on the principal quantum number n. Therefore, all the s orbitals (or all the p, d, f.. orbitals) i.e. orbitals with the same l, have the same form, thus independent of n values. The quantum number m determines the orbital orientation. The number of possible orientations for s orbitals is one, for the p orbital three, for d orbital five and for f orbital seven i.e. (2l + 1). In the absence of any fixed direction, for instance, without electric or magnetic field directions, all the possible orientations of the orbitals are equivalent and correspond to the same energy value. These orbitals are called degenerate. From Figs. (2.8) and (2.9) it is seen that s orbitals are spherical symmetric i.e. they depend on r but are independent of θ and φ angles. Further, only the s orbitals have non-zero electron density at the nucleus while p, d electrons do not come in contact directly with the nucleus. The radial dependence of s orbitals is shown in Fig.2. 8. The p orbitals exist beginning with n = 2 (Table 2.5). Since for l = 1, m can take values 1,0, -1; three p orbitals occur p_1, p_0 and p_{-1} . In Fig.2.9, p orbitals are represented by Y_{lm}^2 as boundary surfaces of dumb bell shape. The d orbitals exist beginning with n = 3 (Table 2.5). Since for l = 2, five possible values for m= 2,1,0, -1, -2 exist, hence five d orbitals occur:$d_2, d_1, d_0, d_{-1}, d_{-2}$. The form of d-orbitals is shown in the Fig.2.9. For all orbitals except 1s there are region in space where $\Psi_{nlm}(r, \theta, \varphi) = 0$ because either $Y_{lm}(\theta, \varphi)$ or $R_{nl}(r)$ is zero. These regions are called nodal surface or simply nodes.

The spherical harmonics can also be written in real form by taking linear combinations of $Y_{lm}(\theta, \varphi)$. The real form of spherical harmonics for m = 0 are

$$Y_{1,\cos\theta} = (\frac{1}{2})^{1/2}(Y_{lm} + Y_{lm}^*) \qquad (2.118)$$

$$Y_{1,\sin\theta} = -i(\frac{1}{2})^{1/2}(Y_{lm} - Y_{lm}^*)$$

For m = 0, $Y_{1,\cos\theta}$ is identical to Y_{10}. The first few spherical harmonics in real form are listed in Table 2.6 The subscript z, x, y, xz, yz etc. are used in the Table2.6 indicate the behaviour of the real spherical harmonics in terms of cartesian coordinates. The designation for example p_z or d_{xy} reflects the orientation of these atomic orbitals along the z axis for p_z or in the xy plane for d_{xy}.

Table 2.6   The first few spherical harmonics in real form

l	m		Spherical Harmonics in Real Form
0	0	s	$= (1/4\pi)^{1/2}$
1	0	p_z	$= (3/4\pi)^{1/2} \cos\theta$
	1	p_x	$= (3/4\pi)^{1/2} \sin\theta \cos\varphi$
		p_y	$= (3/4\pi)^{1/2} \sin\theta \sin\varphi$
2	0	$d_{3z^2 - r^2}$	$= (5/16\pi)^{1/2} (3\cos^2\theta - 1)$
	1	d_{xz}	$= (15/4\pi)^{1/2} \sin\theta \cos\theta \cos\varphi$
		d_{yz}	$= (15/4\pi)^{1/2} \sin\theta \cos\theta \sin\varphi$
	2	$d_{x^2 - y^2}$	$= (15/4\pi)^{1/2} \sin^2\theta \cos 2\varphi$
		d_{xy}	$= (15/4\pi)^{1/2} \sin^2\theta \sin 2\varphi$

Examples

1. Find the quantum number of the Bohr orbit in a hydrogen atom whose radius is 0.0100 nm. What is the energy of a hydrogen atom in this state ?

    We have $r = n^2 a_o$,

    $$r = 0.0100 \text{ nm} = 1 \times 10^{-5} \text{m}.$$

    $$n = (r/a_0)^{1/2} = (\frac{10^{-5} \text{ m}}{5.29 \times 10^{-11} \text{ m}})^{1/2} = 435$$

    $$E_n = -\frac{13.6 \text{eV}}{n^2} = -\frac{13.6 \text{eV}}{(435)^2} = -7.19 \times 10^{-5} \text{eV}$$

2. A beam of electron bombards a sample of hydrogen. Through what potential difference must the electron be accelerated if the first line of the Balmer series is to be emitted.

    For emission of first line of Balmer series, atom must be in n= 3 state. The energy required to raise hydrogen atom from ground state to n = 3 state is

    $$E_n = -\frac{13.6 \text{eV}}{n^2} = -\frac{13.6 \text{eV}}{(3)^2} = -12.1 \text{eV}$$

    Therefore the energy required to raise the atom from ground state to n = 3 state is 12.1 eV.

3. The largest wavelength in the Lyman series is 121.5 nm and the shortest wave- length in the Balmer series is 364.6 nm. Find the longest wavelength of light that could ionise hydrogen.

    The longest wavelength of Lyman series corresponds to transition between n=2 to n=1. The wavelength of 121.5 nm corresponds to 82.3045 x 10^3 cm^{-1}. The shortest wavelength of Balmer series corresponds to transitions from n= ∞ to n=2. The corresponding wavelength of 364.6 nm is equal to 27.4273 x 10^3 cm^{-1}. Thus the total energy required to ionise the hydrogen atom is sum of the above two wavenumbers which is equal to 109.7358 x 10^3 cm^{-1}. This corresponds to a wavelength of 91.13 nm.

4. A positronium atom is a system that consists of a positron and an electron that orbit each other. Compare the wavelengths of the spectral lines of positronium with those of ordinary hydrogen.

    Here the two particles have the same mass m_e ,so the reduced mass $m' = \frac{m_e}{2}$ .The energy levels of the positronium atom are

    $$E_n = \frac{m' E_1}{m_e n^2} = \frac{E_1}{2n^2}$$

    This means that the Rydberg constant for positronium is half as large as it for   ordinary hydrogen. As a result the wavelengths in the positronium spectral lines are all twice those of the corresponding lines in the hydrogen spectrum.

5. Show that the Paschen series in the spectra of hydrogen atom does not overlap the Balmer series and find which series are the first to overlap.

We have for hydrogen atom

$$\frac{1}{\lambda} = \bar{v} = R_\infty \left(\frac{1}{n_f^2} - \frac{1}{n_i^2} \right)$$

Substituting the values of n_i and n_f for various series we obtain the following

Series	n_f	n_i	Range of $\bar{v}$ (in units of R_∞)
Lyman	1	2,3,......, ∞	1-0.75
Balmer	2	3,4,......, ∞	0.25- 0.1388
Paschen	3	4,5,......, ∞	0.111- 0.0486
Brackett	4	5,6,......, ∞	0.0625- 0.0425
Pfund	5	6,7,......, ∞	0.04- 0.0122

From the range of wavenumbers for various series it is seen that Paschen and Balmer series do not overlap and Brackett and Paschen series are first to overlap.

6. Calculate the difference in wavelength between the Balmer H_α lines in atomic hydrogen and deuterium.(R_∞ = 109737 cm^{-1})
We have for Balmer H_α line in atomic hydrogen and deuterium

$$\frac{1}{\lambda_H} = \bar{v}_{II} = R_H \left(\frac{1}{4} - \frac{1}{9} \right) = \frac{5R_H}{36}$$

$$\frac{1}{\lambda_D} = \bar{v}_D = R_D \left(\frac{1}{4} - \frac{1}{9} \right) = \frac{5R_D}{36}$$

and

$$R_H = \frac{R_\infty}{1 + \dfrac{m_e}{M}} = \frac{109737 \text{cm}^{-1}}{1 + \dfrac{1}{1836}} = 109677.3 \text{cm}^{-1}$$

$$R_D = \frac{R_\infty}{1 + \dfrac{m_e}{2M}} = \frac{109737 \text{cm}^{-1}}{1 + \dfrac{1}{2 \times 1836}} = 109707.12 \text{cm}^{-1}$$

Substituting these values of R_H and R_D in the above Equations we have

$$\frac{1}{\lambda_H} = \frac{5R_H}{36} = 15232.95833 \text{cm}^{-1}$$

$$\lambda_H = 656.471 \text{nm}$$

$$\frac{1}{\lambda_D} = \frac{5R_D}{36} = 15237.1 cm^{-1}$$

$$\lambda_D = 656.292 nm$$

The difference in wavelength between the Balmer $H\alpha$ lines in atomic hydrogen and deuterium is Therefore

$$\lambda_H - \lambda_D = (656.471 - 656.292)\ nm = 0.18\ nm$$

7.   Using the infinite nuclear mass approximation in the Bohr model, obtain the wavelength of the resonance line n = 2 to n = 1 for C^{5+} (Z=6 for C).

Rydberg constant for R_C for C is

$$R_C = \frac{R_\infty}{1+\dfrac{m_e}{M}} = \frac{R_\infty}{1+\dfrac{m_e}{M_C}} = \frac{R_\infty}{1+\dfrac{m_e}{12M}} = \frac{109737 cm^{-1}}{1+\dfrac{1}{12\times1836}} = 109732 cm^{-1}$$

From

$$\frac{1}{\lambda} = R_C Z^2 \left(\frac{1}{n_f^2} - \frac{1}{n_i^2}\right)$$

$$\frac{1}{\lambda} = R_C Z^2 \left(1 - \frac{1}{4}\right) = 109732 cm^{-1} \times 36 \times \left(\frac{3}{4}\right)$$

$$\lambda = 3.3752 nm$$

8.   A muon is an unstable elementary particle whose mass is 207 m_e and whose charge is either $+e$ or $-e$. A negative muon can be captured by a nucleus to form a muonic atom A proton captures a negative muon ,(a) find the radius of the first Bohr orbit of this atom (b) find the ionisation energy of this atom.

(a) Here mass m_e is 207 m_e , M = 1836 m_e so the reduced mass m' = 186 m_e . The orbit radius corresponding to n = 1 is given by Eq. (2.11) with mass is replaced by $186m_e$ and radius is

$$r = \frac{a_0}{186} = \frac{5.29\times10^{-11}\,m}{186} = 2.84\times10^{-13}\,m$$

The muon is 186 times closer to the proton than an electron would be.
(b) For n = 1, the energy of hydrogen atom is $E_1 = -13.6$ eV. The energy of the muon E_1' for n= 1 is then

$$E_1' = \frac{m' E_1}{m_e} = 186 E_1 = -13.6 eV \times 186 = -2.53 keV$$

The ionisation energy is therefore 2.53 keV, 186 times that for an ordinary hydrogen atom.

9. An excited hydrogen emits a photon of wavelength λ in returning to the ground state (a) Derive the formula that gives the quantum number n_i of the initial excited state in terms of λ and Rydberg constant R. (b) Use this formula to find n_i for a 102.55 nm photon.

We have

$$\frac{1}{\lambda} = R\left(\frac{1}{n_f^2} - \frac{1}{n_i^2}\right)$$

Putting $n_f = 1$, the above expression gives

$$\frac{1}{\lambda R} = \left(1 - \frac{1}{n_i^2}\right)$$

$$\frac{1}{n_i^2} = 1 - \frac{1}{\lambda R}$$

$$n_i = \left[\frac{\lambda R}{\lambda R - 1}\right]^{1/2}$$

(b)
$$n_i = \left[\frac{102.55 \times 10^{-9}\,\text{m} \times 1.097 \times 10^7\,\text{m}^{-1}}{102.55 \times 10^{-9}\,\text{m} \times 1.097 \times 10^7\,\text{m} - 1}\right]^{1/2} = 3$$

10. When an excited atom emits a photon, the recoil momentum of the atom must balance the linear momentum of the photon. As a result, some of the excited energy of the atom goes into the kinetic energy of its recoil (a) modify $E_i - E_f = h\nu$ to include this effect. (b) Find the ratio between the recoil energy and the photon energy for the n =3 to n =2 transition in hydrogen for which $E_i - E_f = 1.9$ eV. Is this effect a major one?

(a) Conservation of momentum gives

$$Mv = \frac{h\nu}{c}$$

where M is the mass of recoil atom, v is its recoil velocity. Conservation of energy require

$$E_i - E_f = h\nu + \frac{Mv^2}{2}$$

Using the above Equations we have,

$$M^2v^2 = \frac{(h\nu)^2}{c^2}$$

$$\frac{Mv^2}{2} = \frac{(h\nu)^2}{2Mc^2}$$

Hence

$$E_i - E_f = h\nu + \frac{(h\nu)^2}{2Mc^2} = h\nu\left(1 + \frac{h\nu}{2Mc^2}\right)$$

(b)

$$\frac{E_i - E_f}{\dfrac{Mv^2}{2}} = \frac{2Mc^2(E_i - E_f)}{(h\nu)^2} = \frac{2Mc^2}{1.9eV}$$

$$\frac{Mv^2}{2(E_i - E_f)} = \frac{19eV}{2 \times 931 \times 10^6 eV} = 2.04 \times 10^{-9}$$

The effect is negligible.

11.   Show that in atomic units (a.u.) the energy of the ground state of H atom is 1/2 a.u. Show that 1a.u. = 27.2 eV.

We have from Eq. (2.20)

$$E = -\frac{2\pi^2 m_e e^4 Z^2}{(4\pi\varepsilon_0)^2 n^2 h^2}$$

In a.u., a_0 is taken as a unit of length, mass m_e of the electron is used as unit of mass and $h/2\pi$ as unit of angular momentum. The unit of charge is taken to be absolute magnitude e of electronic charge and permittivity of free space ε_o is $1/4\pi$. Therefore, in atomic units

$$m_e = h/2\pi = e = 1, 4\pi\varepsilon_o = 1$$

Eq. (2.20) reduces to

$$E = -(Z^2/2n^2) \text{ a.u.}$$

For ground state of hydrogen atom n = 1 and for hydrogen Z = 1 and above Eq. reduces to

$$E_1 = -(1/2)\text{a.u.}$$

$$1 \text{ a.u.} = -2E_1 = 2 \times 13.6 \text{ eV} = 27.2 \text{ eV.}$$

12.   Show that velocity of light is 137 a.u.

We have velocity of electron in first Bohr orbit

$$v = \frac{e^2}{2h\varepsilon_0} = \alpha c$$

or c = (1/α) a.u.

But $\alpha = 1/137$, therefore c = 137 a.u.

13.   Show that the frequency of the photon emitted by a hydrogen atom in going from the level n+1 to the level n is always intermediate between the frequencies of revolution of the electron in the respective orbits.

From Eqs. (2.17) and (2.22) we have

$$f_n = \frac{2R_\infty c}{n^3}$$

$$f_{n+1} = \frac{2R_\infty c}{(n+1)^3}$$

$$\nu = R_\infty c\left(\frac{1}{n_f^2} - \frac{1}{n_i^2}\right) = R_\infty c\left(\frac{1}{n^2} - \frac{1}{(n+1)^2}\right) = R_\infty c \frac{2n+1}{n^2(n+1)^2}$$

$$\frac{f_n}{\nu} = 1 + \frac{3n+2}{n(2n+1)} > 1,$$

$$\frac{f_{n+1}}{\nu} = 1 - \frac{3n+1}{2n^2+3n+1} < 1$$

14.   Determine the parity of ground state of B (Z= 5) and Oxygen (Z=8).

For multielectron system, parity is even if $\Sigma\ l_i$ is even and odd if $\Sigma\ l_i$ is odd. For B the electronic configuration is $1s^2 2s^2 2p^1$ so $\Sigma\ l_i = 0+0+1 = 1$ i.e. odd so parity of B is odd.

For O the electronic configuration is $1s^2 2s^2 2p^4$ so $\Sigma\ l_i == 0+0+(1+1+1+1) = 4$ (even). Therefore the parity of oxygen is even.

15.   Calculate the average value of <1/r> when the hydrogenic system is in the ground state.

We have

$$<1/r> = \int \psi^*_{100} (1/r)\psi_{100}\, dV = \int \psi^*_{100} (1/r)\psi_{100}\, r^2\, dr\, \sin\theta\, d\,\theta d\,\varphi$$

$$\psi_{100} = (Z^3 / \pi a_0^3)^{1/2} \exp(-Zr/a_0)$$

Substituting this value of ψ_{100} in the expression for <1/r> we have

$$<1/r> = 4(Z/a_0)^3 \int \exp(-2Zr/a_0)\, r^2\, dr$$

$$= Z/a_0$$

16.   Sketch the energy level diagram of hydrogen according to quantum mechanical treatment.

The energy eigenvalues of the hydrogen atom depend only on the principal quantum number n and are therefore degenerate with respect to l and m. For each value of n the orbital quantum number l may take on the values 0, 1,....., n-1. Fig. 2.10 shows an energy level diagram for hydrogen atom. In the fig, the degenerate energy levels with the same n but different l are shown separately. In the figure, the levels are designated by two symbols. The first one gives the principal quantum number n and second one indicates the orbital quantum number l. Thus is the ground state, the first excited state is four fold degenerate and contains a 2s state and three 2p states etc. Transitions between these levels are in accordance with the selection rule $\Delta l = \pm 1$.

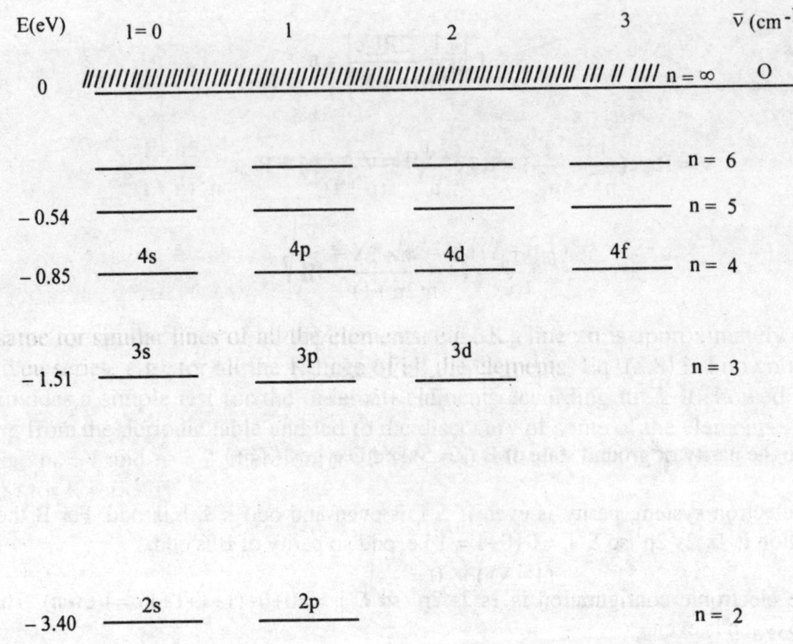

Fig. 2.10 Energy levels of the hydrogen atom, according to the Quantum theory

Problems

2.1   What is the diameter of the hydrogen atom, i.e., the orbit diameter, when the electron is in the orbit n = 30. Find the velocity of the electron in this orbit.

2.2   Assuming the nucleus of the deuteron to have the twice the mass of the hydrogen, compute the wavelength interval between the two lines comprising the second member of the Lyman series.

2.3   In the Bohr model, the electron is in constant motion. How can such an electron have a negative energy ?

2.4   Find the wavelength of the spectral line that corresponds to a transition in the hydrogen from n = 8 to the n = 2 state. In what part of the spectrum is this ?

2.5   How much energy is required to remove an electron in the n = 5 state from a hydrogen atom?

2.6   Find the frequencies of revolution of electrons in n = 1 and n = 2 Bohr orbits. What is the frequency of the photon emitted when an electron in an orbit drops to an n = 1 orbit ?

2.7   An electron typically spends about 10^{-8} s in an excited state before it drops to a lower state by emitting a photon. How many revolutions does an electron in n = 2 Bohr orbit makes in 10^{-8}s?

2.8   When radiation with a continuous spectrum is passed through a volume of hydrogen gas whose atoms are all in the ground state, which spectral series will be present in the resulting absorption spectrum?

2.9   Estimate the temperature of a gas containing hydrogen atom at which the Balmer series lines will be observed in the absorption spectrum.

2.10   Using the infinite nuclear mass approximation in the Bohr model, obtain the wavelength of the first two lines of Paschen series of CVI.

2.11   Using the infinite nuclear mass approximation in the Bohr model, obtain the wavelength of the first line of Lyman series of OVIII.

2.12   Calculate the recoil energy with which a hydrogen atom recoils when emitting a photon in a transition from the n = 5 to the n = 1 level, and show that the recoil energy is negligible in comparison with the energy difference between the two levels.

2.13   Calculate the wavenumber shifts for the first two members of the Balmer series on changing from atomic hydrogen to deuterium.

2.14   What is the wavelength of the first line of Lyman series for a muonic atom?

2.15   Show an electron can circle a nucleus if its orbit contains an integral number of de-Broglie wavelengths.

2.16   What would be the radius of the ground state orbit of positronium ?

2.17   Find the quantum number that characterizes the earth orbit around the sun. The earth's mass is 6×10^{24} Kg, its orbital radius is 1.5×10^{11} m and its orbital speed is 3×10^{4} m/s.

2.18   The wavelength of the first Balmer line of hydrogen is 656.28 nm. Calculate the wavelength of the first line of Balmer series of tritium.

2.19   How many different wavelengths would appear in the spectrum of hydrogen atom initially in the n = 8 state ?

2.20   How many different wavelengths would appear in the spectrum of hydrogen atom with initial state n = 8 and final state n = 2 ?

2.21   Find the shortest wavelength present in the Paschen series of hydrogen.

2.22   A proton and an electron, both at rest initially, combine to form a hydrogen atom in the ground state. A single photon is emitted in this process. What is its wavenumber ?

2.23   Compare the angular momentum of a ground state electron in the Bohr model of the hydrogen atom with its value in quantum theory.

2.24   Sketch Bohr-Sommerfeld orbits for n = 1, 2 and 3.

2.25   Calculate the most probable radius at which an electron will be found when it occupies a 1s orbital of Be^{3+}.

2.26   What is the wavelength of the most energetic photon that can be emitted from a muonic atom with Z = 1.

2.27   Determine the parity of the ground state of N.

2.28   Calculate the average value of distance $<r^2>$ of the electron Li III, when it is in the ground state.

2.29   List the set of quantum number possible for n = 3.

2.30   What are the possible values of the magnetic quantum number m of an atomic electron whose orbital quantum number is l = 3.

2.31   Find the percentage difference between l and the maximum value of l_z for an atomic electron in d orbit.

$$\boxed{3}$$

Vector Representation of Momenta and Vector Coupling Approximations

3.1 Electron Spin

The Schrödinger theory with which we have treated hydrogenic atom (Sec.2.3) so far is inadequate to explain certain details in the spectrum of hydrogen and of other simple atoms. The quantum numbers n, l and m alone cannot explain certain experimental observations

(i) One is the fact that many spectral lines actually consist of two separate lines that are very close together. For example theory predicts a single line of wavelength 656.3 nm for the first line of Balmer series (H_α line). However, two lines 0.14 nm apart were observed for H_α.

(ii) The number of transitions observed when the atoms were placed in an external magnetic field could not be accounted for.

(iii) In 1922, Stern and Gerlach measured the possible values of magnetic dipole moments of silver atoms. They sent a beam of silver atoms in a region in which there was a strong gradient of the magnetic field $\partial B/\partial z$ in z direction transverse to the beam. If the atom has a magnetic moment μ then there is a force on the atom of magnitude $\mu_z \partial B/\partial z$ and transverse deflection of the beam may be observed. Stern and Gerlach found two traces symmetrically displaced on either side of the beam axis with no undeflected trace. The magnetic moment of the electron due to orbital motion is $-\beta_e g_l \, \mathbf{l} \, /\hbar$ (Eq.3.43). The z component of magnetic moment is accordingly proportional to m_l and can take $2l+1$ values (to avoid confusion later, we shall no longer denote the magnetic quantum number related to the orbital angular momentum by m but instead by m_l with subscript l). Since l is an integer, $2l + 1$ is odd. If the magnetic moment of an atom is associated only with the orbital motion, there would be an odd number of discrete traces including the undeflected one because of $m_l = 0$.

From all this it appears that another quantum number is needed for which the number of projections on z axis (quantization axis) is even. To explain these observations Uhlenbeck and Goudsmit in 1925 assigned a new angular momentum to the electron. According to them, the electron in any state spins about its own mechanical axis i.e.. the electron has intrinsic angular momentum denoted by **s** . This spin angular momentum is in addition to the orbital angular momentum. To explain the experimental observations, they assigned to the electron a spin angular momentum $\hbar/2$. For this value of the angular momentum, there are only two orientations of the angular momentum. The notion of electron spin **s** proved to be successful in explaining the various atomic effects.

The magnitude of **s** and z component s_z of the spin angular momentum are related to two quantum numbers s and m_s, by quantization relation which are identical to those for orbital angular momentum. That is

$$s = (h/2\pi)\,[s(s+1)]^{1/2} \qquad\qquad (3.1)$$

$$s_z = m_s\,(h/2\pi) \qquad\qquad (3.2)$$

where $m_s = \pm 1/2$ and is called spin magnetic quantum number. The introduction of electron spin means that a total of four quantum numbers n, l, m_l and m_s is needed to describe each possible state of an atomic electron.

3.2 Spin-Orbit Interaction

Consider an electron of charge –e moving in a Bohr orbit around a nucleus of charge +Ze. Let the velocity of the electron relative to the nucleus be **v** and its position relative to the nucleus be **r**. Since the motion is relative we may consider the electron to be at rest, while the nucleus is going round the electron in a circle of radius r and velocity -**v** (Fig.3.1).

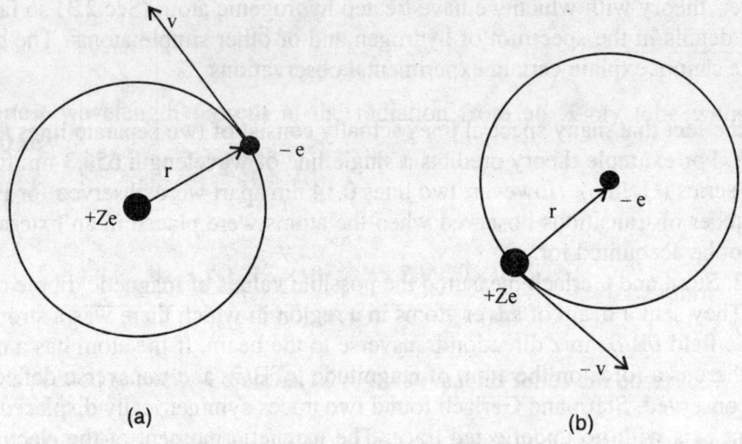

**Fig. 3.1 (a) An electron moves in Bohr's circular orbit, the motion is seen by the nucleus
(b) the same motion, but as seen by the electron. From the point of view of the
electron, the nucleus moves around it**

The motion of the nucleus constitutes a current loop, which produces a magnetic field, and electron is located inside this current loop. The current element **j** because of motion of nucleus is

$$\mathbf{j} = -Ze\mathbf{v} \tag{3.3}$$

According to Ampere's law this produces a magnetic field **B** that at a distance r is

$$\mathbf{B} = \frac{\mu_0}{4\pi}\frac{\mathbf{j}\times\mathbf{r}}{r^3} = -\frac{Ze\mu_0}{4\pi}\frac{\mathbf{v}\times\mathbf{r}}{r^3} \tag{3.4}$$

According to Coulomb law the force between electron of charge –e and nucleus of charge +Ze is given by Eq. (2.1) and electric field is

$$\mathbf{E} = \frac{Ze\mathbf{r}}{4\pi\varepsilon_0\, r^3} \tag{3.5}$$

From Eqs. (3.4) and (3.5), we have

$$\mathbf{B} = -\varepsilon_0 \mu_0 \mathbf{v} \times \mathbf{E} = -\frac{\mathbf{v} \times \mathbf{E}}{c^2} \tag{3.6}$$

Since $c^2 = 1/\mu_0 \varepsilon_0$. The quantity $\mathbf{B}$ is magnetic field strength experienced by the electron when it is moving with velocity $\mathbf{v}$ relative to the nucleus, and therefore through the electric field of strength $\mathbf{E}$ which the nucleus exert on it. The spin-orbit interaction energy ΔE results from the interaction between the magnetic field $\mathbf{B}$ and spin magnetic moment

$$\mu_s = -\frac{g_s e \mathbf{s}}{2m_e} = -\frac{2\pi g_s \beta_e \mathbf{s}}{h} \tag{3.7}$$

The orientational potential energy of spin magnetic moment in this magnetic field is

$$\Delta E = -\mu_s \cdot \mathbf{B} \tag{3.8}$$

Using Eq. (3.7)

$$\Delta E = g_s \beta_e \frac{2\pi \mathbf{s} \cdot \mathbf{B}}{h} \tag{3.9}$$

But the energy has been evaluated in a frame of reference in which the electron is at rest, but what we are really interested in is its value in the frame of reference in which the nucleus is at rest. Transforming back to frame of reference in which nucleus is at rest results in a reduction of the orientational potential energy by a factor of two. Thus, spin-orbit interaction energy given by Eq. (3.9) changes to

$$\Delta E = g_s \beta_e \frac{\pi \mathbf{s} \cdot \mathbf{B}}{h} \tag{3.10}$$

Putting $\mathbf{F} = -e\mathbf{E}$ in Eq.(3.6) we have

$$\mathbf{B} \doteq \frac{\mathbf{v} \times \mathbf{F}}{ec^2} \tag{3.11}$$

Putting $\mathbf{F} = -\dfrac{dV}{dr}\dfrac{\mathbf{r}}{r}$ in Eq.(3.11)

$$\mathbf{B} = -\frac{\mathbf{v} \times \mathbf{r}}{ec^2 r}\frac{dV}{dr} \tag{3.12}$$

Multiplying and dividing Eq. (3.12) by electron mass m_e, we have

$$\mathbf{B} = \frac{\mathbf{r} \times \mathbf{p}}{m_e ec^2 r}\frac{dV}{dr}$$

with $\mathbf{l} = \mathbf{r} \times \mathbf{p}$, the above equation reduces to

$$B = \frac{\mathbf{l}}{m_e e c^2 r} \frac{dV}{dr}$$

(3.13)

Magnetic field experienced by the electron because of its motion about the nucleus with orbital angular momentum $\mathbf{l}$ is proportional to the magnitude of $\mathbf{l}$ and also that $\mathbf{B}$ is in the same direction as $\mathbf{l}$. Substituting the value of $\mathbf{B}$ from Eq. (3.13) in Eq. (3.10), the spin orbit interaction energy is

$$\Delta E = \frac{2\pi \beta_e}{m_e e\, hc^2 r} \frac{dV}{dr} \mathbf{l} \cdot \mathbf{s}$$

(3.14)

energy in wavenumbers is thus

$$\Gamma = \frac{\Delta E}{hc} = \frac{1}{2m_e^2 hc^3 r} \frac{dV}{dr} \mathbf{l} \cdot \mathbf{s} = a\,\mathbf{l} \cdot \mathbf{s}$$

(3.15)

where a is a constant (except for its dependence on r) given by

$$\frac{1}{2m_e^2 hc^3 r} \frac{dV}{dr}$$

3.3 The Vector Model for Atoms

Vector models conveniently picture the relationship between the angular momenta and their components in atoms. In quantum mechanics we have the commutation relation among the components of angular momentum $\mathbf{l}$ as

$$[l_x, l_y] = i\,(h/2\pi)l_z\,;[l_y, l_z] = i\,(h/2\pi)l_x\,;[l_z, l_x] = i\,(h/2\pi)l_y$$

(3.16)

we can say that l_x, l_y and l_z do not commute. If however, one forms the magnitude of $\mathbf{l}$ i.e.

$$l^2 = l_x^2 + l_y^2 + l_z^2$$

then l^2 commute with l_x, l_y and l_z . Thus l^2 and one of its components can be precisely measured simultaneously. In classical mechanics the angular momentum $\mathbf{l}$ is a vector that can have fixed components l_x, l_y and l_z . According to quantum mechanical commutation rules, only one of three components can be fixed. The other two components must vary with time so that l_x and l_y cannot be quantized if l_z is quantized.

The general result for angular momentum is summarized in what is known as the vector model. The vector model diagram of a single angular momentum vector $\mathbf{l}$ is shown in Fig. 3.2. In this figure, $\mathbf{l}$ appears to precess around z-axis, if it did not precess l_x and l_y could be quantized and thus would violate commutation rules. The vector $\mathbf{l}$ has a length given by $\hbar\,[l\,(l+1)]^{1/2}$ where l is an integer and the projection m_l of $\mathbf{l}$ along the z axis can never be perfectly aligned along the z axis.

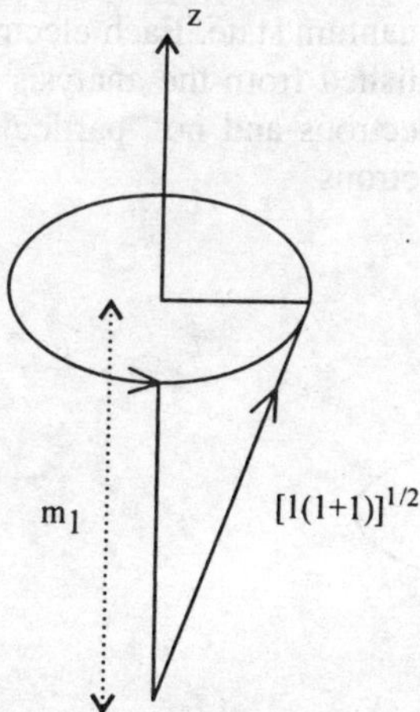

**Fig. 3.2 Precession of orbital angular momentum l
about the axis of quantization z**

If the angular momentum is only due to spins, then **s** having a magnitude $\hbar\,[s(s+1)]^{1/2}$ will precess about the z axis. The projection of **s** along the z-axis is given by m_s (Fig.3.3). Let us consider two angular momenta **l** and **s**, which satisfy the commutation rules. If there is no coupling between **l** and **s**;

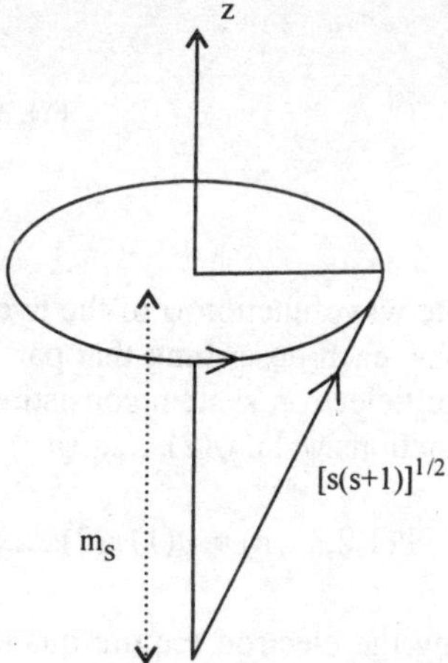

Fig. 3.3 Precession of spin angular momentum s about the axis of quantization z

the vector corresponding to **l** and **s** precess independently about z-axis with their projection m_l and m_s respectively on the z-axis as shown in Fig. 3.4. On the other hand **l** and **s** could precess together and add to form a resultant angular momentum **j**. The only projection along z-axis in this case would be m_j as shown in Fig.3.5.

3.4 Pauli Exclusion Principle

In 1925, W. Pauli discovered the fundamental principle that governs the electron configurations of multielectron atoms. His exclusion principle states that "in a multielectron atom there can never be more than one electron in the same quantum state. Each electron must have a different set of quantum numbers n, l, m_l and m_s". He established from the analysis of experimental data that the exclusion principle represents a property of electrons and not, particularly, of atoms. The exclusion principle operates in any system containing electrons.

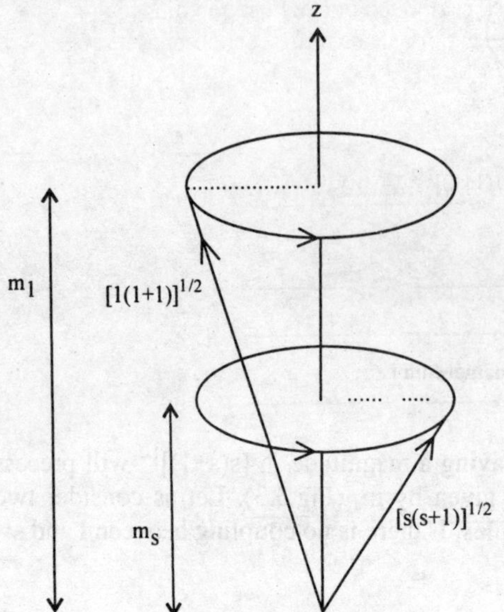

Fig. 3.4 Precession motion of orbital
and spin angular momenta
when they are uncoupled
about the quantization axis z

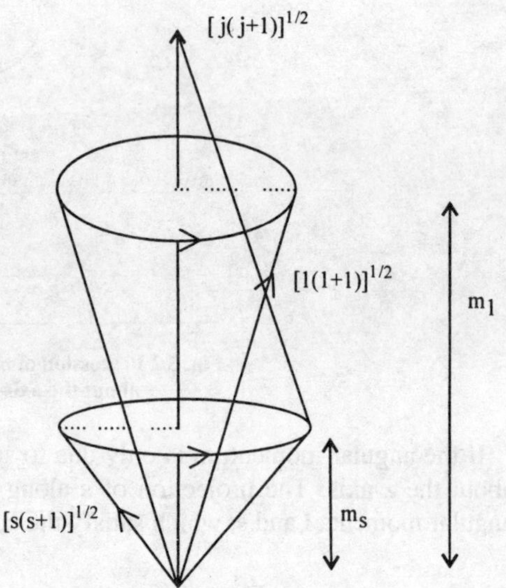

Fig. 3.5 Precession motion of orbital
and spin angular momenta
when they are coupled about
their resultant j

In sec.2.3, it is seen that the complete wavefunction ψ of the hydrogen atom can be expressed as the product of three separate wavefunctions, each describing that part of ψ, which is a function of one of the three, coordinates r, θ and φ. A multielectron system consisting of n non interacting electrons can be expressed as the product of wavefunctions $\psi(1)$, $\psi(2)$, ... $\psi(n)$ of the individual electrons, that is

$$\Psi(1,2,...., n) = \psi(1)\psi(2)...........\psi(n) \tag{3.17}$$

Each of the eigenfunction describing the electron require quantum numbers n, l, m_l to specify the mathematical form of its dependence on the three coordinates. In addition each require one more quantum number m_s to specify the orientation of the spin of the electron. To designate a particular set of four quantum numbers, the symbols such as a, b, c,.. etc are used. Let us consider a wavefunction used to describe a system of two electrons. Suppose electron number 1 is in quantum state a and electron number 2 is in state b. The wavefunction is

$$\Psi_1 = \psi_a(1)\psi_b(2) \tag{3.18}$$

Because the electrons are identical, there is no physical way to distinguish the electron wavefunction given by Eq. (3.18) from the wavefunction

$$\Psi_{II} = \psi_a(2)\psi_b(1) \tag{3.19}$$

in which electron number 2 now has quantum state a, etc. Similarly no conceivable physical experiment could distinguish the six three electron wavefunctions:

$$\psi_a(1)\psi_b(2)\psi_c(3); \psi_a(3)\psi_b(1)\psi_c(2); \psi_a(2)\psi_b(3)\psi_c(1);$$
$$\psi_a(2)\psi_b(1)\psi_c(3); \psi_a(3)\psi_b(2)\psi_c(1); \psi_a(1)\psi_b(3)\psi_c(2) \tag{3.20}$$

Although we might determine that an atom had one electron in state a, another in state b, and a third in state c and so on, but we could not say that a particular electron (say the first electron) was in state a etc.

For two electron system consider linear combinations of Eqs. (3.18) and (3.19)

$$\Psi_S = \sqrt{\frac{1}{2!}}[\psi_a(1)\psi_b(2) + \psi_a(2)\psi_b(1)] \tag{3.21}$$

$$\Psi_A = \sqrt{\frac{1}{2!}}[\psi_a(1)\psi_b(2) - \psi_a(2)\psi_b(1)] \tag{3.22}$$

the factor $(1/2!)^{1/2}$ is needed to normalize Ψ_S and Ψ_A. The function Ψ_S is symmetric with respect to interchange of electron labels i.e.$(1 \rightarrow 2,$ and $2 \rightarrow 1)$ in Eq.(3.21) gives

$$\Psi_S \rightarrow \sqrt{\frac{1}{2!}}[\psi_a(1)\psi_b(2) + \psi_a(2)\psi_b(1)] = \Psi_S \tag{3.23}$$

The function Ψ_A is antisymmetric with respect to interchange of electron labels i.e. $(1 \rightarrow 2,$ and $2 \rightarrow 1)$ in Eq. (3.22) gives

$$\Psi_A \rightarrow \sqrt{\frac{1}{2!}}[\psi_a(1)\psi_b(2) - \psi_a(2)\psi_b(1)] = -\Psi_A \tag{3.24}$$

When the state a is identical to state b, then the antisymmetric wavefuction given by Eq. (3.22) is identically zero i.e.

$$\Psi_A \rightarrow \sqrt{\frac{1}{2!}}[\psi_a(1)\psi_b(2) - \psi_a(2)\psi_b(1)] = 0 \tag{3.25}$$

and symmetric wavefunction given by Eq.(3.21) becomes

$$\Psi_S \rightarrow \sqrt{\frac{1}{2!}}[\psi_a(1)\psi_b(2) + \psi_a(2)\psi_b(1)] = \sqrt{2}\psi_a(1)\psi_a(2) \tag{3.26}$$

Thus antisymmetric wavefunction automatically satisfies Pauli exclusion principle. The Pauli exclusion principle can also be stated as a system containing several electrons must be described by an antisymmetric total wavefunction.

The antisymmetric wavefunction (Eq.3.22) can also be written in Slater determinant form

$$\Psi_A = \sqrt{\frac{1}{2!}} \begin{vmatrix} \psi_a(1) & \psi_a(2) \\ \psi_b(1) & \psi_b(2) \end{vmatrix} \tag{3.27}$$

Similarly, antisymmetric wavefunction for a system of three electrons can be written in Slater determinant form

$$\Psi_A = \sqrt{\frac{1}{3!}} \begin{vmatrix} \psi_a(1) & \psi_a(2) & \psi_a(3) \\ \psi_b(1) & \psi_b(2) & \psi_b(3) \\ \psi_c(1) & \psi_c(2) & \psi_c(3) \end{vmatrix} \tag{3.28}$$

With the help of Pauli's exclusion principle we can assign different quantum states to the electron in a given atom. For an atom, the electrons that have the same principal quantum number n are said to be in the same shell. For a given n, the electrons having the same value of l are said to be in the same subshell. Now we can calculate the maximum number of electrons belonging to the same shell or subshell.

(i) The case of a subshell

For electrons in a subshell we have same value of quantum numbers n and l. These electrons must differ either by the value of quantum number m_l (which can be one of the 2l+1 integral values between –l and +l) or by quantum number m_s (which can take values +1/2 or –1/2). There exists, therefore, 2(2l+1) distinct quantum states corresponding to the same value of n and l and therefore, there can be 2(2l+1) electrons in a subshell of quantum number l. The maximum number of electrons in s, p, d and f subshell are therefore 2, 6, 10 and 14, respectively. A subshell containing 2(2l+1) electrons is said to be complete.

(ii) The case of a shell

For electrons in a shell we have same values of quantum number n but different quantum numbers l, m_l and m_s . The quantum numbers l can have all values from l to n -1. The maximum number of electrons can be obtained by adding the maximum number of electrons in each subshell .

s electron (l=0) +	p electron (l = 1)	+ ....+	Electron with l = n-1
Maximum number	Maximum number		Maximum number
2	6		2(2n-1)

The series may be written as

$$2(2l+1) = 2+6+10+.......+2(2n-1) = 2n^2 \tag{3.29}$$

The maximum number of electrons of quantum number n is therefore $2n^2$. The maximum numbers of electrons in various shells are given in Table 3.1

A shell containing the maximum number of electrons is called a complete shell. We may be tempted to say that electron configuration of any atom follows the general rule

Table 3.1 Number of electrons in various shells

Shell	N	Maximum number of electrons
K	1	2
L	2	8
M	3	18
N	4	32

$$1s^2 2s^2 2p^6 3s^2 3p^6 3d^{10} 4s^2 4p^6 4d^{10} 4f^{14} \qquad (3.30)$$

But this is not true. The actual order in which the levels must be filled so that the resulting energy is minimum (corresponding to a stable atom) is as follows:

$$1s^2 2s^2 2p^6 3s^2 3p^6 4s^2 3d^{10} 4p^6 5s^2 4d^{10} 5p^6 6s^2 4f^{14} 5d^{10} 6p^6 7s^2 6d^{10} \qquad (3.31)$$

A given electron configuration is specified by quantum numbers n and l for each electron but not by m_l and m_s. Therefore, to each single configuration these will correspond a certain number of descriptions, differing in the values of m_l and m_s of each electron. A configuration will therefore have certain degeneracy.

Let us consider a configuration in which a single electron is in each subshell. No electron is characterised by the same quantum number n and l. Thus an electron i may have X_i states according to the values of m_l and m_s. X_i represents the number of places in the subshell with $X_i = 2(2l+1)$. The number of different states corresponding to a configuration will therefore be

$$g = \prod_i X_i \qquad (3.32)$$

Now consider a configuration in which q electrons have the same quantum numbers n and l . For this value of l an electron can have $r = 2(2l+1)$ different states characterised by different values of m_l and m_s. Now we have to put q electrons in r places each containing a maximum of one electron. The number of distinct combinations is

$$g = \frac{r!}{q!(r-q)!} \qquad (3.33)$$

The number g represents the number of different states corresponding to q electrons. The total degeneracy of the configuration is obtained from the product of g and the degeneracy due to electrons in other subshell and is

$$G = \prod_i g_i \qquad (3.34)$$

3.5 Angular Momenta and Magnetic Moment of Atoms

Every electron in an atom has two possible kinds of angular momenta, one due to orbital and other due to its spin motion. The magnitude of the orbital angular momentum vector for a single electron is given by

$$l = (h/2\pi)[l(l+1)]^{1/2} \qquad \text{where } l = 0,1,2,3,\ldots\ldots, n\text{-}1$$

Similarly the magnitude of the spin-angular momentum vector for a single electron is

$$s = (h/2\pi)[s(s+1)]^{1/2} \qquad \text{where } s = 1/2$$

For an electron which has both orbital and spin angular momentum there is a quantum number j associated with the total (orbital plus spin) angular momentum. This is also a vector quantity whose magnitude is given by

$$j = (h/2\pi)[j(j+1)]^{1/2} \tag{3.35}$$

where j can take the values

$$j = l+s, l+s-1, \ldots\ldots\ldots |l-s| \tag{3.36}$$

The number of possible values of j is equal to the smaller of the two numbers 2s+1 and 2l+1. The angular momentum **l** and **s** interact magnetically. If there is no magnetic field, the total angular momentum **j** is conserved in magnitude and direction, and, **l** and **s** precess around the direction of their resultant **j**.

(i) orbital magnetic moment

Consider an electron moving with velocity **v** in circular Bohr orbit of radius r. A charge circulating in a loop constitutes a current of magnitude

$$i = \frac{e}{T} = \frac{ev}{2\pi r} \tag{3.37}$$

where $T = 2\pi r/v$ is the orbital period of the electron. From electromagnetic theory it is known that at large distance from the loop, the magnetic field due to the loop is the same as that of a magnetic dipole located at the centre of the loop. For a current i in a loop of area A, the magnitude of the orbital magnetic dipole moment μ of the equivalent dipole is

$$\mu = iA = \frac{evr}{2} \tag{3.38}$$

The direction of the magnetic dipole moment is perpendicular to the plane of the orbit.

The angular momentum **l** of electron moving in a circular orbit is **r x p**, where **r** is the position vector of the electron and **p** is its momentum. For circular motion **r** and **p** are perpendicular, hence

$$l = rp = m_e rv \tag{3.39}$$

From Eqs. (3.38) and (3.39)

$$\frac{\mu_l}{l} = \frac{e}{2m_e} \tag{3.40}$$

$$\frac{\mu_l}{l} = \frac{2\pi\, g_l \beta_e}{h} \tag{3.41}$$

where $\beta_e = e\,\hbar/2m_e = 0.927 \times 10^{-23}$ amp-m^2 = 0.927×10^{-20} erg /gauss is called the Bohr magneton and $g_l = 1$. g_l is called orbital g factor. The ratio of μ_l to l does not depend on the size of the orbit or on the orbital frequency. Since l is quantized and equal to $\hbar\,[l\,(l+1)]^{1/2}$, hence

$$\mu_l = \beta_e g_l [l(l+1)]^{1/2} \tag{3.42}$$

Since electron charge is negative, the Eq. (3.40) is written as

$$\mu_l = -\frac{el}{2m_e} = -\frac{2\pi\, g_l \beta_e l}{h} \tag{3.43}$$

because of the negative charge of the electron, the direction of μ_l is opposite to that of l.

(ii) spin magnetic moment

The classical picture of an electron spinning on its own axis indicates that there is a magnetic moment associated with this angular momentum also. The magnetic moment due to spin is given by

$$\mu_s = -\frac{2\pi\, g_s \beta_e s}{h} = -g_s \beta_e [s(s+1)]^{1/2} \tag{3.44}$$

where μ_s is the magnetic moment due to electron spin, g_s is simply referred to as the g value of the electron. The value of g_s was through early experiments, to be exactly 2 but its value was found to be 2.0023.

3.6 Coupling of Angular Momenta

The process of combining two angular momenta into a resultant angular momentum is generally called coupling. For one electron the total angular momentum is obtained in just one way by forming the resultant of the spin and orbital angular momentum. A configuration of two electrons with orbital angular momentum l_1, l_2 and spin angular momentum s_1, s_2 allows several way of arriving at the final resultant of the four vectors l_1, s_1, l_2 and s_2. The four vectors can be combined in pairs in six possible ways i.e. (l_1, s_1); (l_2, s_2); (l_1, l_2); (s_1, s_2); (l_1, s_2) and (l_2, s_1). Out of these last two are negligible. The spin-orbit interaction (l_1, s_1) and (l_2, s_2) are magnetic while the interaction (l_1, l_2) and (s_1, s_2) are due mainly to electrostatic force which for (s_1, s_2) are the so-called exchange type. The magnetic spin spin interaction is very small, and its influence is noticeable only in the lightest atoms.

(i) ll coupling

The ll coupling is due to electrostatic interaction between the orbital electrons. Let l_1 and l_2 be the orbital quantum numbers of two electrons with vector $l_1 = [l_1(l_1+1)]^{1/2}$ and $l_2 = [l_2(l_2+1)]^{1/2}$, respectively representing the orbital angular momenta. These vectors add to give a resultant L for the atom with

$$L = l_1 + l_2, l_1 + l_2 - 1, \ldots\ldots\ldots, |\, l_1 - l_2 \,| \qquad (3.45)$$

Consider for example, one electron in p orbit and other in f orbit. For p electron l=1 and for f electron l=3. According to Eq. (3.45) L can have values 4,3 and 2. The length of l_1 and l_2 (in units of $\hbar$) are $(2)^{1/2}$ and $(12)^{1/2}$. The length of L is (in units of $\hbar$): $(20)^{1/2}$, $(12)^{1/2}$ and $(6)^{1/2}$ corresponding to values of L = 4,3 and 2, respectively. l_1, l_2 and L are shown in the vector diagram (Fig.3.6). To construct

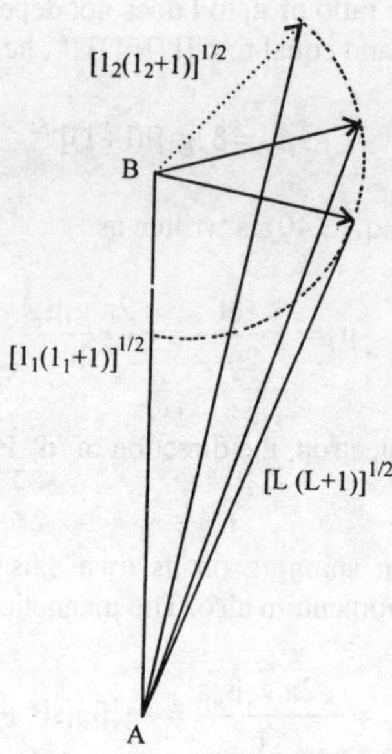

Fig. 3.6 Vector diagram of the space quantization of the orbital motion of the pf configuration

the vector diagram the following procedure is followed. We draw an arrow AB having a length proportional to the length of larger of l i.e.l_2. Now a circle of radius proportional to l_1, i.e. $(2)^{1/2}$ is drawn with B as centre. Now with A as centre the circle of radii $(20)^{1/2}$, $(12)^{1/2}$ and $(6)^{1/2}$, respectively are drawn . The interaction of these circles with one having radius $(2)^{1/2}$ with centre at B are the end of L vector starting from A.

(ii) ss coupling

This coupling arises from electrostatic interaction between the electrons. The **s** vectors add together in a manner analogous to **l** vectors, the only difference being that s has only the value 1/2   having the magnitude $\hbar\,(3/4)^{1/2}$. Two electrons combine to give **S**, the values 0, $\hbar\,(2)^{1/2}$ corresponding to S = 0 and S = 1, respectively. For n electrons, the possible 2S+1 values are n + 1, n − 1, n- 3,...., 2 or 1 depending on whether n is odd or even, respectively, thus giving S either half integer or integral values. In multielectron atoms S values are given by

$$S = s_1 + s_2 + \ldots, s_1 + s_2 + \ldots - 1, s_1 + s_2 + \ldots -2., \ldots\ldots . \qquad (3.46)$$

(iii) LS coupling

The approximation in which the residual electrostatic interaction between electrons is assumed to be large compared with the spin orbit interaction is called LS coupling or Russell-Saunders coupling and is important for light atoms.

For two electrons in LS coupling, l_1 and l_2 form a vector resultant $\mathbf{L}$ about which they are considered to be precessing rapidly; similarly $\mathbf{s}_1$ and $\mathbf{s}_2$ precess rapidly about their resultant $\mathbf{S}$. The precession of $\mathbf{L}$ and $\mathbf{S}$ about their resultant $\mathbf{J}$ is much slower, corresponding to the assumption that the spin-orbit energy splitting is much smaller than the splitting due to electrostatic interaction (with exchange effects). That is to say the classical motion of the orbital system $L = l_1 + l_2$ is nearly independent of that of the spin system $S = s_1 + s_2$ and L and S represent constant of the motion in this approximation. The classical vector model set up in this dynamical way, is concerned with time averages: the component of $l_1(l_2)$ perpendicular to L averages to zero over the many cycles of the rapid precession of l_1 (l_2) about L which take place during one cycle of the slow precession of L about J. Thus only the component of l_1 (l_2) lying along L is taken into consideration. Similarly for s_1 and s_2 about S. The $\mathbf{S}$ and $\mathbf{L}$ couples together to form $\mathbf{J} = \mathbf{L} + \mathbf{S}$ with

$$J = L+S, L+S-1, \dots\dots\dots, |L-S|$$

If several electrons are present in an atom, the LS coupling can be written symbolically as

$$(\mathbf{s}_1, \mathbf{s}_2, \mathbf{s}_3, \dots)(\mathbf{l}_1, \mathbf{l}_2, \mathbf{l}_3, \dots\dots) = (\mathbf{S}, \mathbf{L}) = \mathbf{J}$$

The vector model diagram for a system of two electrons in LS coupling is shown in Fig.3.7. We will now calculate the interaction energy due to combinations (s_1, s_2); (l_1, l_2); (l_1, s_1) and (l_2, s_2). Applying Eq.(3.15) to these four combinations, the four energy relations are:

$$\begin{aligned}
\Gamma_1 &= a_1 \mathbf{s}_1 \cdot \mathbf{s}_2 \\
\Gamma_2 &= a_2 \mathbf{l}_1 \cdot \mathbf{l}_2 \\
\Gamma_3 &= a_3 \mathbf{l}_1 \cdot \mathbf{s}_1 \\
\Gamma_4 &= a_4 \mathbf{l}_2 \cdot \mathbf{s}_2
\end{aligned} \tag{3.47}$$

In LS coupling Γ_1 and Γ_2 predominate. s_1, s_2 and l_1, l_2 precess rapidly around their respective, resultant S and L keeping the angles between each resultant vector and its component constant. However, angles $(\mathbf{l}_1, \mathbf{s}_1)$ and $(\mathbf{l}_2, \mathbf{s}_2)$ are not constant, we therefore, take average value of these angles and hence expressions for Γ_3 and Γ_4 can be written as

$$\Gamma_3 = a_3 l_1 s_1 \overline{\cos\ (\mathbf{l}_1, \mathbf{s}_1)} \tag{3.48}$$

$$\Gamma_4 = a_4 l_2 s_2 \overline{\cos\ (\mathbf{l}_2, \mathbf{s}_2)} \tag{3.49}$$

From $S = \mathbf{s}_1 + \mathbf{s}_2$ we have

$$S^2 = s_1^2 + s_2^2 + 2\mathbf{s}_1 \cdot \mathbf{s}_2$$

$$\mathbf{s}_1 \cdot \mathbf{s}_2 = \frac{1}{2}[S^2 - s_1^2 - s_2^2] = \frac{1}{2}[S(S+1) - s_1(s_1+1) - s_2(s_2+1)] \tag{3.50}$$

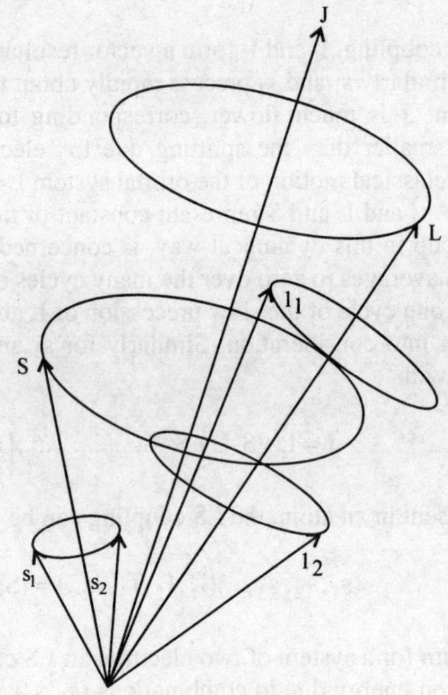

Fig. 3.7 The vector model: L -S coupling

From $L = l_1 + l_2$, we have

$$L^2 = l_1^2 + l_2^2 + 2\, l_1 \cdot l_2$$

$$l_1 \cdot l_2 = \frac{1}{2}[L^2 - l_1^2 - l_2^2] = \frac{1}{2}[L(L+1) - l_1(l_1+1) - l_2(l_2+1)] \tag{3.51}$$

Substituting these values of $s_1 . s_2$ and $l_1 . l_2$ from Eqs.(3.50) and (3.51) in Eq. (3.47) we have

$$\Gamma_1 = \frac{a_1}{2}[S(S+1) - s_1(s_1+1) - s_2(s_2+1)] \tag{3.52}$$

$$\Gamma_2 = \frac{a_2}{2}[L(L+1) - l_1(l_1+1) - l_2(l_2+1)] \tag{3.53}$$

For obtaining the values of Γ_3 and Γ_4 , the average values of the cosine must be evaluated since the angles between the vectors are continuously changing. The average values of their cosine may be written as equal to

$$\overline{\cos\,(l_1, s_1)} = \cos\,(l_1, L) \cos\,(L, S) \cos\,(S, s_1) \tag{3.54}$$

$$\overline{\cos(l_2, s_2)} = \cos(l_2, L)\cos(L, S)\cos(S, s_2) \tag{3.55}$$

Substituting the average values of the cosines from Eqs. (3.54) and (3.55) in Eqs.(3.48) and (3.49) we have

$$\Gamma_3 = a_3 l_1 s_1 \cos(l_1, L)\cos(L, S)\cos(S, s_1) \tag{3.56}$$

$$\Gamma_4 = a_4 l_2 s_2 \cos(l_2, L)\cos(L, S)\cos(S, s_2) \tag{3.57}$$

on adding Eqs. (3.56) and (3.57)

$$\Gamma_3 + \Gamma_4 = [a_3 l_1 \cos(l_1, L)s_1 \cos(S, s_1) + a_4 l_2 \cos(l_2, L)s_2 \cos(S, s_2)]\cos(L, S) \tag{3.58}$$

From $\mathbf{L} = \mathbf{l}_1 + \mathbf{l}_2$ we have $\mathbf{l}_1 = \mathbf{L} - \mathbf{l}_2$ and

$$l_1^2 = L^2 + l_2^2 - 2\mathbf{L} \cdot \mathbf{l}_2$$

$$\cos(L, l_2) = \frac{L^2 + l_2^2 - l_1^2}{2Ll_2} \tag{3.59}$$

Similarly

$$\cos(L, l_1) = \frac{L^2 + l_1^2 - l_2^2}{2Ll_1} \tag{3.60}$$

From $\mathbf{S} = \mathbf{s}_1 + \mathbf{s}_2$ we have

$$s_2^2 = S^2 + s_1^2 - 2\mathbf{S} \cdot \mathbf{s}_1$$

and

$$\cos(S, s_1) = \frac{S^2 + s_1^2 - s_2^2}{2Ss_2} \tag{3.61}$$

Similarly,

$$\cos(S, s_2) = \frac{S^2 + s_2^2 - s_1^2}{2Ss_1} \tag{3.62}$$

Substituting Eqs.(3.61) and (3.62) in Eq.(3.54) we have

$$\Gamma_3 + \Gamma_4 = LS\cos\ (\mathbf{L},\mathbf{S})\left\{\left[a_3\frac{\left(L^2 + l_1^2 - l_2^2\right)\left(S^2 + s_1^2 - s_2^2\right)}{4L^2 S^2}\right] + \left[a_4\frac{\left(L^2 + l_2^2 - l_1^2\right)\left(S^2 + s_2^2 - s_1^2\right)}{4L^2 S^2}\right]\right\} \quad (3.63)$$

$$\Gamma_3 + \Gamma_4 = A\mathbf{L}\cdot\mathbf{S} \quad (3.64)$$

where

$$A = \left\{\left[a_3\frac{\left(L^2 + l_1^2 - l_2^2\right)\left(S^2 + s_1^2 - s_2^2\right)}{4L^2 S^2}\right] + \left[a_4\frac{\left(L^2 + l_2^2 - l_1^2\right)\left(S^2 + s_2^2 - s_1^2\right)}{4L^2 S^2}\right]\right\}$$

$$(3.65)$$

From $\mathbf{J} = \mathbf{L} + \mathbf{S}$ we have

$$J^2 = L^2 + S^2 + 2\mathbf{L}\cdot\mathbf{S}$$

$$\mathbf{L}\cdot\mathbf{S} = \frac{1}{2}[J^2 - L^2 - S^2] = \frac{1}{2}[J(J+1) - L(L+1) - S(S+1)] \quad (3.66)$$

Substituting Eq. (3.66) in Eq. (3.64) we have

$$\Gamma_3 + \Gamma_4 = A\mathbf{L}\cdot\mathbf{S} = \frac{A}{2}[J(J+1) - L(L+1) - S(S+1)] \quad (3.67)$$

The Eq. (3.67) expresses symbolically as Landé interval rule, which states that "the separation between pairs of consecutive levels in a multiplet are proportional to the larger of the J value characterising the pair of levels".

(iv) jj coupling

In this coupling, spin-orbit interaction for an electron is large compared with the electrostatic interaction between electrons On the vector model, this is obtained by coupling together $\mathbf{s}$ and $\mathbf{l}$ for each individual electron to give possible values for their resultant $\mathbf{j}$ and then coupling the $\mathbf{j}$ vectors to give a resultant $\mathbf{J}$. For two electron atom the resultant $\mathbf{J} = \mathbf{j}_1 + \mathbf{j}_2$ such that

$$J = j_1 + j_2, j_1 + j_2 - 1,\ldots\ldots\ldots, |\, j_1 - j_2\,|$$

This coupling is prominent in atoms having large atomic number Z .

In the jj coupling it was assumed that the interaction between spin of each electron and its own orbit is greater than the interaction between two spins and the two orbits, respectively. Since the angles between $(s_1\,,\,s_2)$ and $(l_1,\,l_2)$ are continually changing, the cosines in Γ_1 and Γ_2 must be averaged . Proceeding in exactly the same way as for LS coupling we have terms

$$\Gamma_1 + \Gamma_2 = \frac{A'}{2}[J(J+1) - j_1(j_1+1) - j_2(j_2+1)] \quad (3.68)$$

where

$$A' = \left\{ a_1 \left[\frac{j_1(j_1+1)+s_1(s_1+1)-l_1(l_1+1)}{2j_1(j_1+1)} \right] \left[\frac{j_2(j_2+1)+s_2(s_2+1)-l_2(l_2+1)}{2j_2(j_2+1)} \right] \right\} +$$

$$+ \left\{ a_2 \left[\frac{j_1(j_1+1)+l_1(l_1+1)-s_1(s_1+1)}{2j_1(j_1+1)} \right] \left[\frac{j_2(j_2+1)+l_2(l_2+1)-s_2(s_2+1)}{2j_2(j_2+1)} \right] \right\}$$

and

$$\Gamma_3 = \frac{a_3}{2}[j_1(j_1+1)-l_1(l_1+1)-s_1(s_1+1)] \tag{3.69}$$

$$\Gamma_4 = \frac{a_4}{2}[j_2(j_2+1)-l_2(l_2+1)-s_2(s_2+1)] \tag{3.70}$$

3.7 Term Symbols

Terms are the designation of atomic states described by the L and S values. Term symbols are written as

$$^{2S+1}L$$

where L is total orbital angular quantum number. For each value of L there is a corresponding term designation as follows:

L	0	1	2	3	4	5	6	7	8	9	10
Term	S	P	D	F	G	H	I	K	L	M	N

S is the resultant spin quantum number of the atom, and 2S+1 is the spin multiplicity. The values of 2S+1 are as follows:

2S+1	1	2	3	4	5	6	7
	Singlet	Doublet	Triplet	Quartet	Quintet	Sextet	Septet

For designation of multiplet level we add J value to the term symbol as subscript and term is written as

$$^{2S+1}L_J$$

where J is the total angular momentum quantum number of atom and takes the values from L+S to L-S. For example, ^4F term correspond to the state with L = 3 and 2S+1 = 4 and therefore S=3/2. For ^4F term, L+S = 9/2 and L-S = 3/2. Hence for ^4F term there are four multiplet levels with J = 9/2, 7/2, 5/2, 3/2; that is, $^4F_{9/2}$, $^4F_{7/2}$, $^4F_{5/2}$, $^4F_{3/2}$. The spin multiplicity 2S+1 in the upper left hand corner indicates the number of these multiplet levels.

3.8 Term Derivation from Electron Configuration

In order to determine all the possible terms corresponding to a given configuration, the rules for the addition of angular momenta must be used.

(A) LS coupling

(a) Terms of Non Equivalent Electrons

Non equivalent electrons are electrons belonging to different values of n and l. For example 3s3p or 3s4s. In the former l values are different while in the latter n values are different. The dot between two electrons of the same type indicates that the electron have different total quantum numbers. Thus the non equivalent electrons can be written as e.g. as s.s, 3s4s, 3s3p or sp. Non equivalent electrons can not have the same set of four quantum numbers, therefore Pauli exclusion principle is automatically satisfied. The allowed values of the quantum number L and S are therefore obtained by adding the individual orbital angular momenta l_i of the electrons to form L and spin angular momenta s_i of these electrons to form S according to Eqs.(3.45) and (3.46).

Consider an example of pf electrons. For p electron l=1 and for f electron l = 3, therefore according to Eq.(3.45), L = 4,3,2. For each electron s = ½, therefore according to Eq.(3.46) , S = 0, 1. The possible terms are therefore,

$$^1D, {}^1F, {}^1G, {}^3D, {}^3F, {}^3G \tag{3.71}$$

Let consider the configuration pfd. Adding first the orbital angular momenta and spin of the two electron pf we find the term listed in Eq. (3.71). Now the third electron i.e. d electron has l = 2 and s =1/2. Using Eqs. (3.45) and (3.46) , we see that the addition of the d electron to the term ^1D give terms with L = 4,3,2,1,0 and S = 1/2 (For ^1D term ,l = 2 and s = 0 and for d electron l = 2 and s =1/2) namely ^2S, ^2P, ^2D, ^2F, ^2G. In the same way adding the d electron to the term ^1F yields ^{2}(PDFGH); to the term ^1G yields ^{2}(DFGHI); to ^3D yields ^{2,4}(SPDFG) ; to the term ^3F yields ^{2,4}(PDFGH); to the term ^3G yields ^{2,4}(DFGHI). These results may be summarised by writing that the term we have obtained are

^2S	^2P	^2D	^2F	^2G	^2H	^2I	^4S	^4P	^4D	^4F	^4G	^4H	^4I
2	4	6	6	6	4	2	1	2	3	3	3	2	1

where the number under the term symbol indicates the number of identical terms. The results can also be written as ^{2}[S(2), P(4), D(6), G(6), H(4), I(2)]; ^{4}[S, P(2), D(3), F(3), G(3), H(2), I] where number within parentheses indicates the number of identical terms.

(b) Terms of Equivalent Electrons

If electrons have the same total quantum number n and same quantum number l, they are called equivalent electrons. To obtain the terms, we have to use Pauli exclusion principle as n and l are same . Therefore m_l or m_s must be different. Terms for two equivalent electrons can be easily obtained by using Breit scheme. In this scheme the values of $(m_l)_1$ and $(m_l)_2$ are written down in a row and column as is in the array in Fig. 3.8 (a)

Consider the example of two equivalent p electrons i.e. p^2 .for p electrons l = 1 and m_l = 1,0,-1. To the left of $(m_l)_2 = 1$ and below $(m_l)_1 = 1$, the sum $M_L = 2$ is written. This process continued. Combining $(m_s)_1$ and $(m_s)_2$ to M_S in the same way the array shown in the Fig. 3.8(b) is constructed. If we permit the m_s values to be alike i.e. ($M_S = 1$ or -1), the values of $M_L = 2, 0, -2$ crossed out by the diagonal lines are forbidden (as these come from same values of m_l and hence Pauli exclusion principle is violated) Since the values of M_L in lower left of the array are identical with, and are mirror image of

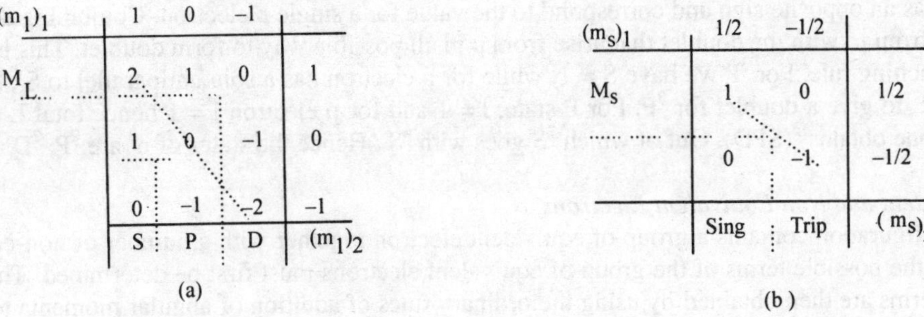

Fig. 3.8 Magnetic quantum numbers for p^2 electron configuration in LS coupling

those in the upper right half , one of these groups must also be eliminated from calculations. Leaving out lower half of the array the remaining values form sums

$$M_L = 1, 0, -1 \quad \text{with} \quad M_S = 1, -1 \tag{3.72}$$

If on the other hand, we permit m_l values to be alike, the values of $M_S = 1, -1$ are forbidden by Pauli principle. Since the two $M_S = 0$ values are identical, one of these must also be excluded. The remaining combinations are

$$M_L = 2, 1, 0, -1, -2; M_S = 0$$
$$M_L = 1, 0, -1; M_S = 0 \tag{3.73}$$
$$M_L = 0; M_S = 0$$

Combining Eqs.(3.72) and (3.73) we have (i) $M_L = 1, 0, -1$ and $M_S = 1, 0, -1$ (ii) $M_L = 2, 0, -2$ and $M_S = 0$ and (iii) $M_L = 0$ and $M_S = 0$. The combination (i) corresponds to $L = 1$ and $S = 1$ giving the term ^3P , combination (ii) corresponds to $L = 2$ and $S = 0$ giving the term ^1D and the combination (iii) corresponds to $L = 0$ and $S = 0$ which correspond to the term ^1S. Thus p^2 configuration gives the terms

$$p^2 \text{ configuration : terms are } ^1S, \, ^1D, \text{ and } ^3P \tag{3.74}$$

For three or more equivalent electrons, Breit scheme cannot be used. However, a method for obtaining terms for three or more equivalent electrons has been given by Gibbs, Wilber and White and also by Russel. Let us apply this method to determine the terms of three equivalent p electrons i.e. p^3 . For p electron m_l values are 1, 0, -1 and $m_s = 1/2$ and $-1/2$ i.e.

m_l	1	0	-1	1	0	-1
m_s	1/2	1/2	1/2	-1/2	-1/2	-1/2

Now we take only those combinations of number three at a time where all three m_s values are plus and those where all three are minus . these are (i) $m_l = 1, 0, -1$ with each having $m_s = 1/2$ (ii) $m_l = 1, 0, -1$ and each $m_s = -1/2$. The combination (i) and (ii) gives $M_L = 0$ with $M_S = -3/2$ and $-3/2$.These corresponds to parts of ^4S term. In obtaining this value we have taken the combinations where all the spins are up or down. Now in the rest of the combinations we have two spins up and one down or one spin up and two down. In each of these combinations , two m_s values are alike are considered alone, they will contain just the combination that gives triplet for two equivalent p electrons i.e. ^3P. The third electron has an opposite sign and correspond to the value for a single p electron. Combining the triplets that arise from p^2 with the doublet that arise from p in all possible way to form doublet. This is done by using branching rule. For ^3P we have S = 1, while for p electron has a spin antiparallel to S , hence net spin is 1/2 to give a doublet for ^3P. For P state, l = 1 and for p electron l = 1 hence total L = 2, 1, 0. Therefor one obtains ^{2}(SPD). Out of which ^2S goes with ^4S . Hence the states of p^3 are ^2P, ^2D and ^4S.

(c) Equivalent and Non Equivalent Electrons

If a configuration contains a group of equivalent electron together with a number of non-equivalent electrons, the possible terms of the group of equivalent electrons must first be determined. The overall possible terms are then obtained by using the ordinary rules of addition of angular momenta to add the nonequivalent electrons. Similarly, if a configuration contains two or more groups of equivalent electrons, the possible terms of each group must first be obtained, and the overall possible terms are then determined by using the ordinary rules for the addition of angular momenta.

The analysis carried out in this section has led to enumeration of the energy levels corresponding to particular configuration, but we have no information as to the position of these levels. Hund's rule partially fulfils this requirement by providing information about the structure of the levels whose energy is the lowest for a ground configuration with only equivalent electrons. These rules are:

(i)   of the terms arising from equivalent electrons those with the highest multiplicity lie lowest in energy.

(ii)   of these, the lowest is that with the highest value of L.

There are two further rules for ground term which tell us whether a multiplet arising from equivalent electron is normal or inverted.

(i)   normal multiplet (the component with the smallest value of J lies lowest in energy) arise from equivalent electrons when an incomplete subshell is less than half filled,

(ii)   inverted multiplet (the component with largest value of J lie lowest in energy) arises from equivalent electrons when an incomplete subshell is more than half filled.

(B) jj coupling

In this coupling the spin s_i of each electron is combined with its own l_i to form a resultant j_i such that j_i takes half integral values only . The two j's are in turn combined to form resultant J such that J takes integral values from $j_1 + j_2$ to$| j_1 - j_2 |$. In the case of jj coupling the notation for the spectral term must specify the quantum numbers $(n_i\ l_i\ j_i)$ of each electron and the total angular quantum number J. The values of the individual j_i's are usually written between parentheses and J as a subscript.

(a) Equivalent Electrons

In the case of equivalent electrons, it is necessary to take into account the effect of Pauli exclusion principle. The statement of Pauli exclusion principle for jj coupling is "no two electrons can have all the four quantum numbers n, l, j and m_j same". Let us consider two equivalent p electrons. The j values of two electrons are j_1 = 1/2, 3/2 and j_2= 1/2 and 3/2. Now for the same value of j, m_j cannot be same.

We can use Breit scheme for determining the states. We combine each of the two m_{j1} with each of two m_{j2} to obtain all possible combinations. We ignore those combinations having same values of m_{j1} and m_{j2} (restriction by Pauli exclusion principle). Of the four resultant arrays there will two like that lower right one in the Fig.3.9. Since the electrons are equivalent, one of these must be entirely excluded. In the lower left hand array, the diagonal and either upper right or lower left of the array must be excluded. The same is true for the lower centre array. The remaining combinations are just sufficient to form five terms $(3/2,3/2)_{2,0}$, $(1/2,1/2)_0$; $(1/2,3/2)_2$ and $(1/2,3/2)_2$. This corresponds to to the five term $^1S_0$, $^1D_2$, $^3P_0$, $^3P_1$ and $^3P_2$ in LS coupling. Total number of levels having a given value of J for a given electron configuration must be the same in LS and jj coupling.

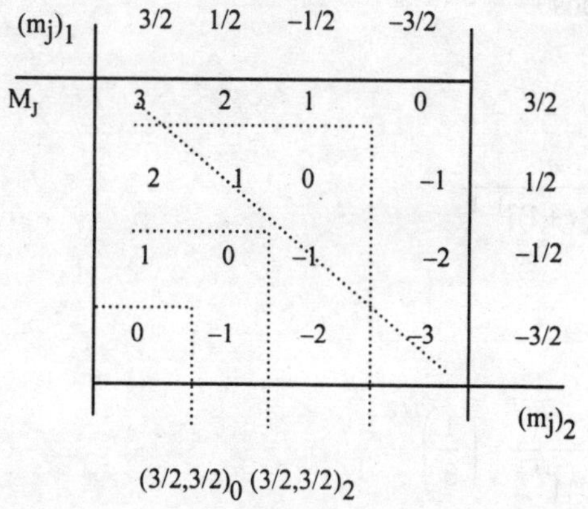

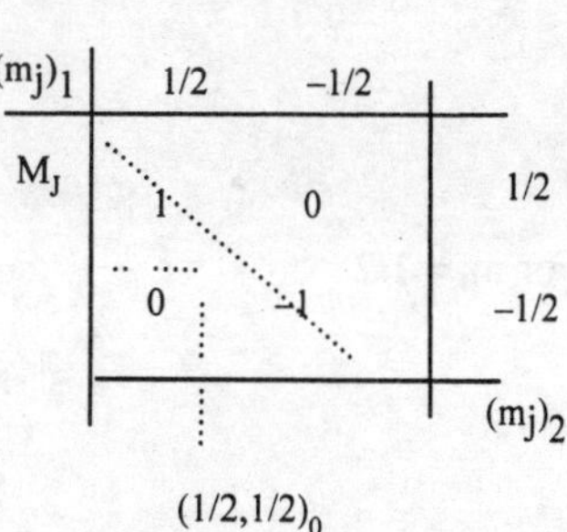

Fig. 3.9 Magnetic quantum number for p^2 electrons in jj coupling

(b) Nonequivalent Electrons

Let us consider pd configuration. Electrons p and d are designated as 1 and 2. For p electron $l_1=1$ and $s_1 =1/2$ and therefore $j_1 = 3/2, 1/2$. For d electron $l_2 = 2$ and $s_2 = 1/2$ and therefore $j_2 = 5/2$ and $3/2$. The j_1 and j_2 can combined in the following ways:

j_1	j_2	J	Term symbol
1/2	3/2	2,1	$(1/2, 3/2)_{2,1}$
1/2	5/2	3,2,1	$(1/2, 5/2)_{3,2,1}$
3/2	3/2	3,2,1,0	$(3/2, 7/2)_{3,2,1,0}$
3/2	5/2	4,3,2,1	$(3/2, 5/2)_{4,3,2,1}$

Examples

1.   Calculate the two possible orientations of the spin vector **s** with respect to a quantization axis.

Let z axis is the quantization axis. We have $\mathbf{s} = \hbar\,[s(s+1)]^{1/2}$ where $s = 1/2$ and

$$s_z = m_s\,\hbar \text{ with } m_s = \pm 1/2.$$

$$s_z = s\,\cos\theta$$

hence

$$\cos\theta = \frac{s_z}{s} = \frac{m_s}{[s(s+1)]^{1/2}}$$

For $m_s = 1/2$

$$\cos\theta_1 = \frac{\dfrac{1}{2}}{\left[\dfrac{1}{2}\left(\dfrac{1}{2}+1\right)\right]^{1/2}} = \left(\frac{1}{3}\right)^{1/2}$$

$$\theta_1 = 54.73^0$$

For $m_s = -1/2$

$$\cos\theta_2 = -\left(\frac{1}{3}\right)^{1/2}$$

$$\theta_2 = 125.25^0$$

m_s makes angle θ_1 and θ_2 with the z axis.

2.   Consider a p electron in a one-electron system. Calculate values of l, s, j and possible angle between **l** and **s**.

For p electron, l = 1, s = 1/2 and therefore j = 1/2, 3/2. The corresponding states are designated by $^2P_{3/2,1/2}$.

We have

$$\mathbf{j} = \mathbf{l} + \mathbf{s}$$

and

$$j^2 = (\mathbf{l}+\mathbf{s})\cdot(\mathbf{l}+\mathbf{s}) = l^2 + s^2 + 2\,\mathbf{l}\cdot\mathbf{s}$$

$$\cos\,(\mathbf{l},\mathbf{s}) = \frac{j^2 - l^2 - s^2}{2} = \frac{j(j+1) - l(l+1) - s(s+1)}{2[l(l+1)]^{1/2}[s(s+1)]^{1/2}}$$

for j = 3/2, l=1, s=1/2 we have

$$\cos\theta = \cos\,(\mathbf{l},\mathbf{s}) = \frac{\dfrac{15}{4} - 2 - \dfrac{3}{4}}{2(2)^{1/2}(3/4)^{1/2}} = \left(\frac{1}{6}\right)^{1/2}$$

$$\theta = 66^0$$

for j = 1/2, l=1, s=1/2 we have

$$\cos\theta = -2\left(\frac{1}{6}\right)^{1/2}$$

$$\theta = 145^0$$

3. Consider the $D_{5/2}$ state of the electron. Calculate (a) the possible values of m_j and j_z (b) the different possible orientation of j vector in space.

(a) For $D_{5/2}$, l = 2, s =1/2, j =5/2. Therefore mj = 5/2, 3/2, 1/2, -1/2, -3/2, -5/2. and $j_z = m_j\,\hbar = \hbar\,(\pm$ 5/2, ± 3/2, ± 1/2) i.e there are six possible values of m_j for j =5/2.

(b) Now j_z is the projection of j on the z axis. If angle between z axis and j is θ then

$$j_z = j\cos\theta$$

$$m_j(h/2\pi) = (h/2\pi)\,[j(j+1)]^{1/2}\cos\theta$$

$$\cos\theta = \frac{m_j}{[j(j+1)]^{1/2}}$$

for j = 5/2, we have

$$[j(j+1)]^{1/2} = (35/4)^{1/2}$$

for m_j = 5/2,

$$\cos\theta = \frac{(5/2)}{(35/4)^{1/2}} = 0.85$$

$$\theta = 32^0$$

Similarly

m_j	$\cos\theta$	θ
3/2	0.51	59.5^0
1/2	0.17	79.8^0
-1/2	-0.17	100.2^0
-3/2	-0.51	120.5^0
-5/2	-0.85	148^0

4. What is the degeneracy of the ground state of carbon $1s^2 2s^2 2p^2$.

Using Eq. (3.33) we have

(a) for the set of two 1s electron: q= 2, r = 2, g = 1
(b) for the set of two 2s electron: q= 2, r = 2, g = 1

(c) for the set of two 2p electron: q= 2, r = 6, $g = \dfrac{6!}{4!2!} = 15$

From Eq. (3.34) the product of three we have degeneracy G = 15.

5. What are the configurations of the electrons of atoms of the following elements (a) Si (Z=14) (b) Br (Z= 35).

According to Eq. (3.31) the electronic configuration of

(a) Si (Z=14) is $1s^2 2s^2 2p^6 3s^2 3p^2$ and (b) Br (Z=35) is $1s^2 2s^2 2p^6 3s^2 3p^6 4s^2 3d^{10} 4p^5$.

6. Why is it impossible for $2^2D_{3/2}$ state to exist?

For the given state, principal quantum number n = 2, L = 2, S= 1/2 and J=3/2. Now l can have values only 0,1,.. , n-1. Hence for n = 2 we can have l = 0 and 1 which correspond to S and P state respectively. Hence $^2D_{3/2}$ with n = 2 cannot exist.

7. Explain on the basis of Hund's rules why the ground state of carbon is $^3P_0$ and that of Oxygen is $^3P_2$.

The electronic configuration of C (Z=6) is $1s^2 2s^2 2p^2$. For two equivalent p electrons   we have from Eq. (3.74) the terms ^1S, ^1D and ^3P. According to Hund's rule the state of maximum multiplicity lie lowest. Off the three allowed states, P state has the maximum multiplicity. For ^3P, L=1, S=1. Therefore J= 2,1,0. Hence we have $^3P_{2,1,0}$. According to Hund's rule $^3P_0$ is the ground state as the shell is less than half filled. In case of oxygen (Z=8) the electronic configuration is $1s^2 2s^2 2p^4$. Since the terms of p^n are same as that of p^{6-n}. Therefore, the terms of p^4 are ^1S, ^1D and ^3P. According to Hund's rule the state of maximum multiplicity lie lowest. Off the three allowed states, P state has the maximum multiplicity. For ^3P, L=1, S=1. Therefore J= 2,1,0. Hence we have $^3P_{2,1,0}$. According to Hund's rule $^3P_2$ is the ground state as the shell is more than half filled.

8. A level multiplet is observed with spacing in the ratio of 3:5:7. What are the possible values of S and L?

From Eq. (3.67) we have

$$\Gamma_J = \Gamma_3 + \Gamma_4 = A\mathbf{L}\cdot\mathbf{S} = \frac{A}{2}[J(J+1)-L(L+1)-S(S+1)]$$

$$\Gamma_{J+1} = A\mathbf{L}\cdot\mathbf{S} = \frac{A}{2}[(J+1)(J+2)-L(L+1)-S(S+1)]$$

Subtracting the above two equations

$$\Gamma_{J+1} - \Gamma_J = (\Gamma_3 + \Gamma_4)_{J+1} - (\Gamma_3 + \Gamma_4)_J = A(J+1)$$

Similarly,

$$\Gamma_J - \Gamma_{J-1} = AJ$$

$$\frac{\Gamma_{J+1} - \Gamma_J}{\Gamma_J - \Gamma_{J-1}} = \frac{A(J+1)}{AJ} = \frac{7}{5}$$

from this, J = 5/2 . Therefore, the highest level has J = 7/2 i.e. L+S = 7/2, and other levels are therefore J= 5/2, 3/2 and 1/2 .The minimum value of J = L-S =1/2.
From L+S = 7/2 and L-S=1/2 we have L = 2 and S = 3/2. The state is therefore ^4D .

9. Find the possible terms for f^2 configuration.

For f electron, l=3 and m_l = 3, 2, 1, 0, -1, -2, -3 and s = 1/2 and m_s = 1/2 and –1/2. To obtain the terms, we have to use Pauli exclusion principle, as n and l are same. Therefore m_l or m_s must be different. Terms for two equivalent electrons can be easily obtained by using Breit Scheme. Following the procedure given in Sec.3.8, the followings tables are constructed

$(m_l)_1$	3	2	1	0	−1	−2	−3	$(m_l)_2$
$M_L =$	6	5	4	3	2	1	0	3
	5	4	3	2	1	0	−1	2
	4	3	2	1	0	−1	−2	1
	3	2	1	0	−1	−2	−3	0
	2	1	0	−1	−2	−3	−4	−1
	1	0	−1	−2	−3	−4	−5	−2
	0	−1	−2	−3	−4	−5	−6	−3

If we permit the m_s values to be alike i.e.($M_S = 1$ or -1), the values of $M_L = 6, 5, 4, 3, 2, 1, 0, -1, -2, -3, -4, -5, -6$ crossed out by the diagonal lines are forbidden (as these come from same values of m_l and hence Pauli exclusion principle is violated) Since the values of M_L in lower left of the array are identical with, and are mirror image of those in the upper right half , one of these groups must also be eliminated from

$$(m_S)_1 \quad 1/2 \qquad -1/2$$

$M_S =$	1	0	1/2
	0	-1	-1/2

$$(m_S)_2$$

calculations. Leaving out lower half of the array the remaining values form sums

M_L	M_S
5,4,3,2,1,0,-1,-2,-3,-4,-5	1,-1
3,2,1,0,-1,-2,-3	1,-1
1,0,-1	1,-1

If on the other hand, we permit m_l values to be alike, the values of $M_S = 1, -1$ are forbidden by Pauli Principle. Since the two $M_S = 0$ values are identical, one of these must also be excluded. The remaining combinations are

M_L	M_S
6,5,4,3,2,1,0,-1,-2,-3,-4,-5, -6	0
5,4,3,2,1,0,-1,-2,-3,-4,-5	0
4,3,2,1,0,-1,-2,-3,-4	0
3,2,1,0,-1,-2,-3,-4	0
2,1,0,-1,-2	0
1,0,-1	0
0	0

On combining the above we have

M_L	M_S	L	S	Term
6,5,4,3,2,1,0,-1,-2,-3,-4,-5, -6	0	6	0	^1I
5,4,3,2,1,0,-1,-2,-3,-4,-5	1,0,-1	5	1	^3H
4,3,2,1,0,-1,-2,-3,-4	0	4	0	^1G
3,2,1,0,-1,-2,-3	1,0,-1	3	1	^3F
2,1,0,-1,-2	0	1	0	^1D
1,0,-1	1,0,-1	3	1	^1P
0	0	0	0	^1S

Terms from f^2 configurations are ^{1}(SDGI) ^{3}(PFH) a total of 91 independent states.

10. Show for a ps electron configuration that the total ^{3}P separation is same in both LS and jj coupling.

For ps configuration we have $l_1 = 1, l_2 = 0$, $s_1 = 1/2$ and $s_2 = 1/2$ .In LS coupling, for ^{3}P we have L= 1, S=1 and J= 2,1,0.Substituting these values in Eq. (3.67) and using Eq.(3.65) we have

$$\Gamma \text{ (for } ^3P_2) = (A/2)[6- 2 - 2] = A = a_3/2$$

$$\Gamma \text{ (for } ^3P_0) = (A/2)[0- 2 - 2] = -2 A = a_3$$

Therefore

$$\Gamma (^3P_2 - {}^3P_0) = 3A = 3a_3/2$$

In jj coupling we have for p, $l_1=1$, $s_1=1/2$ and $j_1 =3/2,1/2$ and for s electron $l_2=0$, $s_2=1/2$ and j_2. Therefore J vales are J= $j_1 + j_2$, ..., $j_1 - j_2$. Thus J values are 2, 1 and 0. Using these values in Eqs.(3.68), (3.69) and (3.70) we have

$$\Gamma \text{ (for } j_1=3/2)_{J=2} = \frac{a_3}{2}\left[\frac{15}{4}-2-\frac{3}{4}\right] = \frac{a_3}{2}$$

$$\Gamma \text{ (for } j_1=1/2)_{J=0} = \frac{a_3}{2}\left[\frac{3}{4}-2-\frac{3}{4}\right] = -a_3$$

and

$$\Gamma \text{ (for } j_1=3/2)_{J=2} - \Gamma \text{ (for } j_1=1/2)_{J=0} = 3a_3/2$$

The interval is therefore same in both LS coupling and jj coupling.

11. Derive all the terms arising from the electron configuration fg.

For f electron, $l= l_1= 3$ and $s=s_1=1/2$ for g electron $l = l_2 = 4$ and $s = s_2 = 1/2$. Since fg electrons are nonequivalent, therefore the terms can be obtained using Eqs. (3.45) and (3.46) as follows:

$$L= 4 + 3, 4 + 3 - 1,................, 4 - 3 = 7, 6, 5, 4, 3, 2, 1$$

and

$$S= 1/2 + 1/2 , \text{ and } 1/2 - 1/2 = 1, 0.$$

The multiplicity is therefore 3 (triplet) and 1(singlet). The terms corresponding to L values are represented by P(L=1), D(L=2), F(L=3), G(L=4), H(L=5), I(L=6), K(L=7).Thus the terms are

$$^{1,3}(P,D,F,G,H,I,K).$$

12. Construct vector diagram for $^2F_{5/2}$.

For $^2F_{5/2}$, the multiplicity $2S+1 = 2$ or $S=1/2$. For F state $L= 3$ and $J =5/2$ is given. To construct the vector diagram the following procedure is followed. We draw an arrow AB having a length proportional to length of L i.e. $k[3(3+1)]^{1/2} = k(12)^{1/2}$. Now a circle of radius corresponding to length proportional to S i.e $k\sqrt{\frac{1}{2}\left(\frac{1}{2}+1\right)} = k\sqrt{\frac{3}{4}}$ is drawn with B as centre. Now with A as centre the circle of radius corresponding to $k\sqrt{\frac{5}{2}\left(\frac{5}{2}+1\right)} = k\sqrt{\frac{35}{4}}$ is drawn . The interaction of these circles with one having radius $k(3/4)^{1/2}$ with centre at B is the end of **J** vector starting from A (Fig.3.10)

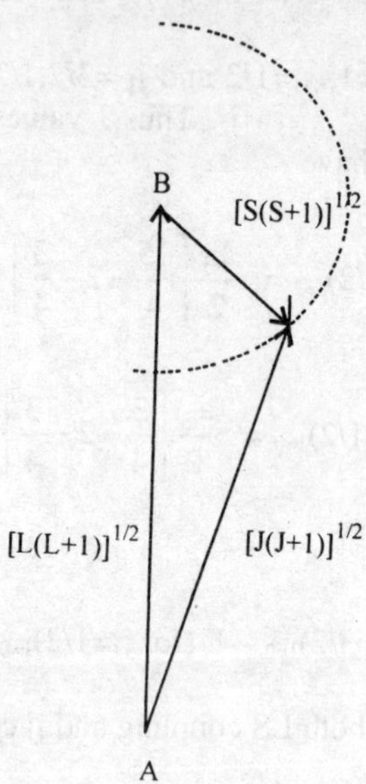

Fig. 3.10 Vector diagram for $^2F_{5/2}$

13. Suppose that the electron in a hydrogen atom is moving in n th Bohr orbi. (a) calculate the electric current (b) the magnetic dipole moment.

(a) Since

$$i = \frac{e}{T} = \frac{ev}{2\pi r}$$

Substituting the values of r and v from Eqs. (2.11) and (2.14) ,respectively in the above expression

$$i = \frac{4\pi^2 m_e Z^2 e^5}{(4\pi \varepsilon_0)^2 h^3 n^3}$$

Putting

Z = 1
m_e = 9.1095 x 10^{-31} Kg
e = 1.6022 x 10^{-19} C
$(h/2\pi)$ = 1.055 x 10^{-34} J sec
$(1/4\pi\varepsilon_o)$ = 9 x 10^9 N-m^2/coul^2

$$i \approx \frac{1.056 \times 10^{-3}}{n^3}\ \text{amp.}$$

(b) The magnetic moment due to orbital motion is given by

$$(\mu_l)_n = \frac{e}{2m_e}\,l$$

$$(\mu_l)_n = \frac{e}{2m_e}\,r\,m_e\,v = \frac{e\,m_e\,rv}{2n}$$

Substituting the values of r and v from Eqs. (2.11) and (2.14) in the above we have

$$(\mu_l)_n = \frac{n\,e\,h}{4\pi\,m_e} = \beta_e\,n$$

14. Derive the terms arising from electron configuration 3d4p5p.

   Since the electrons of the configuration 3d4p5p are nonequivalent, therefore application of Pauli exclusion principle is not required. Let us first calculate the terms arising from 3d4p. For d electron l = 2, s = 1/2 and for p electron l = 1 and s = 1/2. Combining the l and s values according to Eqs. (3.45) and (3.46) we obtain L = 3,2,1 and S = 1,0. Now consider 5p electron, its l value is 1 and s value is 1/2. Combining these values of l and s with L and S values obtained for 3d4p using Eqs. (3.45) and (3.46) we obtain

L	l	L+l, -----,L-l	S	s	S+s, ...,S-s	Term
3	1	4,3,2	1	1/2	3/2, 1/2	^{2,4}(D,F,G)
3	1	4,3,2	0	1/2	1/2	^{2}(D,F,G)
2	1	3,2,1	1	1/2	3/2, 1/2	^{2,4}(P,D,F)
2	1	3,2,1	0	1/2	1/2	^{2}(P,D,F)
1	1	2,1,0	1	1/2	3/2, 1/2	^{2,4}(S,P,D)
1	1	2,1,0	0	1/2	1/2	^{2}(S,P,D)

The terms from 3d4p5p configuration are:

$$^2S(2),\ ^2P(4),\ ^2D(6),\ ^2F(4),\ ^2G(2),\ ^4S(1),\ ^4P(2),\ ^4D(3),\ ^4F(2),\ ^4G(1)$$

15. Derive the terms arising from electron configuration p^3sd.

The terms corresponding to p^3 configuration are obtained as the procedure given in Sec.3.8 and the terms are ^2P, ^2D and ^4S. Terms corresponding sd is obtained as follows:

For s electron $l_1 = 0$, $s_1 = 1/2$ for d electron $l_2 = 2$ and $s_2 = 1/2$. Following the procedure given in example 3.11, the terms obtained are: ^1D and ^3D. To obtain the terms from the configuration p^3sd , first take the term ^2P of p^3 and ^1D of sd and combine them together. For ^2P the values of L and S are 1 and 1, respectively. For ^1D, L = 2 and S = 0.

Therefore $L_{resultant} = 3,2,1$ and $S_{resultant} = 1/2$ and therefore the corresponding states are $^2(PDF)$. Similarly,

Combination of ^2P with ^3D gives : $^2P, ^2D , ^2F, ^4P, ^4D, ^4F$
Combination of ^2D with ^1D gives : $^2S, ^2P, ^2D , ^2F, ^2G,$
Combination of ^2D with ^3D gives : $^2S, ^2P, ^2D , ^2F, ^2G, ^4S, ^4P, ^4D, ^4F, ^4G,$
Combination of ^4S with ^1D gives : $^4D,$
Combination of ^4S with ^3D gives : $^2D , ^4D, ^6D$

There are 28 terms: $^2S(2), ^2P(4), ^2D(5), ^2F(4), ^2G(2), ^4S, ^4P(2), ^4D(4), ^4F(2), ^4G, ^6D$.

16. Show that

$$\sum_{J=|L-S|}^{J=|L+S|}(2J+1) = (2L+1)(2S+1)$$

The possible values of J are equal to the smaller of two numbers 2S+1 and 2L+1. Let 2S+1 is smaller than 2L+1. The number of J values is then 2S+1. The first term in the above expression is 2(L-S)+1, Common difference is 1. Therefore, the value of the sum using AP series is

$$\frac{2S+1}{2}\{2[2(L-S)+1]+[(2S+1)-1]\} = (2S+1)(2L+1)$$

Let 2L+1 is smaller than 2S+1. The number of J values is then 2L+1. The first term in the above expression is 2(S-L)+1, Common difference is 1. Therefore, the value of the sum using AP series is

$$\frac{2L+1}{2}\{2[2(S-L)+1]+[(2L+1)-1]\} = (2S+1)(2L+1)$$

17. Give a Schematic representation of the interaction energy for sp configuration in LS coupling.

Position of the energy levels are given by Eqs. (3.52), (3.53) and (3.67). For sp configuration we have $l_1 = 0$ and $s_1 = 1/2$ for s electron and $l_2 = 1$ and $s_2 = 1/2$ for p electron. Therefore L = 1, S = 1, 0. The terms for ps configuration are $^1P_1$ and $^3P_{2,1,0}$. The values of various terms are given in the following Table.

Spectral Term	S	L	J	Γ_1	Γ_2	$\Gamma_3 + \Gamma_4$
$^1P_1$	0	1	1	$-(3/4)a_1$	0	0
$^3P_0$	1	1	0	$a_1/4$	0	$-2A$
$^3P_1$	1	1	1	$a_1/4$	0	$-A$
$^3P_2$	1	1	2	$a_1/4$	0	A

The effects of various interactions are shown in the above table and in Fig.3.11. Starting at the left with only the energy attributed to the n and l values of the two electrons. Then the effect of Γ_1 is shown. This causes a splitting into singlet and triplet. The contribution of Γ_2 is zero. On the extreme right the combined effect of $\Gamma_3 + \Gamma_4$ is shown. The terms $\Gamma_3 + \Gamma_4$ vanishes for singlet .The term value can be written as $T = T_0 - (\Gamma_1 + \Gamma_2 + \Gamma_3 + \Gamma_4)$ where T_0 is a hypothetical term value for the centre of gravity of the entire electron configuration.

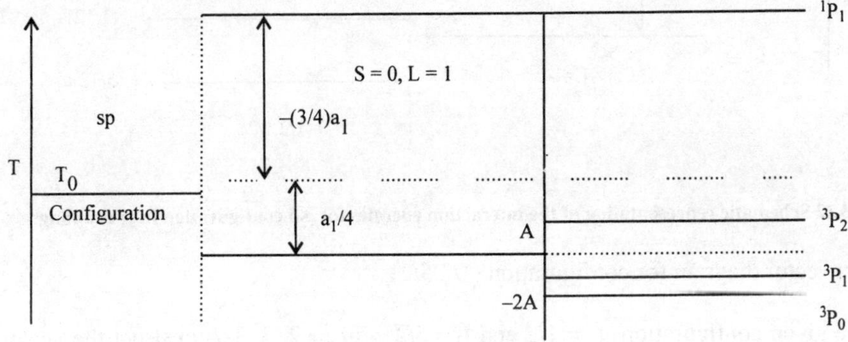

Fig. 3.11 Schematic representation of the interaction energies for sp configuration

18. Give a Schematic representation of the interaction energy for sp configuration in jj coupling.

Position of the energy levels are given by Eqs. (3.68) - (3.70) .For sp configuration we have $l_1 = 0$ and $s_1 = 1/2$ for s electron and $l_2 = 1$ and $s_2 = 1/2$ for p electron. Therefore $j_1 = 1/2$, $j_2 = 3/2$, 1/2. The terms for ps configuration are $(1/2, 1/2)_{1,0}$, $(1/2, 3/2)_{2,1}$. The values of various terms are given in the following Table.

j_1	j_2	Γ_3	Γ_4	$\Gamma_1 + \Gamma_2$	J
1/2	3/2	0	$a_4/2$	$-(5/12)a_1$	$(3/2, 1/2)_1$
				$(3/12)a_1$	$(3/2, 1/2)_2$
1/2	1/2	0	$-a_4$	$-(1/12)a_1$	$(1/2, 1/2)_1$
				$(3/12)a_1$	$(1/2, 1/2)_0$

The effects of various interactions are shown in the above table and in Fig.3.12. Starting at the left with only the energy attributed to the n and l values of the two electrons. Then the effect of Γ_4 is shown. This causes a splitting into singlet and triplet. The contribution of Γ_3 is zero. On the extreme right the combined effect of $\Gamma_1 + \Gamma_2$ is shown. The term value can be written as $T = T_0 - (\Gamma_1 + \Gamma_2 + \Gamma_3 + \Gamma_4)$ where T_0 is a hypothetical term value for the centre of gravity of the entire electron configuration.

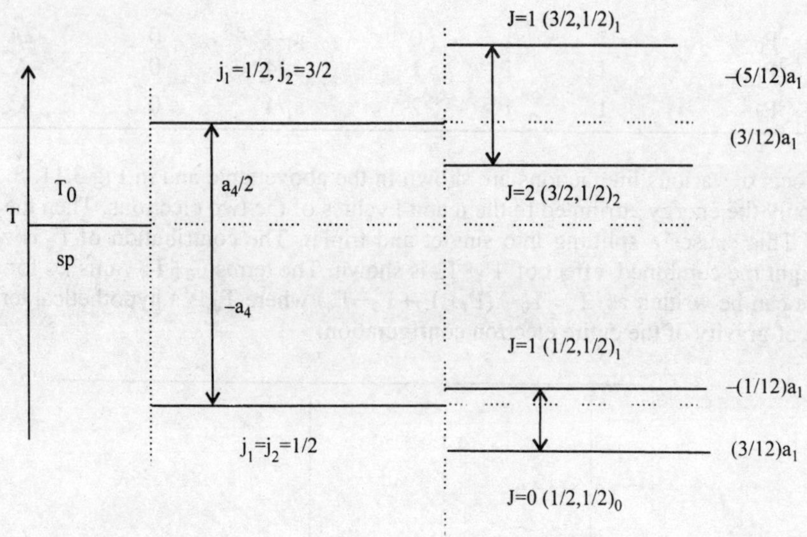

Fig. 3.12 Schematic representation of the interaction energies for sp configuration in jj coupling

19. Construct vector diagram for configuration $(1/2, 5/2)_{2,3}$

   For the given configuration $j_1 = 1/2$ and $j_2 = 5/2$ with J= 2, 3. To construct the vector diagram the following procedure is followed. We draw an arrow AB having a length proportional to length of j_2 i.e. $k\sqrt{\dfrac{5}{2}\left(\dfrac{5}{2}+1\right)} = k\sqrt{\dfrac{35}{4}}$. Now a circle of radius corresponding to length proportional to j_1

i.e. $k\sqrt{\dfrac{1}{2}\left(\dfrac{1}{2}+1\right)} = k\sqrt{\dfrac{3}{4}}$ is drawn with B as centre. Now with A as centre the circle of radii

corresponding to $k[3(3+1)]^{1/2} = k\,(12)^{1/2}$ and $k[2(2+1)]^{1/2} = k\,(6)^{1/2}$ are drawn . The interaction of these circles with one having radius $k(3/4)^{1/2}$ with centre at B are the end of J vectors starting from A (Fig.3.13).

20. Evaluate the ratio of the orbital magnetic dipole moment to the orbital angular momentum for an electron moving in an elliptical orbit of the Bohr-Sommerfeld atom.

   By Kepler's second law, the rate dA/dt at which the radius r drawn from the nucleus to the electron describes area is such that

$$\frac{dA}{dt} = \frac{1}{2r^2}\frac{d\theta}{dt}$$

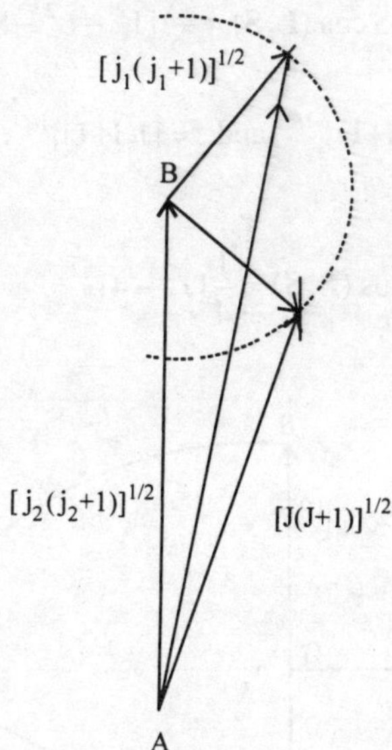

Fig. 3.13 Vector Diagram of $(1/2 , 5/2)_{3,2}$

$$2m_e \frac{dA}{dt} = m_e r^2 \frac{d\theta}{dt} = L$$

where θ is the azimuthal angle and L the angular momentum . Integrating for one complete period T we get

$$2m_e A = LT$$

$$A = \frac{LT}{2m_e}$$

The electron passes a given point on its orbit 1/T times a second and therefore, is equivalent to a current i = –e/T. Now if a current flows in a closed circuit of area A, then at a great distance it produces a field practically identical with that of magnet having a moment

$$\mu = iA,$$

$$\mu = -\frac{e}{T} \frac{iT}{2m_e} = -\frac{e}{2m_e} L$$

21. Graphically represent the Landé interval rule for a ^{3}P term.

For ^{3}P state L = 1, S = 1 and J= 2,1,0.Lande's Interval rule given by Eq. (3.67) is

$$ALS \cos(\mathbf{L,S}) = \frac{A}{2}[J^2 - L^2 - S^2]$$

where L = $[L(L+1)]^{1/2}$, S = $[S(S+1)]^{1/2}$ and J = $[J(J+1)]^{1/2}$. Substituting the values of L and S in Eq. (3.67) we obtain

$$\cos(\mathbf{L,S}) = \frac{1}{4}[J^2 - 4]$$

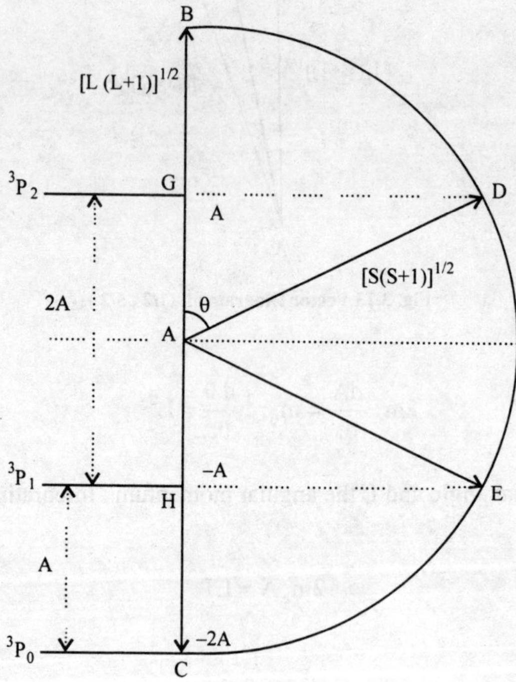

Fig. 3.14 Graphical representation of the Landé interval rule for a ^3P term

For J= 2,

$$\cos(L,S) = \cos\theta = \frac{1}{4}[6-4] = \frac{1}{2}$$

$$\theta = 60^0$$

For J= 1,

$$\theta = 120^0$$

For J=0,

$$\theta = 180^0$$

To represent Landé interval rule graphically, draw a vertical line BC of length $= 2k[L\,(L+1)]^{1/2}$ $= 2k(2)^{1/2}$ with centre at A. k is some constant. With A as centre, draw a circle of radius k $[S\,(S+1)]^{1/2}$ which cuts line BC. With A as origin, draw lines at angles 60^0, 120^0, and 180^0 to AB. These lines cut the circle at D, E and C, respectively (Fig.3.14). Draw perpendicular from D, E and C on the lines BC that cut BC at G, H and C. The interval GH and HC are in the ratio 2:1.

Problems

3.1   Consider an atom with an electron configuration of $1s^2 2s^2 2p^1$. What are the values of l, s, and j, and the corresponding values of the angular momenta L,S, and J, respectively.

3.2   If atom could contain electrons with principal quantum numbers up to and including n = 7, how many elements would there be ?

3.3   Calculate the angles between the total and orbital angular momentum vector for $^2D_{3/2}$ state.

3.4   What is the degeneracy of (a)an excited state of nitrogen of configuration$1s^2 2s^2 2p^2 3p^1$. (b) the configuration 2p3d4f (c) the configuration 1s2s2p.

3.5   What are the configurations of the electrons of atoms of the following elements (a) Ca (Z=20) (b) Cl (Z= 17).

3.6   Calculate the possible angles between **L** and **S** vectors for the f electron in an atom.

3.7   A level multiplet is observed with spacing in the ratio of (a) 1:2:3:4,(b) 3:5. What are the possible values of S and L ?

3.8   Find the possible terms for d^2 configuration.

3.9   Find the possible terms for d^3 configuration.

3.10   Show for a ds electron configuration that the total ^3D separation is same in both LS and jj coupling .

3.11   Derive all the terms arising from the electron configuration np n'p n"p.

3.12   Derive all the terms arising from the electron configuration pdf.

3.13   Derive the terms arising from electron configuration p^2sd

3.14   Draw vector diagrams to show the possible terms for two non-equivalent p electrons.

3.15   Draw vector diagrams for the two term arising from an s and a d electron in both LS and jj coupling.

3.16   Give a Schematic representation of the interaction energy for pd configuration in LS coupling.

3.17   Assuming jj coupling holds, list the possible terms $(j_1,j_2)_J$ of the (a) $(nl\ 5/2)^2$ (b) $(nl\ 5/2)^3$ (c) $(nl\ 3/2)^3$ (d) $(nl\ 1/2)^2$ and (e) $(nl\ 5/2)^5$.

3.18   Graphically represent the Landé interval rule for (a) ^3D (b) ^4P (c) ^4D and (d) ^5F terms.

4

Atomic Spectra

4.1 Spectrum of Hydrogen Atom

In hydrogen atom, the single electron moves in a Coulomb field, free from the effect of inter electron repulsion. The Schrödinger equation for hydrogen atom (Sec.2.3), assuming nucleus to be of infinite mass is

$$H_0 \Psi_{nlm} = \left[\frac{p^2}{2m_e} + V(r) \right] \Psi_{nlm} = E_n \Psi_{nlm} \qquad (4.1)$$

The Schrödinger equation for hydrogen atom is exactly soluble and energy eigenvalues E_n are given by Eq. (2.102) and eigenfunctions are written as [Eq. (2.107)]

$$\Psi_{nlm}(r, \theta, \varphi) = R_{nl}(r) Y_{lm}(\theta, \varphi) \qquad (4.2)$$

These solutions correctly predict the series of spectral lines. Figs. 2.2 and 2.10 show the energy levels according to Bohr theory and Schrödinger theory, respectively. Fig. 2.2 shows the first few of an infinite number of series of transitions in the hydrogen atom, all smoothly converging to the ionization limit n = ∞. Spectroscopic observations with apparatus of high resolution show that different lines such as H_α (n = 3 → n = 2 transition), for example, consists of several closely spaced wavelength components indicating that the levels have a structure. The Bohr theory and Schrödinger theory is however, inadequate to explain the fine structure of the levels.

In Bohr treatment and Schrödinger treatment of hydrogen atom it is assumed that electron velocities to be much smaller than the velocity of light. This is not correct for electron moving close to the nucleus. Therefore, requirements of theory of relativity cannot be ignored. Dirac has given a relativistic equation, which in the limit of low electron velocities can be written, in the form of a Schrödinger equation with the additional term from the electron spin and other relativistic effect.

The Hamiltonian H with additional terms from electron spin and relativistic effects is

$$H = H_0 + H_1 + H_2 + H_3$$

$$H = H_0 - \frac{cp^4}{(2m_0 c)^3} + \frac{1}{2m_0^2 c^2} \left(\frac{1}{r} \frac{dV}{dr} \right) \mathbf{l} \cdot \mathbf{s} - \frac{h^2 e}{8\pi^2 m_0^2 c^2} \mathbf{E} \cdot \nabla \qquad (4.3)$$

H_0 is given by Eq. (4.1) and the third term is given by Eq. (3.14). The last three terms in Eq. (4.3) provide the perturbation H'. The second term is relativistic correction to the kinetic energy, third is spin-orbit interaction term and fourth is Darwin term. m_0 is electron rest mass .Because H involves the spin operator s, therefore we must include a spin function as a factor in the solution. We take this spin function χ to be an eigenfunction of s^2 and s_z for spin 1/2 with the property

$$s^2\chi = s(s+1)\chi = \frac{3}{4}\chi \tag{4.4}$$

$$s_z\chi = m_s\chi = \pm\frac{1}{2}\chi \tag{4.5}$$

with no interaction between spin and orbit, the eigenfunction of H can be expressed as

$$\Psi_{nlmm_s} = \Psi_{nlm}\chi(m_s) \tag{4.6}$$

Ψ_{nlmm_s} are functions of space and spin coordinates. The Eq. (4.3) can be treated by first-order perturbation theory using function given by Eq. (4.6) as zeroth –order function.

(a) relativistic correction to the kinetic energy

According to theory of relativity

$$m_e = \frac{m_0}{\sqrt{1-(v^2/c^2)}}$$

where m_0 is electron rest mass, v is the velocity of electron. The kinetic energy of the electron is

$$T = \frac{p^2}{2m_e} = \frac{p^2}{2m_0}\left[1-\frac{v^2}{c^2}\right]^{-1/2} = \frac{p^2}{2m_0}\left[1-\frac{1}{2}\left(\frac{v}{c}\right)^2 - \frac{1}{8}\left(\frac{v}{c}\right)^4 - - - - \right]$$

with the approximation $p^2 = (m_0 v)^2$ and retaining only the term up to $(v/c)^2$

$$T = \frac{p^2}{2m_0} - \frac{cp^4}{(2m_0c)^3} \tag{4.7}$$

$$T = T_0 - \frac{T_0^2}{2m_0c^2} \tag{4.8}$$

where $T_0 = (p^2/2m_0c^2)$. The Hamiltonian H_0 is

$$H_0 = T_0 + V(r)$$

$$T_0 = H_0 - V(r)$$

and

$$H_1 = -\frac{T_0^2}{2m_0c^2} = -\frac{1}{2m_0c^2}[H_0 - V(r)]^2 \tag{4.9}$$

The above term is modification to the non-relativistic kinetic energy T_0. Using the Schrödinger function, the energy shift in first-order perturbation theory is

$$\Delta E_1 = -\frac{1}{2m_0c^2}\left\langle [H_0 - V(r)]^2 \right\rangle_{nlmm_s} = -\frac{1}{2m_0c^2}\left\langle [H_0 + \frac{Ze^2}{4\pi\,\varepsilon_0 r}]^2 \right\rangle_{nlmm_s} \tag{4.10}$$

$$\Delta E_1 = -\frac{1}{2m_0c^2}\left[E_n^2 + \left(\frac{Ze^2}{4\pi\,\varepsilon_0}\right)^2 \left\langle \frac{1}{r^2} \right\rangle_{nlm_l} + \frac{2E_n Ze^2}{4\pi\,\varepsilon_0}\left\langle \frac{1}{r} \right\rangle_{nlm_l} \right] \tag{4.11}$$

Substituting the values of $<1/r>_{nlm_l}$ and $<1/r^2>_{nlm_l}$ from Table 2.4

$$\Delta E_1 = E_n\alpha^2 Z^2 \left[-\frac{1}{4n^2} + \frac{1}{n^2} - \frac{1}{n(1+\frac{1}{2})} \right] = \frac{\alpha^2 Z^2}{n^2}E_n\left(\frac{3}{4} - \frac{n}{1+\frac{1}{2}} \right) \tag{4.12}$$

where α is fine structure constant given by Eq.(2.15). This term depends on l as well as on n. It is small than E_n by a factor $\alpha^2 Z^2 \approx v^2/c^2$.

(b) spin-orbit term

From Eq. (4.3) the spin-orbit term is

$$H_2 = \frac{1}{2m_0^2 c^2}(\frac{1}{r}\frac{dV}{dr})\mathbf{l}\cdot\mathbf{s} \tag{4.13}$$

The total angular momentum $\mathbf{j}$ is

$$\mathbf{j} = \mathbf{l} + \mathbf{s} \tag{4.14}$$

Taking dot product of this equality times itself and on simplifying

$$\mathbf{l}\cdot\mathbf{s} = \frac{1}{2}[j^2 - l^2 - s^2] \tag{4.15}$$

The functions $\Psi_{nlm_l m_s}$ which are simultaneous eigenfunction of the operator H_0, l^2, s^2, l_z and s_z are not adequate because $\mathbf{l}.\mathbf{s}$ does not commute with l_z or s_z. However, zero order wavefunction Ψ_{nljm_j} can be formed by linear combinations of $\Psi_{nlm_l m_s}$. This new wave function forms a satisfactory basis set in which the operator $\mathbf{l}.\mathbf{s}$ is diagonal.

From Eqs. (4.13) and (4.15)

$$H_2 = \frac{1}{2m_0^2 c^2} \left(\frac{1}{r} \frac{dV}{dr} \right) \frac{j^2 - l^2 - s^2}{2} \qquad (4.16)$$

The spin-orbit energy for the state is just the expectation value of H_2. Therefore, energy arising from spin-orbit interaction is

$$\Delta E_2 = \frac{1}{2m_0^2 c^2} \left\langle \left[\frac{j^2 - l^2 - s^2}{2} \right] \left(\frac{1}{r} \frac{dV}{dr} \right) \right\rangle_{nljm_j} = \frac{h^2}{8\pi^2 m_0^2 c^2} \frac{[j(j+1) - l(l+1) - s(s+1)]}{2} \frac{Ze^2}{4\pi\,\varepsilon_0} < \frac{1}{r^3} > \qquad (4.17)$$

Using the value of $<1/r^3>$ from Table 2.4 in Eq. (4.17)

$$\Delta E_2 = -\frac{\alpha^2 Z^2}{n^2} E_n \frac{n}{l(1+\frac{1}{2})(l+1)} \frac{[j(j+1) - l(l+1) - s(s+1)]}{2} \qquad (4.18)$$

For $j = l- (1/2)$, $s = 1/2$

$$\Delta E_2 = \frac{\alpha^2 Z^2}{n} E_n \frac{1}{l(2l+1)} \qquad (4.19)$$

For $j = l+(1/2)$, $s = 1/2$

$$\Delta E_2 = -\frac{\alpha^2 Z^2}{n} E_n \frac{1}{(l+1)(2l+1)} \qquad (4.20)$$

For $l = 0$, the spin orbit interaction vanishes, and $\Delta E_2 = 0$.

(c) Darwin term

The Darwin term which applies only to $l = 0$ is

$$H_3 = -\frac{h^2 e}{16\pi^2 m_0^2 c^2} E \cdot \nabla \qquad (4.21)$$

The energy shift derived from it

$$\Delta E_3 = \frac{Ze^2 h^2}{8\pi\, m_0^2 c^2} |\psi(0)|^2 \qquad (4.22)$$

Using the value of $|\psi(0)|^2$ from Eq.(2.107)

$$\Delta E_3 = -\frac{\alpha^2 Z^2 E_n}{n} \tag{4.23}$$

Thus total energy shift is

$$\Delta E = \Delta E_1 + \Delta E_2 + \Delta E_3 = E_n \frac{(Z\alpha)^2}{n^2}\left(\frac{n}{j+(1/2)} - \frac{3}{4}\right) \tag{4.24}$$

each value of j correspond to two possible values of l given by l = j ± (1/2). Adding ΔE to the non - relativistic energy E_n,

$$E_{nj} = E_n\left[1 + \frac{\alpha^2 Z^2}{n^2}\left(\frac{n}{j+(1/2)} - \frac{3}{4}\right)\right] \tag{4.25}$$

The $|E_{nj}|$ of the electron is slightly increased with respect to non-relativistic value $|E_n|$. Using Eqs. (2.20) and (2.22) in Eq.(4.25)

$$E_{nj} = -(hcR)\frac{Z^2}{n^2}\left[1 + \frac{\alpha^2 Z^2}{n^2}\left(\frac{n}{j+(1/2)} - \frac{3}{4}\right)\right] \tag{4.26}$$

$$T_{nj} = -\frac{E_{nj}}{hc} = -\frac{RZ^2}{n^2}\left[1 + \frac{\alpha^2 Z^2}{n^2}\left(\frac{n}{j+(1/2)} - \frac{3}{4}\right)\right] \tag{4.27}$$

Eq. (4.26) was first obtained by Sommerfeld by supplimenting Bohr's quantization postulates with postulates of relativistic mechanics. Sommerfeld employed a quantum number $n_\theta = j +(1/2)$. However, his derivation omitted the electron spin and so his result must be considered fortuitous.

From Eq. (4.26) it is seen that the energy depends on n and j but not on l: $2p_{1/2}$ orbital has the same energy as the $2s_{1/2}$. For a given value of n, the level with largest j value lies higher in energy. The interval between adjacent levels with the same n is

$$E_{n,j+1} - E_{n,j} = \frac{hcR\alpha^2 Z^4}{n^3(j+\frac{1}{2})(j+\frac{3}{2})} \tag{4.28}$$

The spacing therefore decreases rapidly as one proceeds to higher principal quantum number. Substituting the value R and α^2 and simplifying

$$\Delta T_{nj} = -\frac{E_{n,j+1} - E_{n,j}}{hc} = -\frac{5.84Z^4}{n^3(j+\frac{1}{2})(j+\frac{3}{2})}\,cm^{-1} \tag{4.29}$$

The energy separation between two levels corresponding to j = l + (1/2) and j = l − (1/2) (l≠ 0) by Eq.(4.24) is

$$\delta E = \Delta E_{j=l+1/2} - \Delta E_{j=l-1/2} = \frac{R\alpha^2 Z^4}{2n^3 l\,(l+1)}\,cm^{-1} \tag{4.30}$$

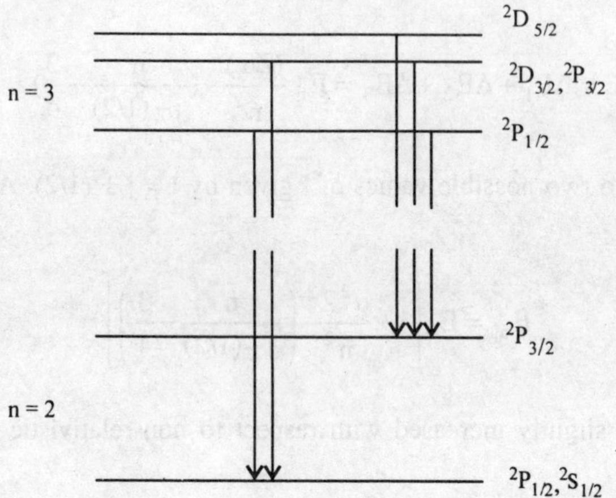

Fig. 4.1 Fine structure of the n = 2 and n = 3 levels of hydrogen according to Dirac theory

Fig.4.1 show the structure of n = 2 and n = 3 levels of hydrogen atom according to the Eq. (4.29)

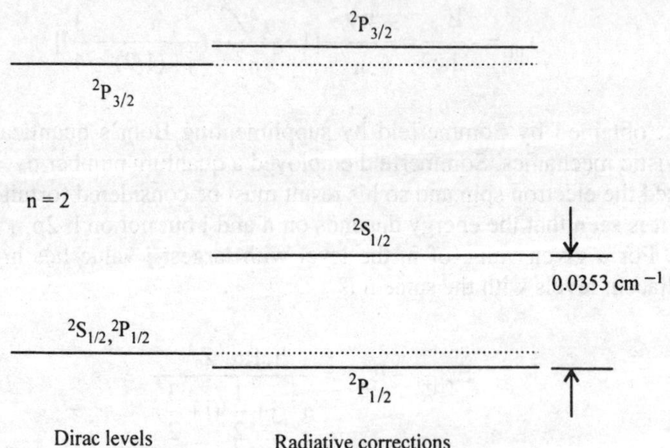

Fig. 4.2 A diagram (not to scale) of the Lamb shift of the n = 2 levels of atomic hydrogen

considering spin other relativistic effects. Transitions between n = 3 → n = 2 are observed according to the selection rules $\Delta l = \pm 1$, $\Delta j = 0, \pm 1$. The fine structure as shown in Fig.4.1 does not quite agree with experimental results. Levels with same j but different l are not coincident. The separation of levels having same j but different l is quite small. Lamb and Rutherford found that the $^2S_{1/2}$ level lies above $^2P_{1/2}$ level by 0.0353 cm^{-1} . This shift of S term is called Lamb shift. Dirac theory had to be modified to account for the Lamb shift and this theory is called quantum electrodynamics. This takes into account the interaction neglected in Dirac theory between the electron and the radiation field. The Lamb shift is proportional to n^{-3} and therefore decreases rapidly with n.

4.2 Spectra of Alkali Metal Atoms

The simplest series observed in spectra of neutral atoms apart from those in hydrogen occurs in the spectra of alkali atoms lithium (Li), sodium (Na), potassium (K), rubidium (Rb) and cesium (Cs). The normal configurations of alkali atoms are given in Table 4.1.

Table 4.1 Electronic configurations of alkali metals

Li (Z = 3)	$1s^2 2s$
Na (Z = 11)	$1s^2 2s^2 2p^6 3s$
K (Z = 19)	$1s^2 2s^2 2p^6 3s^2 3p^6 4s$
Rb (Z = 37)	$1s^2 2s^2 2p^6 3s^2 3p^6 3d^{10} 4s^2 4p^6 5s$
Cs (Z = 55)	$1s^2 2s^2 2p^6 3s^2 3p^6 3d^{10} 4s^2 4p^6 4d^{10} 5s^2 5p^6 6s$

An alkali atom in its ground state consists of one or more closed shell of electron plus single (valence) electron in a new shell in ns orbit (n = 2 for Li, 3 for Na, 4 for K,.. etc.). The valence electron is responsible for the quite simple spectra of alkali atoms. The closed shell have zero total orbital angular momentum and zero spin angular momentum, designated as $^1S_0$ and are essentially unchanged in transition between the two state of valence electron. The atomic spectra of alkali atoms is explained by considering the excited state of alkali atom which contains the same closed shells with only valence electron is excited. The valence electron can be excited to various s, p, d, f,.. orbits resulting in doublet terms as given in Table 4.2. Fig. 4.3 shows the energies of the ground state and first few excited states

Table 4.2 Terms of Rb atom

Normal Configuration	Excited Configuration	n	Term
$[Kr]^{36} 5s$	$[Kr]^{36} ns$	6, 7, ...	$n^2 S_{1/2}$
$[Kr]^{36} 5s$	$[Kr]^{36} np$	6, 7, ...	$n^2 P_{1/2, 3/2}$
$[Kr]^{36} 5s$	$[Kr]^{36} nd$	4, 5, ...	$n^2 D_{3/2, 5/2}$
$[Kr]^{36} 5s$	$[Kr]^{36} nf$	4, 5, ...	$n^2 F_{5/2, 7/2}$

of Rb atom , obtained from an analysis of the optical line spectra of Rb. For the P, D and F states, each level consists of two closely spaced multiplet components of different J value whose separation on the scale of the figure is too small to be shown. In comparison with n = 5 level of hydrogen, the ground state 5^2S of Rb is about 3.6 eV and first excited state 5^2P is about 2 eV more negative. On the other hand second excited state 5^2D of Rb is about 1.12 eV more negative than n = 4 state of hydrogen while 4^2F of Rb have nearly the same energy as that of n = 4 of hydrogen. For a given n, the energy is most negative for the smallest value of l. This is because electron spends more time near the centre of atom, where it experiences the full nuclear charge. For Rb, the l dependence makes 6^2S more negative than 5^2D or 4^2F . However, for large radii subshell with large value of n the l dependence become less important and the energy of valence electron become very close to the energy levels of an electron in H atom. Qualitatively, the same situation persists in other alkali atom energy levels.

In alkali atoms the effective potential V(r) is radically different from a Coulomb potential especially for large value of Z. At small distances the potential is more attractive than due to Coulomb force, so that the binding energy of the orbital ns is always greater than of the ns level of hydrogen. Therefore,

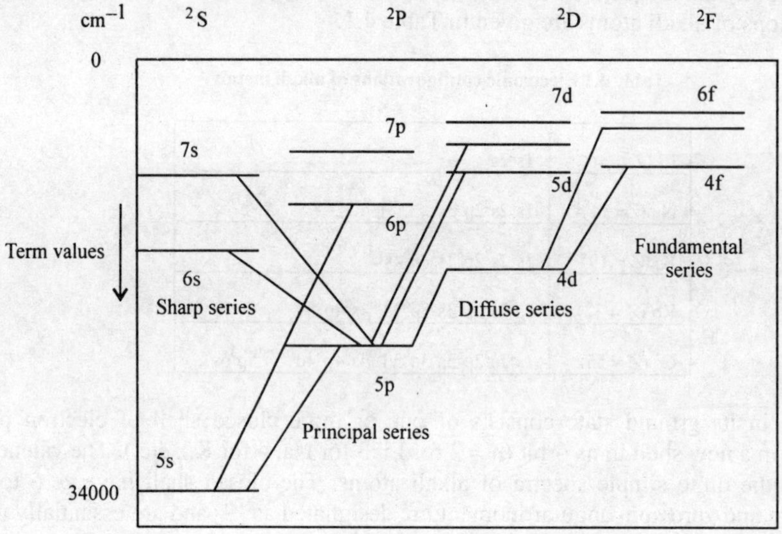

Fig. 4.3 Term diagram of Rb

the energy levels are lower than the hydrogenic value for a given n. Moreover, the energy depends strongly on l. For small l the electron orbit is highly eccentric; the electron penetrate the core and experiences a strong attraction towards the nucleus corresponding to a state of low energy. For larger l the penetration become less and energy levels become more hydrogen like. The non-penetrating orbits are defined if the term value are hydrogen like and that orbits for which term values are different is defined as penetrating orbits. For distances larger than the radii of core of alkali atoms, the nuclear charge Ze is shielded by – (Z-1)e charge of electrons of the core. The valence electron experiences the same Coulomb potential due to a single charge +e as an electron in a hydrogen atom. Thus for large value of n energy levels of alkali atom approaches that of hydrogen.

The transition of electron from one energy level to another is governed by the selection rules $\Delta l = \pm 1$ and are shown in Fig.4.3. allowed transitions between different energy levels give rise to a large number of lines which can be grouped in to four series:

(i) Principal Series: This series arises due to transition from ^{2}P state to normal ^{2}S state. It can be observed in emission as well as in absorption. The lines of this series are intense and are usually observed in absorption spectrum as at normal temperature, most of the atoms are in ground state. The lines of this series lie in ultraviolet region except one, which lies in the visible region. In alkali metal the strongest line in this series corresponds to transition from lowest ^{2}P state to lowest ^{2}S states is called a resonance line (for Rb transition 5 ^{2}P state to 5 ^{2}S state).

(ii) Sharp series: This series arise because of transition from ^{2}S states to lowest ^{2}P state. The lines of this series lie in visible and infrared regions. Lines of this series are quite narrow.

(iii) Diffuse series: This series arises from transition from ^{2}D state to lowest ^{2}P state. The lines of this series lie in the visible and infrared region. The lines of this series are diffuse on one end or both the end of the line.

(iv) Fundamental or Bergmann Series: This series arises from transitions from ^{2}F state to the lowest ^{2}D state. The lines of this series lie in the visible and infrared regions.

The lines of sharp, diffuse and fundamental series occur in emission.

From the energy level diagram (Fig.4.3) it is clear that (a) sharp and diffuse series have common limit (b) the difference between the limit of the principal series and common limit of the sharp and diffuse series is equal to the first member of the principal series. This is known as Rydberg-Schuster law and (c) the difference between the limit of the diffuse and fundamental series is equal to the first line of diffuse series.

Each series in alkali atoms converge towards shorter wavelength, as do the series of hydrogen spectrum. However, the wavenumber of lines fit the formula for hydrogen

$$\bar{v} = R_A Z^2 \left(\frac{1}{n_e'^2} - \frac{1}{n_e''^2} \right)$$
(4.31)

and spectral term

$$T = \frac{R_A Z^2}{n_e^2}$$
(4.32)

if non-integral quantum number n_e are used. The effective quantum number $n_e = n - \mu$ (l) where n is an integer and μ, called quantum defect, is a fraction and is a function of l. R_A is Rydberg constant for element A. According to Rydberg-Ritz

$$n_e = n - \mu - \frac{\delta(l)}{n^2}$$
(4.33)

$n - n_e$ is almost constant within one term series, but slightly larger for the lowest term owing to the Ritz correction $\delta(l)/n^2$. μ (l) is very nearly independent of n for a given l. For Rb, the quantum defects for various configurations are given in Table 4.3. The larger departure of μ (S) from zero and to a lesser extent of μ (P) and μ (D) indicates deep penetration of the core. On the other hand; f configuration are practically hydrogen like. The small dependence of μ (l) on n may be caused by the polarisation of charge cloud of the core by the valence electron. This polarisation gives rise to a further attractive force towards the centre. The valence electron on the average is closer to the core for small values of n.

Table 4.3 Experimental values of the Quantum defect in Rb.

Term	n = 4	n = 5	n = 6	n = 7
S	-	3.195	3.154	3.144
P	-	2.713	2.676	2.664
D	1.233	1.293	1.32	1.33
F	0.011	0.016	0.02	-

The quantum defect is used to explain the increased binding energy of the electron. However, increased binding energy can be explained on the basis of shielding by the core electrons. The net charge seen by the valence electron depends upon the shielding by the core electrons. If the shielding is complete, the valence electron sees only +e charge. On the other hand incomplete shielding increases the charge seen by the valence electron. Therefore, in place of assuming defect in principal quantum number one can write in term values the effective nuclear charge but without quantum defect. The term value is

$$T = \frac{R_A Z_{eff}^2}{n^2} \tag{4.34}$$

Fine structure

Careful examination of alkali spectra reveals that the lines of various series have a structure known as fine structure. It is observed that:

(a) Each member of principal series consists of a pair of transitions known as simple doublets. The difference of the wavenumbers of the components decreases rapidly as we pass towards ultraviolet.

(b) The sharp series members are all simple doublets. The wave number differences of sharp series doublet remain constant.

(c) All members of diffuse series consist of compound doublet.

(d) The wavenumber difference of the first doublet in the principal series is equal to the sharp series doublet separation .

The splitting of the levels and hence splitting of the lines is due to spin-orbit interaction acting on the valence electron. Other relativistic effects, which are important for hydrogen atom is generally quite negligible for the valence electron of the alkali atoms. Assuming Bohr model, the average value of $v/c \sim 10^{-2}$ for the ground state. The associated relativistic effects for valence electron is of same magnitude while spin- orbit interaction increases in magnitude in going from H to elements further up the periodic table, so it dominates the other relativistic effects. All the energy levels of the valence electron, in an alkali atom, except $l = 0$ are split into two. One level corresponding to a total angular momentum $j = l + (1/2)$ and other to $j = l - (1/2)$ with $j = l - (1/2)$ lying lower than $j = l + (1/2)$.

Using Eqs. (2.20), (2.22), (4.19) and (4.20) the term value is

$$\Delta \bar{v} = (\Delta T)_{j=l+1/2} - (\Delta T)_{j=l-1/2} = \frac{\Delta E_1}{hc} - \frac{\Delta E_2}{hc} = \frac{R_A \alpha^2 Z^4}{n^3 l\,(l+1)} \tag{4.35}$$

For a given element $\Delta \bar{v}$ falls off rapidly with increasing n and l. Further, it increases with increase in atomic number. The above expression is derived assuming Coulombic field. However, in alkali atoms, the valence electron is in central field, which is not a Coulombic field because of electrostatic screening by the electrons of the core of the atom. To take account of the penetration of the core, Landé modified the expression by replacing Z^4 by $Z_i^2 Z_0^2$ and n by n_e, where $Z_0 e$ is equal to the charge of the nucleus plus closed shell of electrons and $Z_i e$ is an effective inner charge which must be treated as parameter. As a very rough guide $Z_i \approx Z - 4$ for p electrons and $Z_i \approx Z - 11$ for d electrons. The Eq.(4.35) is known as Landé formula for the doublet splitting.

Fig. 4.4 show the transitions giving rise to the principal doublet arising from the allowed selection rules $\Delta S = 0$, $\Delta l = \pm 1$, $\Delta j = 0, \pm 1$. For the ^2P level, $l = 1$ and $s = 1/2$, therefore $j = 3/2$ and $1/2$. For ^2S level $l = 0$, $s = 1/2$ and $j = 1/2$. The doublet P level split into two while there is no splitting of doublet S level because of spin-orbit interaction. The difference in wavenumber spacing between the doublet reflects the splitting of doublet P level. As the value of n increases, the splitting of ^2P levels decreases. Therefore, for very large value of n the splitting of P level is negligible and in the series limit only one line is observed. For the same value of n, the doublet separation is large in heavy elements. For example the doublet separation for Na and Rb are 2.49 cm^{-1} and 237.6 cm^{-1}, respectively for n = 5.

Fig.4.5 show the fine structure transitions for sharp series, which take place from different excited

^{2}S levels to lowest ^{2}P level. The spacing of the doublet is governed by the splitting of the lowest ^{2}P level. Since the sharp series doublet separation is produced solely by the lowest ^{2}P level on which all the transitions terminate, therefore, it remain constant.

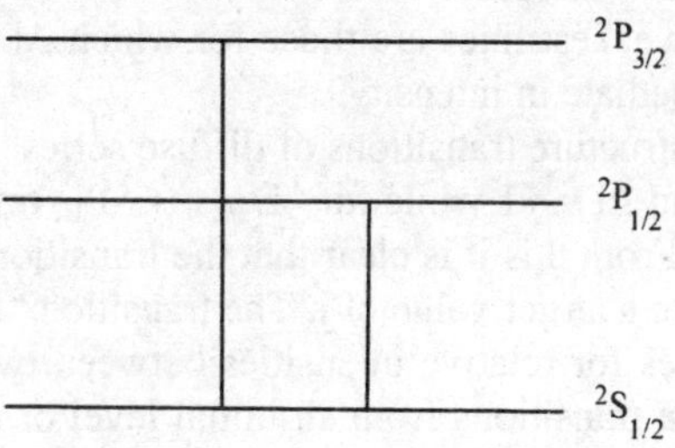

Fig. 4.4 Origin of principal doublet

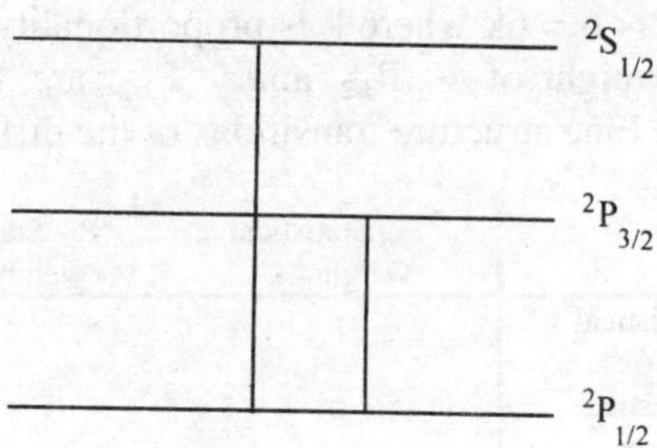

Fig. 4.5 Origin of sharp doublet

Fig. 4.6 shows the fine structure transitions for diffuse series, which take place from different excited ^{2}D level to lowest ^{2}P level. Each of D and P level split into two by spin-orbit interaction. As a result of allowed selection rules, three transitions are observed. For large value of n, the splitting of D levels may be too small for close pair of transitions ($^2D_{5/2} \rightarrow \ ^2P_{3/2}$ and $^2D_{3/2} \rightarrow \ ^2P_{3/2}$) to be resolved. It is for this reason that the set of three transitions has become known as compound doublet rather than triplet. Same situation exists for the fine structure of lines of fundamental series. The spacing of the compound doublet remains constant for large values of n in the case of diffuse and fundamental series. This is because of negligible splitting of excited D and F levels and then splitting is governed by splitting of lowest P and D levels which is constant.

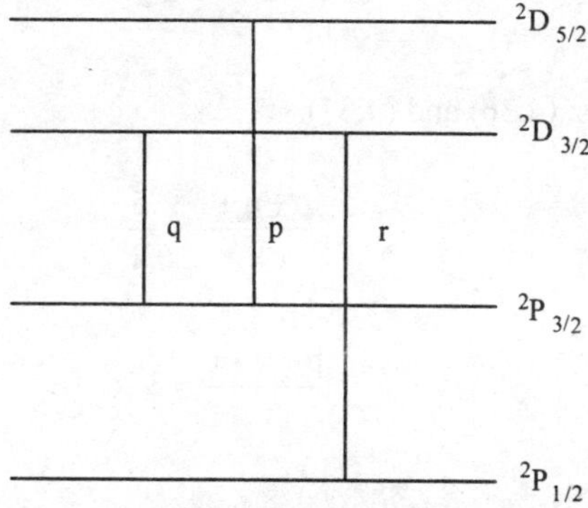

Fig. 4.6 Compound doublet of diffuse series

Intensity rules

(i) For a change Δl ($\neq 0$) in the transition from one term to another in LS coupling the strongest lines are those for which Δj has the same sign as Δl. Of these, the intensities of the lines increase as the magnitude of j increases.

(ii) For a change $\Delta l \neq 0$, The weakest lines are those for which Δj and Δl have opposite sign. The lines with $\Delta j = 0$ are intermediate in intensity.

Let us apply these rules to fine structure transitions of diffuse series. From Fig. 4.6 it is clear that for transition $^2D_{5/2} \rightarrow {}^2P_{3/2}$, $\Delta j = 0$ and $\Delta l = +1$ while for $^2D_{3/2} \rightarrow {}^2P_{1/2}$ transition $\Delta j = 1$ and $\Delta l = +1$, that is, Δj and Δl have the same sign. From this it is clear that the transition $^2D_{5/2} \rightarrow {}^2P_{3/2}$ is stronger than $^2D_{3/2} \rightarrow {}^2P_{1/2}$ as the former involve a larger value of j. The transition $^2D_{3/2} \rightarrow {}^2P_{3/2}$ is weakest as $\Delta j = 0$ though $\Delta l = 1$. Quantitative rules for relative intensities between two multiplets in LS coupling is: the sum of the intensities of all the transitions from an initial level or to a final level of a multiplet is proportional to the statistical weight, $2j + 1$, of that level. The constant of proportionality is common to all levels of a given multiplet.

Let us consider again the fine structure transitions of diffuse series. It involves three spectral lines. The statistical weight of $^2D_{5/2}$ is $\propto 6 = 6k$ where k is proportionality constant. The statistical weight of $^2D_{3/2} = 4k$. The statistical weight of $^2P_{3/2}$ and $^2P_{1/2}$ are 4k' and 2k', respectively. k' is proportionality constant of P level. Fine structure transitions of the diffuse series can be written as

	$^2P_{3/2}$ Statistical weight 4k'	$^2P_{1/2}$ Statistical weight 2k'
$^2D_{5/2}$ Statistical weight 6k	p	-
$^2D_{3/2}$ Statistical weight 4k	q	r

Let p, q and r represent unknown intensities of the three allowed transitions and zero for the forbidden transition (Fig. 4.6). From the sum rules: the sum of lines starting from $^2D_{5/2}$ is to the sum starting from $^2D_{3/2}$ is

$$\frac{p}{q+r} = \frac{6k}{4k} = \frac{3}{2} \tag{4.36}$$

Sum of the lines ending on $^2P_{3/2}$ to the sum ending on $^2P_{1/2}$ is

$$\frac{p+q}{r} = \frac{4k'}{2k'} = 2 \tag{4.37}$$

Adding 1 to both sides of Eqs. (4.36) and (4.37)

$$\frac{p+q+r}{q+r} = \frac{5}{2} \tag{4.38}$$

$$\frac{p+q+r}{r} = 3 \tag{4.39}$$

Dividing Eqs. (4.38) and (4.39)

$$\frac{r}{q+r}=\frac{5}{6} \tag{4.40}$$

From Eq. (4.40) we have, r = 5q. Putting this value of r in Eqs. (4.36) and (4.37) and on solving we obtain p = 9q. Thus p: q: r = 9: 1: 5.

4.3 Spectra of the He Atom

The helium (He) has electronic configuration $1s^2$. These two electrons in the configuration are responsible for the chemical valence of two and general characteristic of the optical spectra. Helium in its ground state have two 1s electrons. For this system, both the electrons have l = 0. Since n, l and m_l are same therefore, according to the Pauli exclusion principle their spins are antiparallel. Thus for $1s^2$, S and L are zero and hence J = 0. The ground state is accordingly $^1S_0$. Let us assume that one of the two electrons is excited to the 2 p orbits yielding a configuration 1s2p (n = 1 for the s electron and n = 2 for the p electron). Pauli exclusion principle does not apply, as the two electrons are non-equivalent. For the s and p electrons, l = 0 and 1, respectively. This leads to L = 1. The state is therefore, represented by the symbol P. The spins can either be parallel or antiparallel giving S = 1 and 0, respectively. In the first case multiplicity is 2S+1 = 3 (triplet) and for S = 0 the multiplicity is 1(singlet). The J values for triplet are 2,1,0 and for singlet J = 1. The possible terms are therefore, $^3P_{210}$ and $^1P_1$. Similarly, one of the two electrons is promoted to various excited configuration and resulting states are given in Table 4.4.

Table 4.4 LS terms for some configurations

Normal Configuration	Excited Configuration	n	Terms
$1s^2$	-		$^1S_0$
$1s^2$	1sns	2, 3, 4, ...	$^1S_0$, $^3S_1$
$1s^2$	1snp	2, 3, 4, ...	$^1P_1$, $^3P_{210}$
$1s^2$	1snd	3, 4, 5, ...	$^1D_2$, $^3D_{321}$
$1s^2$	1snf	4, 5, 6, ...	$^1F_3$, $^3F_{432}$

The ground state configuration leads only to singlet whereas each excited configuration arising from the promotion of an electron gives rise to singlet and triplet states. According to Hund's rule, the triplet states lie lower than the corresponding singlet states. Fig. 4.7 show the energies of the ground state and first few excited states of helium atom obtained from an analysis of the optical line spectra. For the same n, the energy of the term increases in the order of nS, nP, nD, nF, ...From the Fig.4.7 it is seen that there is no $1\,^3S$ state in helium. The lowest triplet state is $2\,^3S$, although the lowest singlet state is $1\,^1S$. Because of Pauli's exclusion principle the state 1^3S is not allowed. The energy difference between the 1^1S and 2^3S is quite large (19.72 eV) reflecting the tight binding closed shell electrons. Helium atoms in singlet states (antiparallel spins) constitute parahelium and those in triplet states (parallel spins) constitute orthohelium. The lowest state of orthohelium, that is, 2^3S does not combine with the ground state 1^1S. Those terms, which cannot go to a lower state with the emission of radiation and correspondingly cannot be reached from a lower state by absorption, are called metastable states. The 2^1S state is also metastable, since the selection rule $\Delta l = \pm 1$ does not allow any transition to 1^1S. The metastability of the 2^3S state is however, stronger than that of 2^1S, since the transition $2^3S \rightarrow 1^1S$ would contradict the prohibition of an ortho-para transition as well as $\Delta L = \pm 1$. However, $2^3S \rightarrow 1^1S$ transitions can be caused by collision with other atoms or with the walls of a containing vessel, which distorts the electronic structure of the atom, and allow a change of spin to occur.

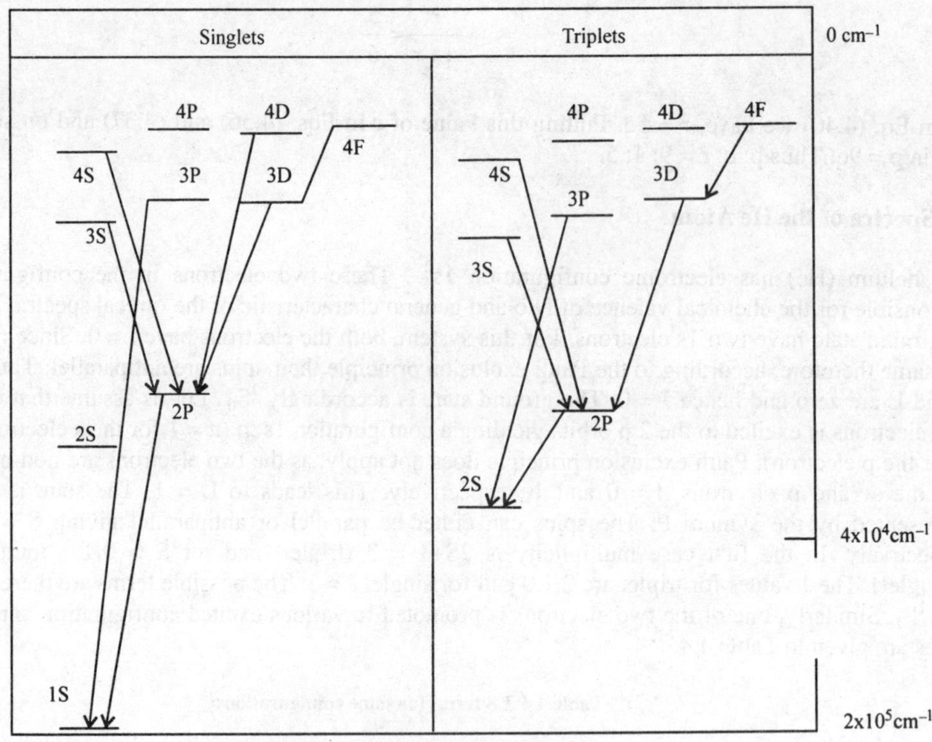

Fig. 4.7 Energy level diagram of He

The transition from one energy state to another is governed by the selection rule $\Delta L = \pm 1$, $\Delta S = 0$, $\Delta J = 0, \pm 1$. Allowed transitions between different energy levels gives rise to a large number of lines which can be put into two distinct system of lines: a system of closely spaced triplets and a system of singlets. In the emission spectrum of helium atom, lines of singlet and triplet systems may be grouped into four series each:

(i)   Principal series: Two principal series for singlet are observed one is from higher P state to normal 1^1S_0 state and member of this series lie in far ultraviolet region. This series also exists in absorption. Besides this there exists another principal series whose spectral lines arise from transitions from higher ^1P states to the 2^1S state. This principal series lie from infrared to near ultraviolet region. The members of the triplet principal series are due to transitions from higher ^3P state to 2^3S state. The triplet principal series lie from infrared to ultraviolet region.

(ii) Sharp series: The sharp series arise due to transitions from S states to the lowest P state. The lines of sharp series of singlet and triplet system lie in visible and ultraviolet region.

(iii) Diffuse series: The diffuse series arise from transition between higher D states to the lowest P state. The lines of this series for both singlet and triplet system lie in visible and ultraviolet region.

(iv) Fundamental series: This series arise from transitions between higher F states to the lowest D state. The lines of this series for singlet and triplet system lie in infrared region.

Under high resolution, the lines of parahelium system are single and those of orthohelium show closely spaced triplet structure, which is very difficult to resolve. The triple fine structure is due to spin-orbit interaction. Fig 4.8(a) shows $^3P \rightarrow {}^3S$ transition. ^3S state has one component $^3S_1$ while ^3P has three components $^3P_2$, $^3P_1$ and $^3P_0$. Three lines are observed according to allowed selection rules, $\Delta L = \pm 1$, $\Delta J = 0, \pm 1$ and $\Delta S = 0$, between these states. Fig.4.8 (b) show the fine structure of triplet diffuse series. For triplet D we have S = 1, L = 2 and hence J = 3,2,1. According to allowed selection rules, six transitions are possible. The complete spectrum should consist of six lines. Normally, the very close spacing is not resolved and only three lines are observed. For this reasons the spectrum is referred to as compound triplet. For He, the component separations $^3P_0 - ^3P_1 = 0.996$ cm^{-1} and $^3P_1 - ^3P_2 = 0.078$ cm^1. The terms of 2^3P of He are inverted with $^3P_0$ component lying highest in energy and $^3P_2$ the lowest. Further there is breakdown of Landé interval rule. This is due to spin-spin interaction.

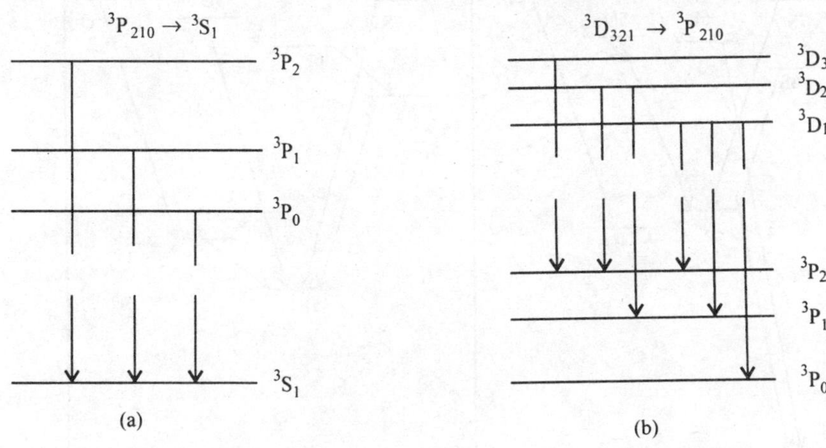

Fig. 4.8 Fine structure splitting of (a) $^3P_{210} \rightarrow ^3S_1$, (b) $^3D_{321} \rightarrow ^3P_{210}$ transitions. The component of each triplet levels are shown in normal order . For He the order is inverted in both P and D state

4.4 Spectra of Two Valence Electron Systems

The normal electronic configuration of neutral atoms of group II elements (alkaline elements) are given in Table 4.5 .

Table 4.5 Ground configuration of group II elements

Element	Atomic number Z	Ground state configuration
Be	4	$1s^22s^2$
Mg	12	$1s^22s^22p^63s^2$
Ca	20	$1s^22s^22p^63s^23p^64s^2$
Zn	30	$1s^22s^22p^63s^23p^63d^{10}4s^2$
Sr	38	$1s^22s^22p^63s^23p^63d^{10}4s^24p^65s^2$
Cd	48	$1s^22s^22p^63s^23p^63d^{10}4s^24p^64d^{10}5s^2$
Ba	56	$1s^22s^22p^63s^23p^63d^{10}4s^24p^64d^{10}5s^25p^66s^2$
Hg	80	$1s^22s^22p^63s^23p^63d^{10}4s^24p^64d^{10}4f^{14}5s^25p^65d^{10}6s^2$

It is the last two equivalent electrons in the configuration for the group II elements that are responsible for the general characteristic of the optical spectra. These elements in their ground state have two equivalent s electrons in the nth (n = 2, 3, 4, 5 and 6) orbit and therefore the ground state is n^1S_0. On exciting one of these two electrons to higher s, p, d or f orbit one obtains a system of two non-equivalent electrons. These system gives singlet and triplet S, P, D and F states as described in the case of helium.

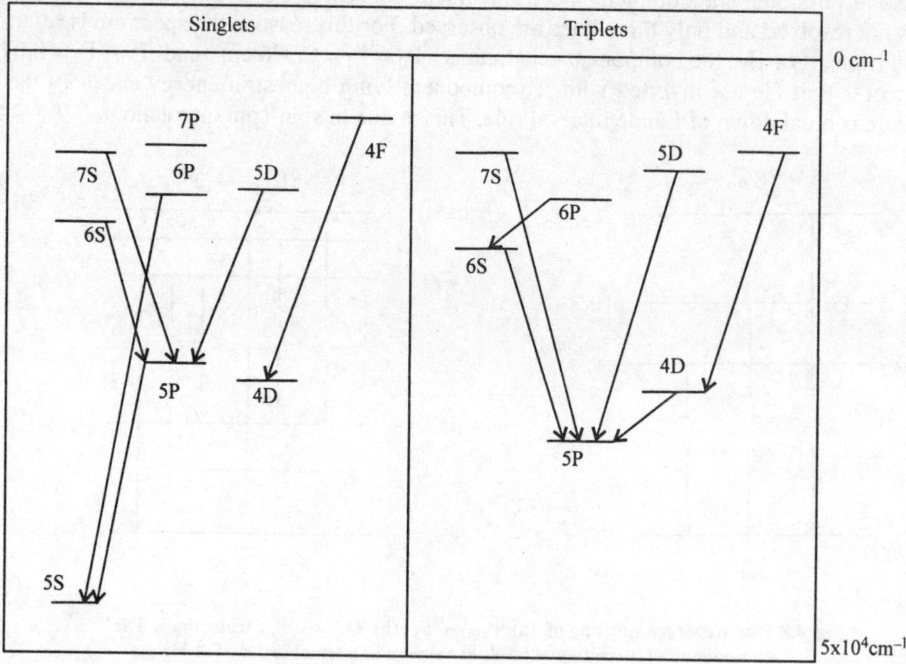

Fig. 4.9 Energy level diagram of Sr

The ground state configuration leads only to singlet states whereas each excited configuration arising from the promotion of an electron gives rise to singlet and triplet states with later lying lower than the corresponding singlet state. Fig. 4.9 shows the energies of the ground states and first few excited states of Sr atom obtained from the analysis of the optical line spectra. This figure also shows some of the transitions allowed by selection rules given in Sec.4.3.

In the Fig.4.9, the fine structure splitting is not shown. Allowed transitions between different energy levels give rise to a large number of lines which can arise from transitions (a) between closely spaced triplets and (b) between a systems of singlet. In the spectra of alkaline earth metals, the lines of triplet system and of singlet system can be further classified into four chief series. These are:

(i)   Principal series: This series arise from transitions from higher P states to the lowest S state.

(ii) Sharp series: This series arise from transitions from higher S states to the lowest P state

(iii) Diffuse series: This series arise from transitions from excited D states to the lowest P state.

(iv) Fundamental series: This series arise because of transitions from excited F states to lowest D state.

The regions in which the lines of these four series of singlet and triplet system lie for various elements are given in Tables 4.6 and 4.7, respectively. Each series converge towards shorter wavelength to fit the wavenumber of the spectral line to hydrogen like formula

$$\bar{\nu} = RZ^2 \left(\frac{1}{n_e''^2} - \frac{1}{n_e'^2} \right) \tag{4.41}$$

Table 4.6 Spectral region of singlet series of II A group elements

Element	Principal series	Sharp series	Diffuse series	Fundamental series
Be	UV	IR-UV	Visible, UV	IR
Mg	UV	IR, Visible	IR, Visible	IR, Visible
Ca	Visible, UV	IR, Visible	Visible	Visible, UV
Zn	UV	IR-UV	Visible, UV	IR, Visible
Sr	Visible, UV	IR, Visible	IR, Visible	Visible
Cd	UV	IR-UV	Visible, UV	IR, Visible
Ba	Visible, UV	IR, Visible	IR, Visible	IR, Visible
Hg	UV	IR-UV	IR, Visible	IR, Visible

Table 4.7 Spectral region of triplet series of II A group elements

Element	Principal series	Sharp series	Diffuse series	Fundamental series
Be	IR, Visible	UV	UV	IR
Mg	IR, Visible	Visible, UV	Visible, UV	IR, Visible
Ca	IR, Visible	Visible, UV	IR - UV	Visible, UV
Zn	IR, Visible	Visible, UV	UV	IR
Sr	IR, Visible	Visible, UV	IR - UV	Visible, UV
Cd	IR, Visible	Visible, UV	UV	IR
Ba	IR, Visible	Visible, UV	Visible, UV	Visible, UV
Hg	IR, Visible	Visible, UV	UV	IR

are successful only if non integral quantum numbers are used. For each spectral series the term $T'' = R Z^2 / n_e''^2$ is constant and is the wavenumber of series limit. The effective quantum number is $n_e = n - \mu$, where n is an integer and μ is a fraction called quantum defect which is zero for non-penetrating orbits. As in case of alkali atoms, the penetration of various possible electron orbits into the atom core is measured by the deviation of the term values from those of hydrogen. Two methods are used to measure the penetration (i) by modifying the nuclear charge in the term value expression for hydrogen, (ii) modifying the principal quantum number. These leads to term values for (i) as

$$T = \frac{R(Z - \sigma)^2}{n^2} \tag{4.42}$$

and for (ii) as

$$T = \frac{RZ_0^2}{(n - \mu)^2} \tag{4.43}$$

where σ is a screening constant, Z_0 is the effective nuclear charge when the electron is well outside the core. From the spectrum, ignoring fine structure, it is observed that:

(i)   Triplet sharp and diffuse series have one common limit and singlet sharp and diffuse another.

(ii) The frequency difference between common limit of triplet sharp and diffuse series and triplet principal series is equal to the frequency of the first member of the principal series (Rydberg-Schuster law).

(iii) The frequency difference between the common limit of the triplet sharp and diffuse series and the triplet fundamental series is equal to the frequency of the first member of the triplet diffuse series (Runge law).

(iv) The same is true even if fine structure is taken into account.

(v) (i), (ii), and (iii) holds for singlet series also.

Fine structure

The examination of alkaline spectra reveals that lines of various triplets' series have a structure known as fine structure. It is observed that:

(i)   All members of the principal series are composed of three lines with decreasing separation with increasing frequency, and approach a single limit.

(ii) All members of sharp series are composed of three lines with the same separation, and approach a triple limit.

(iii) All members of diffuse and fundamental series contain six lines, three strong and three satellites and approach a triple limit.

The fine structure in triplet series is due to splitting of various energy levels by spin-orbit interaction as described previously in case of alkali atoms. In case of alkaline metals, in triplet levels we have $S = 1$ and L can take values 0, 1, 2 and 3 for S, P, D and F states, respectively. For S state, $S = 1$ and $L = 0$, therefore, $J = 1$. For $^3S_1$ there is no splitting of the level. For P state, $L = S = 1$, therefore $J = 2,1,0$ and P state is split into three components with different J values. Similarly, D and F states, represented by $^3D_{321}$ and $^3F_{432}$, respectively splits into three components each.

In the triplet principal series transitions take place from $^3P_{210}$ to lowest $^3S_1$ level as shown in the Fig. 4.10. According to selection rules

$$\Delta S = 0, \Delta L = \pm 1, \Delta J = 0, \pm 1$$

three transitions take place. The spacing between the transitions depends on the splitting of P level which decreases with increasing n [Eq. (4.35)]. For large value of n, the splitting of the P level is negligible and as a result, one observes a single transition in series limit.

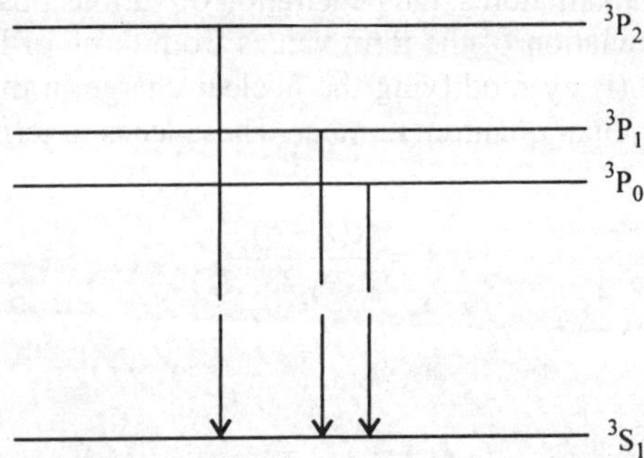

Fig. 4.10 Structure of simple triplet $^3S - ^3P$

Fig.4.11 show the fine structure transitions for sharp series, which take place from different excited

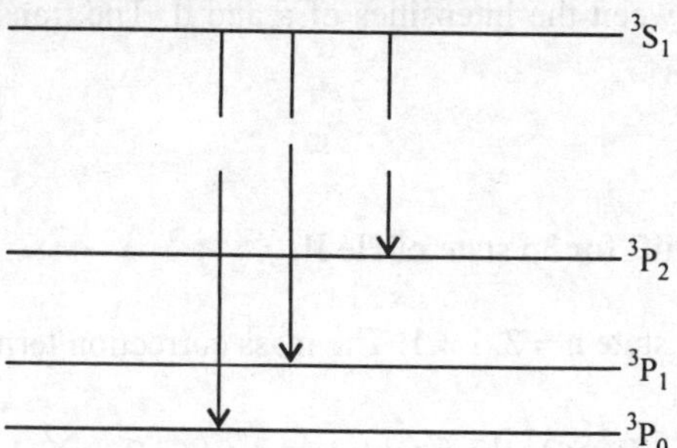

Fig. 4.11 Fine structure transitions for sharp series

triplet S states to lowest triplet P state. As a result of allowed transitions, three lines are observed. The spacing between the lines of triplet is governed by the splitting of the lowest triplet P state. Since the sharp series triplet separation is produced by the splitting of the lower $^3P_{210}$ level on which all the transitions terminates, therefore it remains constant.

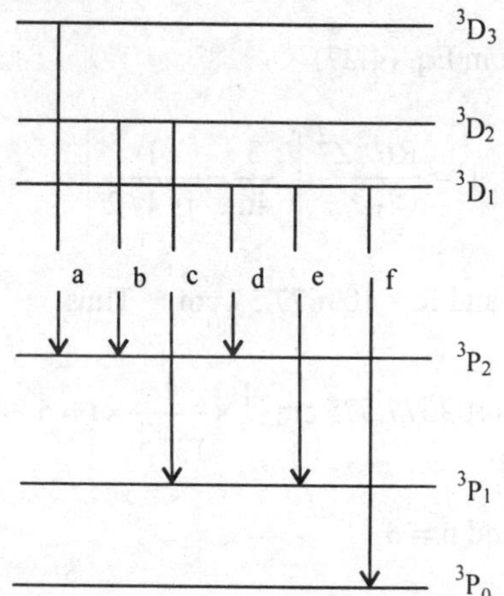

Fig. 4.12 Structure of composite triplet $^3P - ^3D$

Fig.4.12 shows the fine structure transitions for the triplet diffuse series, which take place from different excited ^3D levels to lowest ^3P level. Each of ^3P and ^3D level splits into three levels. As a result of allowed selection rules, six transitions are possible. The spacing between the transitions depends on the splitting of ^3P and ^3D levels. For large value of n, the spacing between $^3D_{321}$ levels diminishes and therefore $^3D_{321}$ levels are not resolved. As a result of this only three transitions are observed whose spacing depends on the splitting of $^3P_{210}$ levels. Same situation exists for the fundamental series. However, now for large value of n, the spacing between the transitions depends on the splitting of $^3D_{321}$ levels, which is constant. Since for sharp and diffuse series the transitions terminate at lowest ^3P level therefore, for large value of n, the spacing between transitions is same for both sharp and diffuse series.

Qualitatively, the relative intensity can be obtained by rules given for alkali atom spectra. Let us apply those rules for the members of diffuse series. For transitions (a) $\Delta J = +1$ with $J = 3$, (b) $\Delta J = 0$, $J = 2$, (c) $\Delta J = +1$, $J = 2$, (d) $\Delta J = -1$, $J = 1$, (e) $\Delta J = 0$, $J = 1$and (f) $\Delta J = +1$, $J = 1$.For all these transitions

ΔL has the same value. Thus transition a is the strongest and d is the weakest. The intensities of other transitions are intermediate between the intensities of a and d. The transitions in order of decreasing intensities are a >c >f >b >e >d.

Examples

1.   Evaluate mass correction shift for 3p state of He II.

   For He II, Z = 2 and for 3p state n = 3, l = 1. The mass correction term is given by Eq. (4.12)

$$\Delta E_1 = \frac{(1/137)^2 \times 2^2}{3^2}\, \frac{13.6\,\text{eV}}{3^2}\left(\frac{3}{4}-2\right)=\left(\frac{2}{9\times137}\right)^2 \frac{13.6\times5}{4}\,\text{eV}$$

$$\Delta E_1 = 4.472834\times10^{-5}\ \text{eV}$$

2.   Find the relativistic energy-shifts of 3s, 3p and 3d for H (R = 109677,57 cm^{-1}, α = 1/137)

   The relativistic energy shift from Eq. (4.27)

$$\Gamma = \frac{R\alpha^2 Z^4}{n^3}\left[\frac{3}{4n}-\frac{1}{j+1/2}\right]$$

   For hydrogen Z= 1, α = 1/137 and R = 109677.57 cm^{-1} Thus

$$R\alpha^2 Z^4 = 109677.575\ \text{cm}^{-1}\times\frac{1}{137^2}\times1 = 5.84\ \text{cm}^{-1}$$

   For 3s, l = 0, s = 1/2 , j = 3/2 and n = 3

$$\Gamma = \frac{5.84\ \text{cm}^{-1}}{3^3}\left[\frac{3}{12}-1\right] = \frac{5.84\ \text{cm}^{-1}}{27}\left(-\frac{3}{4}\right) = -0.1622\ \text{cm}^{-1}$$

   For 3p, n = 3, l = 1, s = 1/2 and j = 1/2, 3/2

$$\Gamma_{3,1/2} = \frac{5.84\ \text{cm}^{-1}}{27}\left[\frac{1}{4}-1\right] = -0.1622\ \text{cm}^{-1}$$

$$\Gamma_{3,3/2} = \frac{5.84\ \text{cm}^{-1}}{27}\left[\frac{1}{4}-\frac{1}{2}\right] = -\frac{5.84\ \text{cm}^{-1}}{27\times4} = -0.054\ \text{cm}^{-1}$$

   For 3d, n = 3, l =2, s = 1/2 and j = 3/2 , 5/2

$$\Gamma_{3,3/2} = -0.054 \text{ cm}^{-1}$$

$$\Gamma_{3,5/2} = \frac{5.84 \text{ cm}^{-1}}{27}\left[\frac{1}{4} - \frac{1}{3}\right] = -\frac{5.84 \text{ cm}^{-1}}{27 \times 12} = -0.0180 \text{ cm}^{-1}$$

3. Find the relativistic energy shifts of 3s, 3p and 3d for Be IV and compare the result with that of hydrogen atom.

Here Z = 4 for Be IV, therefore

$$R\alpha^2 Z^4 = 5.84 \text{ cm}^{-1} \times 4^4 = 5.84 \text{ cm}^{-1} \times 256$$

so every value for 3s, 3p, 3d for Be IV would be 256 times 3s, 3p and 3d for H as evaluated in example 2.

4. If the doublet splitting of the first excited state $2^2P_{3/2}$ of HeII is 5.84 cm^{-1}, calculate the corresponding separation of H.

From Eq. (4.27) $\Delta\Gamma \propto Z^4$

$$\frac{\Delta\Gamma_H}{\Delta\Gamma_{He}} = \frac{(Z)^4_H}{(Z)^4_{He}} \times 5.84 \text{ cm}^{-1} = \frac{5.84 \text{ cm}^{-1}}{16} = 0.365 \text{ cm}^{-1}$$

5. For the lowest ^2P term for Na, the doublet splitting is 17.2 cm^{-1} and quantum defect is 0.884. Determine the value of Z_i using Landé formula.

The Landé formula for doublet splitting is given by Eq. (4.35)

$$\Delta\bar{v} = \frac{5.84 \text{ cm}^{-1} Z_i^2 Z_0^2}{n_e^3 \, l \, (l+1)}$$

for p state, $l = 1$ and lowest ^2P term has n = 3 and therefore $n_e = 3 - 0.884 = 2.116$ and $Z_0 = 1$.

$$\Delta\bar{v} = 17.2 \text{ cm}^{-1} = \frac{5.84 \text{ cm}^{-1} Z_i^2}{(2.116)^3 \times 2}$$

$$Z_i^2 = \frac{17.2 \times (2.116)^3 \times 2}{5.84} = 55.80$$

$$Z_i = 7.47 \approx 8$$

6. If the valence electron in sodium is excited to the 4^2D state, what are different routes open for the electron in returning to the normal state.

An electron in a 4^2D state returns to the normal state 3^2S by following the selection rules $\Delta l = \pm 1$. It has the following possible routes open along which it may return to the normal state:

1. $4^2D \rightarrow 4^2P \rightarrow 3^2D \rightarrow 3^2P \rightarrow 3^2S$
2. $4^2D \rightarrow 4^2P \rightarrow 3^2S$
3. $4^2D \rightarrow 3^2P \rightarrow 3^2S$
4. $4^2D \rightarrow 4^2P \rightarrow 4^2S \rightarrow 3^2P \rightarrow 3^2S$

7. Calculate the Rydberg denominators for the first term value of the principal series of sodium. The wavenumber of the transition is 16973.7 cm^{-1}. The $R_{Na} = 109734$ cm^{-1}.

Spectral term is given by

$$T = -\frac{R_{Na}}{(n-\mu)^2} = \frac{R_{Na}}{n_e^2}$$

$$n_e^2 = \frac{109734 \text{ cm}^{-1}}{16973.7 \text{ cm}^{-1}} = 6.4649$$

$$n_e = 2.54$$

8. The effective quantum number for the ground state of Rb is 1.805. Determine the ionization potential of Rb ($R_{Rb} = 109737$ cm^{-1}).

From Eq. (4.32) term value of the ground state is

$$T = \frac{R}{n_e^2} = \frac{109737 \text{ cm}^{-1}}{(1.805)^2} = 33682.06 \text{ cm}^{-1}$$

the ionization energy I is equal to remove the electron from the ground state and is equal to energy corresponding to the ground state term

$$I = \frac{33682.06}{8065} \text{ eV} = 4.176 \text{ eV}$$

9. The first ionization potential of sodium is two fifth that of hydrogen. Calculate the effective nuclear charge of the sodium atom as far as the 3s electron is concerned.

For hydrogen

$$E_n = -\frac{RZ^2}{n^2}$$

For H, Z = 1, n = 1. $E_1 = E_H = -R$ cm^{-1} = -13.6 eV. The ionization energy of hydrogen is thus 13.6 eV.

$$\frac{I_H}{I_{Na}} = \frac{(RZ_{eff}^2 / n^2)}{[RZ_{eff}^2 / n^2]_{Na}}$$

For Na, n = 3 and for hydrogen Z_{eff} = 1. Putting (I_{Na}/I_H) = 2/5 in the above equation

$$\frac{5}{2} = \frac{9}{Z_{eff}^2}$$

$$Z_{eff}^2 = \frac{18}{5} = 3.6 \text{ and } Z_{eff} = 1.897$$

10. The principal and sharp series for the Li atom converge to continua at 43487 and 28583 cm^{-1}, respectively. Calculate the quantum defect for the common term in each of the series (R_{Li} = 109729 cm^{-1}).

The common term for sharp and principal series is 2^2P term. The wavenumber position of this term is 28535 cm^{-1} and

$$n_e^2 = \frac{R_{Li}}{T} = \frac{109729 \text{ cm}^{-1}}{28583 \text{ cm}^{-1}} = 3.83896$$

$$n_e = 1.9593$$
$$\mu = n - n_e = 2 - 1.9593 = 0.0407$$

11. Six lines lying close together in the spectrum of Sr. Their wavenumbers are 20109, 20124, 20147, 20503, 20518 and 20690 cm^{-1}. Identify the energy levels connected by this group of transitions.

Strontium has two valence electrons giving singlet and triplet terms. We assume LS coupling is valid and that the six lines arise from transitions between two triplet terms. We therefore look for an interval rule, which will show up as the ratio of small integral numbers between some of the frequency differences. The frequency differences are 0, 15, 38, 394, 409 and 581 cm^{-1} . It is found that

$$\frac{38-15}{15-0} = \frac{23}{15} \approx \frac{3}{2}$$

and

$$\frac{394-0}{581-394} = \frac{394}{187} \approx \frac{2}{1}$$

we therefore suspect that according to the interval rule, the level J = 3, 2, 1 occur in one term, which must therefore be a ^{3}D term, and the level J = 2, 1, 0 occur in the other, which must be a ^{3}P term. The energy level diagram as shown in the Fig. 4.13 can be constructed from this information.

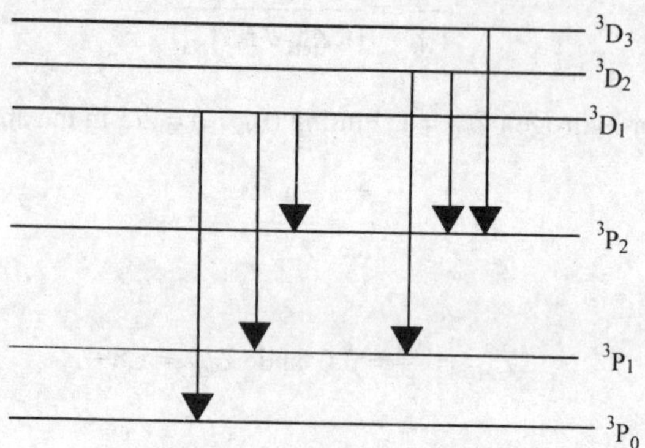

Fig. 4.13 Energy level diagram

12. The principal and sharp series for the sodium atom converge to continua at 41450 and 24477 cm^{-1}, respectively. Calculate the ionization potential of Na.

The wavenumber corresponding to series limit of principal series give the position of normal ground state 3^2S of Na. Hence ionization energy corresponds to 41450 cm^{-1}. Since 1 eV $\approx$ 8065 cm^{-1}. Thus 41450 cm^{-1} is equal to (41450/8065) eV = 5.139 eV.

13. Compute value of $Z - \sigma$ and quantum defect μ for the electron configuration 4s4p of Ca. The term value is 33989 cm^{-1} and Rydberg constant is 109737 cm^{-1}.

From Eq. (4.42)

$$(Z-\sigma)^2 = \frac{Tn^2}{R} = \frac{33989 \text{ cm}^{-1} \times 4^2}{109737 \text{ cm}^{-1}} = 4.9557$$

$$Z - \sigma = 2.226$$

From Eq. (4.43)

$$(n-\mu)^2 = \frac{RZ_0^2}{T} = \frac{109737 \text{ cm}^{-1}}{33989 \text{ cm}^{-1}} =$$

$$n - \mu = 1.7968$$
$$\mu = 2.2032$$

14. Determine theoretical intensity ratio for the doublet trasnsition $^2F - {}^2G$.

   The fine structure transitions $^2F - {}^2G$ involves three spectral lines. The statistical weight of $^2F_{5/2}$ is $\propto 6 = 6k$ where k is proportionality constant. The statistical weight of $^2F_{7/2} = 8k$. The statistical weight of $^2G_{7/2}$ and $^2G_{9/2}$ are 8k' and 10k', respectively. k' is proportionality constant of G level. Fine structure transitions of $^2F - {}^2G$ can be written as

	$^2F_{7/2}$ Statistical weight 8k	$^2F_{5/2}$ Statistical weight 6k
$^2G_{9/2}$ Statistical weight 10k'	p	-
$^2G_{7/2}$ Statistical weight 8k'	q	r

the numbers directly below and to the right of the term symbols are the statistical weight. Let p, q and r represent unknown intensities of the three allowed transitions and zero for the forbidden transition (Fig. 4.14). From the sum rules: the sum of lines starting from $^2G_{9/2}$ is to the sum starting from $^2G_{7/2}$ is

$$\frac{p}{q+r} = \frac{10k'}{8k'} = \frac{5}{4}$$

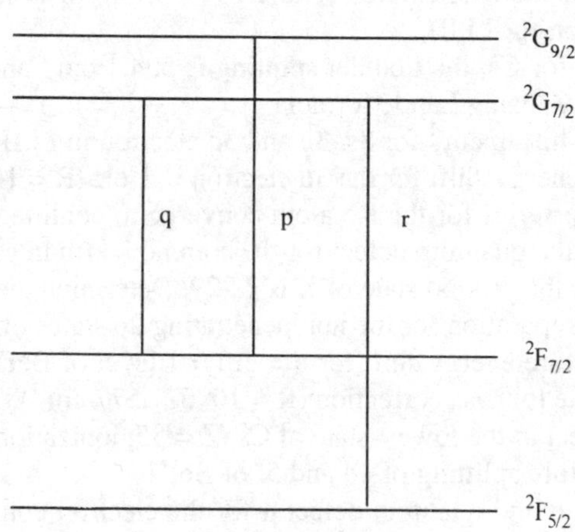

Fig. 4.14 Composite doublet $^2G - {}^2F$

Sum of the lines ending on $^2F_{7/2}$ to the sum ending on $^2F_{5/2}$ is

$$\frac{p+q}{r} = \frac{8k}{6k} = \frac{4}{3}$$

Adding 1 to both sides of above equations

$$\frac{p+q+r}{q+r} = \frac{18}{8} = \frac{9}{4}$$

$$\frac{p+q+r}{r} = \frac{7}{3}$$

On dividing

$$\frac{r}{q+r} = \frac{27}{28}$$

From the above equation we have, $r = 27q$. Using this value of r we obtain $p = 35\,q$. The intensity ratio is thus $p:q:r = 35:1:27$.

Problems

4.1   If the valence electron in sodium is excited to the 4^2P state, what are different routes open for the electron in returning to the normal state ?

4.2   If the valence electron in sodium is excited to the 3^2D state, what are different routes open for the electron in returning to the normal state ?

4.3   If the valence electron in Lithium is excited to 3^2D state, what are different routes open for the electron in returning to the normal state 2^2S.

4.4   Calculate the fine structure splitting of 2p, 3p and 4p of Boron like hydrogen.

4.5   The doublet splitting of the first excited state $2^2P_{3/2} - 2^2P_{1/2}$ of H is 0.365 cm^{-1}. Calculate the corresponding separation for Li III.

4.6   For the lowest ^2P term for Cs, the doublet splitting is 554.1 cm^{-1} and quantum defect is 3.623. Determine the value of Z_i using Landé formula.

4.7   Find the fine structure shift in cm^{-1} for 3s, 3p and 3d electron in Li III (R = 109677.57 cm^{-1}).

4.8   Find the fine structure energy shift for the 3d electron in HeII (R = 109677.57 cm^{-1}).

4.9   The principal and sharp series for the Rb atom converge to continua at 20874 and 33691 cm^{-1}, respectively. Calculate the quantum defect for the common term in each of the series.

4.10   The quantum defect for the ground state of K is 2.229. Determine the ionization energy of K.

4.11   Compute doublet term separation for the non penetrating 2p states of Li.

4.12   Calculate the fine structure energy shift for n = 3, l = 1 level of BeIV in cm^{-1} and compare this with the energy shift due to mass correction (R = 109677.577 cm^{-1}).

4.13   Calculate quantum defect in the lowest state of Cs (Z = 55, ionization potential = 3.89 eV).

4.14   Calculate the fine structure splitting of 4d and 5f of BeIV.

4.15   Compute value of $Z - \sigma$ and quantum defect μ for the electron configuration 4s5d of Ca. The term value is 6557 cm^{-1} and Rydberg constant is 109737 cm^{-1}.

4.16   Determine theoretical intensity ratio for the doublet trasnsition $^2F - {}^2D$.

4.17   In the high-resolution emission spectrum of the cesium atom, the first member of the principal series has fine structure components at 11178.7 and 11732.7 cm^{-1}. The first member of the fundamental series has components at 9971 and 9873 cm^{-1}. The multiplet splitting is negligibly small for the ^2F term. Calculate the relative separation of the components of the first member of the diffuse series.

4.18   The term values of some of the singlet states of Mg are:

State	Term values (cm^{-1})
3S	61669
3P	26618
4S	18166
3D	15266

Calculate the wavelengths of the allowed transitions.

4.19   The term values of some of the states of Na are as follow:

State	Term values (cm^{-1})
3S	41449.7
$3P_{1/2}$	24493
$3P_{3/2}$	24476
4S	15710.2
3D	12277
4F	6861

Sketch these states on an energy level diagram and calculate the wavelengths of allowed transitions

5

X - Ray Spectroscopy

5.1 The Origin of X-Rays

X-rays are radiations in the electromagnetic spectrum of wavelength between ~10 pm and ~ 10 nm. X-rays are characterized by index of refraction very close to 1 for all materials. Fig. 5.1 shows a diagram of an X-ray tube. Electrons are emitted thermally from the heated cathode C and are accelerated

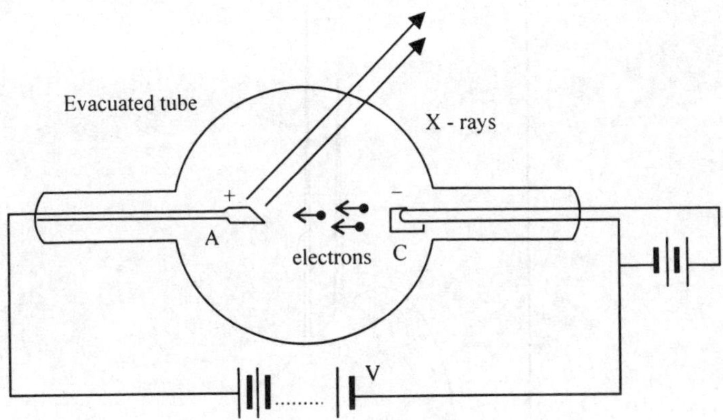

Fig. 5.1 An X – ray tube

towards the anode target A , made of a metal, by the applied potential V. X-rays are produced when accelerated electrons strike the target. The face of the target is at an angle relative to the direction of electron beam and the X-rays that leave the anode target pass through the side of the tube. The tube is evacuated to allow electrons reach the target without interruption.

The X-rays are produced both by deceleration of electrons in the metal target and by the excitation of the core electrons in the atom of the target. The first process gives a broad continuous spectrum and the second gives sharp lines. When a moving electron is stopped suddenly, all its energy appears as photon of frequency ν of X-rays. The energy of an electron of charge e in dropping through a potential difference V is eV and

$$E = h\nu = \frac{hc}{\lambda} = eV$$

$$\lambda = \frac{hc}{E} = \frac{hc}{eV} \tag{5.1}$$

An electron will not lose all its energy in this way; it will have a number of glancing collisions with the atoms that it collides and causing them to vibrate. As a result of this the temperature of the target increases. Eq. (5.1), therefore, gives the minimum value λ can possibly have and accounts for the short wavelength cut off. Larger wavelengths are more probable and so the rapid increase in the intensity. The intensity falls off gradually indicating that there is no upper limit. Fig. 5.2 shows the X-ray spectrum that result when molybdenum target is bombarded by electron at 35 keV. The electron beam on striking the target not only get decelerated but also a small fraction of electrons of the beam strike the target and ejects the inner shells electrons. The atom is then unstable, and outer shell electrons in the same atom will drop into the hole (vacancy) caused by the ejection of the electron. In doing so it loses energy and a photon is emitted. If E is the energy lost we have

$$\lambda = \frac{hc}{E} \tag{5.2}$$

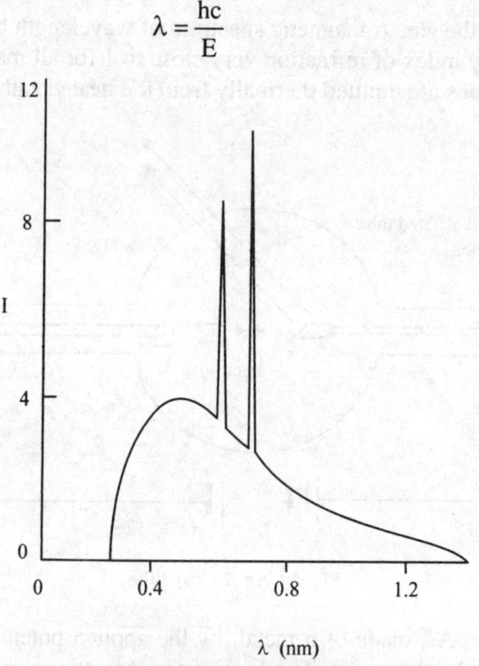

Fig. 5.2 X – ray spectra of molybdenum at 35 kV accelerating potential

E is a definite quantity associated with the electron energy change in the atom. Therefore, the wavelength concerned is specific. Several wavelengths are possible and they constitute the characteristic X-ray line spectrum shown as peaks in Fig.5.2. The energy of the characteristic X-ray produced is very weakly dependent on the chemical structure in which the atom is bound indicating that non-bonding shells of atoms are the characteristic X-ray source. The resulting characteristic spectrum is superimposed on the continuum. An atom remains ionized for a very short time ($\sim 10^{-14}$ s) and thus the incident electrons that arrive about every $\sim 10^{-17}$ s can repeatedly ionize an atom. However, not all outer electrons can fall into holes to provide X-rays.

   Suppose an electron from K shell is ejected and a hole is created in the shell. Subsequently an electron from a higher shell L, M, N,... will make a radiative transition filling the hole. The energy of the photon emitted will be in the range from a few keV to a few hundred keV and thus lie in the X-ray region. The emission spectrum, which results in a line spectrum, forms a simple series. The lines originating from a K shell vacancy are called K_α , K_β , K_γ , ... lines and corresponding to L→K, M→K, N→K, ... transition. K_α is the strongest. The emission of K series is accompanied by other series. As a result of lines of K series, the vacancies are created in L, M, N, ... shells which are filled by

M, N, ... shell electrons. Thus K series may be accompanied by L, M, N, ... and so on. These transitions can be shown on an energy level diagram which is different from that used in the atomic spectra of valence electrons. The most important difference is that X-ray diagram gives the energy of the atom when one electron of the quantum number n, l, j is missing, that is, diagram describes the energy levels of a hole, with quantum number n, l, j that jumps from one subshell to the next when the atom emits X-ray line spectrum. As a hole representation, the absence of an electron of negative energy, the energy associated with the hole is positive. Thus the energy of all the levels of X-ray diagram is positive. The energy levels are shown in Fig.5.3. These levels are also identified by capital letters K, L, M, N, corresponding to n = 1, 2, 3, 4, respectively.

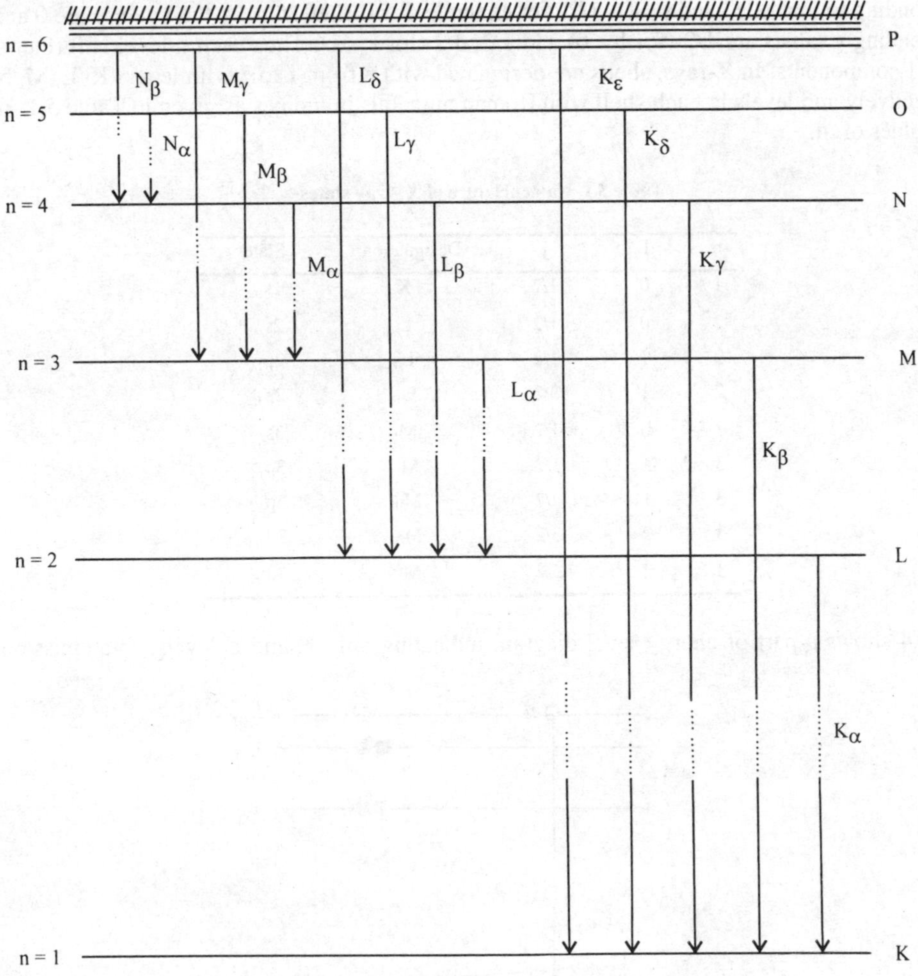

Fig. 5.3 The origin of X-ray spectra

According to the wavelength, X-rays can be classified as hard or soft. The K X-rays are called hard while L radiation is soft and M, N, O X-rays are very soft. M and L radiation arising from heavy elements are harder (i.e. of shorter wavelengths) than K radiation of lighter elements. K radiation of an element is more penetrating than that of L, M, N, O etc.

5.2 X-Ray Emission Spectra

For every filled subshell or shell, spin, orbital and total angular momenta are zero. If one electron is removed from the closed subshell, the allowed values of spin, orbital and total angular momenta for the rest of the electrons in the subshell are identical with those of the single electron, which can complete the subshell. Thus the states of np^1, nd^1 .... are same as that of np^5, nd^9, .. , respectively. Hence the quantum numbers n, l and j of subshell with one electron missing are identical with those of an electron, which can complete the subshell. Since only an electron is missing from the subshell, therefore s = S = 1/2 and multiplicity is 2. In Fig.5.3 the energy levels are drawn for various values of n. However, for each value of n, there are n-1 values of l and for each value of l, there are two j values corresponding to j = l +1/2 and j = l – 1/2. For example, for L shell (n = 2), the l values are 0 and 1 and corresponding j values are 1/2 (for l = 0) and 1/2, 3/2 (for l = 1). Thus each n level is further split up into 2n-1 components. In X-rays, shells are designated with n from 1 to 7 with letters K, L, M, N, O, P, Q, respectively and levels in each shell with Roman numerals in indices as given in Table 5.1 for some of the values of n.

Table 5.1 Nomenclature of X – ray states

n	l	j	Designation	State
1	0	1/2	K	$1s_{1/2}$
2	0	1/2	L_I	$2s_{1/2}$
2	1	1/2	L_{II}	$2p_{1/2}$
2	1	3/2	L_{III}	$2p_{3/2}$
3	0	1/2	M_I	$3s_{1/2}$
3	1	1/2	M_{II}	$3p_{1/2}$
3	1	3/2	M_{III}	$3p_{3/2}$
3	2	3/2	M_{IV}	$3d_{3/2}$
3	2	5/2	M_V	$3d_{5/2}$

Fig. 5.4 shows a part of energy level diagram indicating only K and L levels. The emission of an

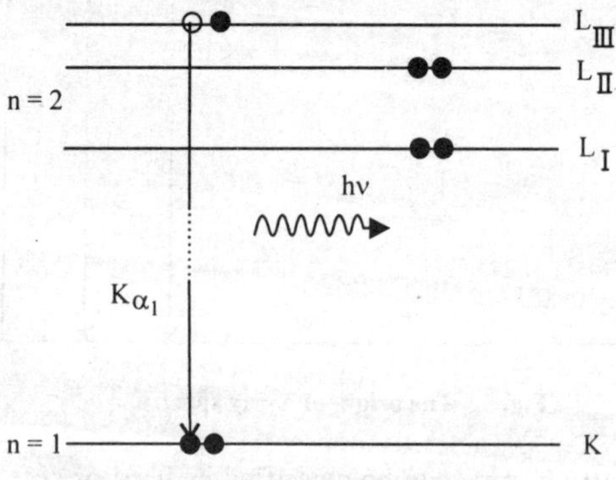

Fig. 5.4 X-ray emission spectrum

X-ray line occurs as a result of transition from one excited state to another, that is, both the initial and final states are excited. For example, emission of K line implies transition of an electron from level L to that of K accompanied by emission of a photon whose energy is equal to the difference in energy between these levels. The selection rules for the transitions are

$$\Delta l = \pm 1, \Delta j = 0, \pm 1 \tag{5.3}$$

According to the selection rules, all the transitions between K and L levels are not allowed. For example, $L_I \rightarrow K$ is not allowed because $\Delta l = 0$ is forbidden. One of the two allowed transitions is shown in Fig.5.4. On the energy level diagram transitions are denoted in a number of ways: by an arrow pointing in the direction of electron transition with emission of X-ray photon e.g. from level L_{III} to level K or by an arrow indicating the transition of the atom from level K to level L_{III}. The two allowed transitions between L and K levels are further given the name $K_{\alpha 1}$ and $K_{\alpha 2}$. The name $K_{\alpha 1}$ is given to the intense line and weaker is $K_{\alpha 2}$. Similarly, $K_{\beta 1}$ and $K_{\beta 2}$ are in order of decreasing intensity. A similar system is used for γ lines. The $K_{\alpha 1}$ line arises from a K-L_{III} ($1s_{1/2} \rightarrow 2p_{3/2}$) transition for the atom corresponding to the electron from L_{III} filling the vacancy in the K shell. The $K_{\alpha 2}$ line corresponds to electron from L_{II} filling the vacancy in the K shell. The statistical factor of L_{III} is $2j+1=4$ while that of L_{II} is 2, therefore, one can expect $K_{\alpha 1}$ to be twice as intense as the $K_{\alpha 2}$ line. Similarly, $K_{\beta 1}$ (K-M_{III} transition) is twice as strong as $K_{\beta 2}$ (K-M_{II} transition). However, it does not mean that $K_{\alpha 1}$ and $K_{\beta 1}$ are equally intense. The rule of proportionality to the statistical weight is good only for the intensity of lines of very nearly the same energy.

Apart from the transitions corresponding to selection rules given by Eq. (5.3), sometimes-weak lines arise, which are forbidden by the selection rules. These lines are due to electric quadrupole $\Delta l = 0, \pm 2$; $\Delta j = 0, \pm 1, \pm 2$ or magnetic dipole $\Delta l = 0, \Delta j = 0, \pm 1$ transitions.

5.3 Dependence of Position of Emission Lines on the Atomic Number

In 1913, Moseley observed that the wavelength of characteristic X-ray lines shift continuously with the change in the element atomic number Z. He observed that with increasing Z, the wavelength of X-ray decreases and hence the frequency of X-ray increased. A plot of square root of frequency versus Z gives a straight line. This increase in frequency with increase of Z is due to increasing binding energy of the electron with increase of number of protons in the nucleus.

The electron energy E_{nl} for a given n and l, according to Eq. (4.42) is

$$E_{nl} = -\frac{hcRZ*^2}{n^2}$$

where $Z*= Z - \sigma_{nl}$ is the effective charge of the nucleus, σ_{nl} is value of screening constant. The term value is

$$T = -\frac{E_{nl}}{hc} = \frac{RZ*^2}{n^2}$$

For an emission line, the frequency (in cm^{-1})

$$\bar{v} = R\left(\frac{Z*_f^2}{n_f^2} - \frac{Z*_i^2}{n_i^2}\right) \tag{5.4}$$

with $n_i > n_f$ and σ is approximately constant for initial and final state, that is, $Z^*_i = Z^*_f = Z^*$

$$\bar{\nu} = RZ^{*2}\left(\frac{1}{n_f^2} - \frac{1}{n_i^2}\right) \tag{5.5}$$

Putting

$$a = \left(\frac{1}{n_f^2} - \frac{1}{n_i^2}\right) \tag{5.6}$$

$$\bar{\nu} = RZ^{*2}\,a \tag{5.7}$$

$$\sqrt{\frac{\bar{\nu}}{R}} = Z^*\sqrt{a} = \sqrt{a}\,(Z - \sigma_{nl}) \tag{5.8}$$

where a is same for similar lines of all the elements, e.g., $K_{\alpha1}$ line . σ is approximately equal for all the lines of a given series, e.g., for all the K lines of all the elements. Eq. (5.8) is known as Moseley law. This law provides a simple test for the order of elements according to Z. It showed where elements were missing from the periodic table and led to the discovery of some of the elements.

For K_α line, $n_f = 1$ and $n_i = 2$, therefore a= 0.866. σ is ≈ 1 and 7.4 for K and L shells, respectively. The Eq. (5.8) for K series is

$$\sqrt{\frac{\bar{\nu}}{R}} = 0.866(Z-1) \tag{5.9}$$

$$\bar{\nu} = (0.866)^2 \times 109737 \times (Z-1)^2\,\mathrm{cm}^{-1} \tag{5.10}$$

$$\lambda = \frac{1}{\bar{\nu}} = \frac{10^{-4}\,\mathrm{cm}}{8.23 \times (Z-1)^2} \tag{5.11}$$

The wavelength for K series is given by Eq. (5.11) and for L series it can be obtained from Eq. (5.8) by putting the appropriate values of Z^* and a.

5.4 X-ray Emission (Doublet) Spectra

The separation between adjacent X-ray terms of the same principal quantum number is often referred to as an X-ray doublet. The intervals that correspond to differences between levels with the same n, l and s but different j are called spin doublets or regular doublets. The frequency interval remains constant all through the spectrum of the same element. However increases very slowly with rising value of Z. For example spin doublets are L_{II}-L_{III}, M_{II}-M_{III} or M_{IV}-M_V etc.

The screening doublets (irregular doublets) are pair of energy levels having equal n and j but l quantum number differs by 1. The separation between lines remains constant for all the elements. The examples of screening doublets are: L_I - L_{II} ($2s_{1/2}$ – $2p_{1/2}$); M_{III} - M_{IV} ($3p_{3/2}$ - $3d_{3/2}$); M_I - M_{II}($3s_{1/2}$ -$3p_{1/2}$) etc.

The general features of the doublet separation can be explained as follows. The energy of a given level due to the Coulomb interaction of an electron with the nucleus and the energy of a level in a first approximation is given by

$$E_n = -\frac{hcRZ^{*2}}{n^2} = -\frac{hcR(Z-\sigma_1)^2}{n^2}$$

(5.12)

where σ_1 is screening constant depending on the value of l. Taking into account the correction due to relativistic effect and spin-orbit interaction, the position of the energy level is given by a similar expression to that of hydrogen given by Eq.(4.26) with Z replaced by Z*

$$E_{nlj} = -\frac{hcR(Z-\sigma_1)^2}{n^2} + \frac{hcR(Z-\sigma_2)^2}{n^3}\left(\frac{3}{4n} - \frac{1}{j+1/2}\right)$$

Since the screening constant represents an average over ranges of effective Z value, therefore we expect different screening constants for expression with different power of Z. The screening constant σ_2 is less than σ_1 .

The term value is

$$T_{nlj} = -\frac{E_{nlj}}{hc} = \frac{R(Z-\sigma_1)^2}{n^2} - \frac{R(Z-\sigma_2)^4}{n^3}\left(\frac{3}{4n} - \frac{1}{j+1/2}\right)$$

(5.13)

The doublet separation

$$\Delta T = \frac{R\alpha^2(Z-\sigma_2)^4}{n^3 l(l+1)}$$

(5.14)

Thus the energy difference ΔT (cm^{-1}) between regular doublets is proportional to the fourth power of (Z - σ_2). The value of σ_2 is independent of Z but depend on the orbit.σ_2 increases as one goes to subshells farther from the nucleus. It is shown that $\sigma_2 = 3.5$ for L_{II} , L_{III} ; 8.5 for M_{II} , M_{III} ; 13 for M_{IV} and M_V etc.

The first of the two terms in Eq. (5.13) predominates. Therefore ignoring the term in α^2

$$T = \frac{R(Z-\sigma_1)^2}{n^2}$$

(5.15)

For the two levels a and b

$$\sqrt{T_a} = \sqrt{R}\,\frac{(Z-\sigma_1)_a}{n}$$

(5.16)

$$\sqrt{T_b} = \sqrt{R}\ \frac{(Z-\sigma_1)_b}{n} \tag{5.17}$$

Since a and b involve different orbits σ_1 is different in the two case

$$\Delta\sqrt{T} = \sqrt{T_a} - \sqrt{T_b} = \frac{\sqrt{R}}{n}\left[(\sigma_1)_b - (\sigma_1)_a\right] = \frac{\sqrt{R}}{n}\Delta\sigma_1 \tag{5.18}$$

where $\Delta\sigma_1$ is the difference in screening constant. The screening constant σ_1 increases with Z.

5.5 Satellites

The lines in the X-ray spectra are often found to be accompanied by faint satellites. These lines do not fit into conventional energy level diagram and are therefore, called non-diagram lines. They occur usually on the short wavelength side of the diagram lines. The origin of satellites is related to transitions between the states of double and multiple ionization. For example, suppose incident electron beam striking the target of X-ray tube has sufficiently high energy, to eject an electron from K shell as well as from L shell. This results in holes in K and L shells. Such a state may undergo a radiative transition into any one of a number of other double ionization, for example, KL $\rightarrow$ LL (an electron dropping from L shell into K shell). The loss of energy due to emission of line in the transition KL $\rightarrow$ LL is greater than the normal K $\rightarrow$ L transition which gives rise to K_α line. Another important source of satellite line is the Auger process. In this process when an electron from an outer orbit make a transition to a hole in the core orbital, a photon is emitted which may be passed to another electron of outer orbit leading to its ionization. Thus a doubly ionized state is produced. For example, an electron drops from L shell into the vacancy in K shell and the photon released is used for ejecting an electron from one of the higher shells. In this way Auger effect produces a doubly ionized state, which can then give rise to the emission of satellite of L series.

5.6 Continuous X-Ray Emission

When X-rays are excited by electron, then in addition to characteristic spectrum a continuous spectrum of a different nature arises. The continuous part of the spectrum is due to slowed down movement of an electron striking the target in the Coulomb field of the nucleus. The continuous spectrum has a sharp limit λ_0 on the lower wavelength side whose value depend on the X-ray tube voltage and is independent of atomic number Z of the target element. The relationship between wavelength and X-ray tube voltage is given by Eq. (5.1). Below this short wavelength limit no X-ray radiation is observed. Wavelength of maximum intensity λ_m is approximately one and half time λ_0 and

$$\lambda_m V^{1/2} = \ \text{constant} \tag{5.19}$$

Intensity of continuous X-ray is nearly proportional to the square of voltage and to the first power of atomic number that is

$$I = k\,Z\,V^2 \tag{5.20}$$

k is constant of proportionality.

The position of characteristic spectrum is independent of voltage of X-ray tube. By raising voltage the limit of continuous X-ray spectrum shifts towards shorter wavelength side.

5.7 X-Ray Absorption Spectra

When a beam of X-ray passes through a sample, the intensity of rays become weaker because of absorption and scattering. The absorption of X-ray emission energy occurs as a result of a single process. The X-ray photon knocks out electron of a shell and the energy of the absorbed photon is thus transformed into kinetic energy of the electron plus the potential energy of the excited atom that equals binding energy of the electron. The X-ray emission line yields information on the difference in binding energies between two electronic states. However, to determine absolute binding energies, absorption of X-ray has to be studied. For such a purpose an X-ray continuum is used and sample absorption as a function of wavelength is measured.

The X-rays of the longest wavelength force out electrons from the outer shell. With increasing energy of X-rays (decreasing wavelength) a small part of it is required to knock out the electron from the given shell. This causes reduced absorption. This decrease continue till the X-ray energy is sufficient to knock out electron from the next deeper lying inner shell and give rise to a sharp increase in absorption as shown in Fig.5.5. It causes a discontinuous behaviour of absorption μ with λ. The discontinuous behaviour of absorption and wavelength or frequency is called absorption edge and corresponds to photon critical energy. The absorption edges are labelled as K edge, L edge, M edge etc. From the observed absorption edge an approximate binding energy is obtained. For increasing energy of X-ray, the threshold for photoemission of deeper and deeper shells are reached and additional contributions to the total absorption are obtained resulting in a number of edges. Apart from the edge there is a general fall off in absorption due to λ^3 dependence of the absorption coefficient given by

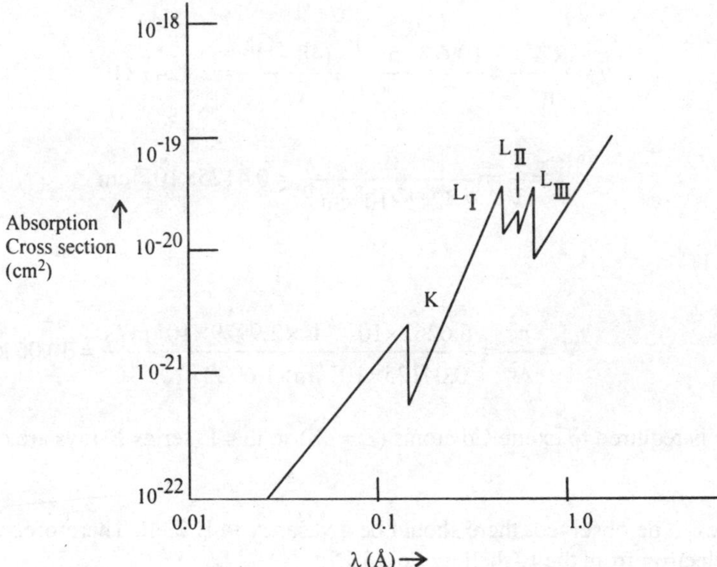

Fig. 5.5 X-ray absorption spectrum with absorption edges

$$\mu = k \, \rho \, \lambda^3 \, Z^3$$

where k is a constant and ρ is the density. The soft X-rays are abosrbed strongly in comparison with hard X-rays. The wavelength of the absorption edge is defined in terms of the orbital energy

$$\lambda_k = \frac{hc}{E_k} \qquad\qquad (5.21)$$

The absorption edge exhibit fine structure corresponding to the fine structure of the core state. However, K edge also exhibits a structure, which is due to discrete levels close to the series limit.

Examples

1. Find the shortest wavelength present in the radiation from an X-ray tube whose accelerating potential is 50 keV.

    From Eq. (5.1)

$$\lambda_{min} = \frac{hc}{eV} = \frac{6.6262 \times 10^{-34}\,Js \times 2.9979 \times 10^{8}\,m/s}{1.6022 \times 10^{-19}\,C \times 50000V} = 0.0248nm$$

2. What energy is needed to excite Cd atoms (Z = 48) so that all series are observed?

    To observe all series, it is necessary to have a vacancy in K shell. The energy required to remove a K electron is from Eq. (5.15) is

$$\bar{v} = \frac{RZ^{*2}}{n^2} = \frac{109737cm^{-1} \times (48-1)^2}{1} = 2.4241 \times 10^{8}\,cm^{-1}$$

$$\lambda = \frac{1}{\bar{v}} = \frac{1}{2.4241 \times 10^{8}\,cm^{-1}} = 0.4125 \times 10^{-8}\,cm$$

    From Eq. (5.1)

$$V = \frac{hc}{\lambda e} = \frac{6.6262 \times 10^{-34}\,Js \times 2.9979 \times 10^{8}\,m/s}{0.04125 \times 10^{-9}\,m \times 1.6022 \times 10^{-19}\,C} = 30.06\,keV$$

3. What energy is required to excite Cd atoms (Z = 48) so that L-series X-rays are observed (σ is 7.4 for L series)

    For L series to be observed, there should be a vacancy in L shell. Therefore, energy required to remove the electron from the L-shell by Eq. (5.15)

$$\bar{v} = \frac{R(Z-7.4)^2}{4} = \frac{109737 \text{cm}^{-1} \times (40.6)^2}{4} = 4.522 \times 10^7 \text{cm}^{-1}$$

Using 1 eV = 8065.48 cm^{-1}

$$V = \frac{4.522 \times 10^7}{8065.48} = 5.607 \text{keV}$$

4. Which element has K_α X-ray line whose wavelength is 1541.23 XU ?

1000 XU = 1.00202 Å. The wavelength λ = 1.54123 x 1.00202 Å = 1.5443 Å

From Eq. (5.11)

$$(Z-1)^2 = \frac{10^{-4}}{8.23 \times 1.5443 \times 10^{-8}} = 786.8$$

$$Z - 1 \approx 28.05$$

$$Z = 29$$

The element with atomic number Z = 29 is Cu.

5. If K and L energy levels of an element lie at roughly 78 keV and 12 keV, respectively , compute the approximate wavelength of K_α line. What minimum potential difference across an X-ray tube is required to excite this line? At approximately what wavelength is the K absorption edge?

For observing K_α line, a vacancy has to be created in the K shell. The K level lie at 78 keV , therefore, this much potential difference has to be applied to remove the electron. The K_α line is due to transition between K and L shells. The energy difference between these two is (78 – 12) keV= 66 keV. The wavelength corresponding to this much of potential difference from Eq. (5.1) is 0.185 Å. The K edge from Eq. (5.21) is

$$\lambda_k = \frac{hc}{E_k} = \frac{6.6262 \times 10^{-34} \text{Js} \times 2.9979 \times 10^8 \text{m/s}}{1.6022 \times 10^{-19} \text{C} \times 78 \times 10^3 \text{V}} \approx 0.01589 \text{nm}$$

6. The L_{II} - L_{III} regular doublet separation for Ag (Z=47) is 0.173 keV. Evaluate the $L_{II} - L_{III}$ regular doublet separation in In (Z = 49) (σ_2 = 3.5 for L_{II}, L_{III}).

From Eq. (5.14)

$$(\Delta T)_{Ag} = \frac{R\alpha^2 (47 - 3.5)^4}{n^3 l(l+1)} = 0.173 \text{ keV}$$

$$(\Delta T)_{In} = \frac{R\alpha^2 (49-3.5)^4}{n^3 l(l+1)}$$

Dividing the above two equations

$$(\Delta T)_{In} = \frac{(45.5)^4}{(43.5)^4} \times 0.173 \text{ keV} \approx 0.207 \text{keV}$$

Problems

5.1   Which element has K_α X-ray line whose wavelength is 1796.37 XU ?

5.2   Calculate the longest wavelength of K series of Cu $(Z = 29)$ with $\sigma = 1$.

5.3   Which element has K_α X-ray line whose wavelength is 0.05396 nm ?

5.4   The $M_{II} - M_{III}$ regular doublet separation for Mo $(Z = 42)$ is 0.017 keV. Evaluate the $M_{II} - M_{III}$ regular doublet separation in Ag $(Z = 47)$ and $\sigma_2 = 8.5$ for M_{II}, M_{III}.

6

Atoms in External Fields

6.1 Zeeman Effect

In 1876 P. Zeeman observed that when a light source is brought into a magnetic field, each spectral line is split into a number of components. This means that in the presence of a magnetic field the energy levels of the atom, which are involved in the transition, must be split into several components. This splitting results from the interaction of the electronic magnetic moment of the atom with the magnetic field.

According to the classical theory, the ratio of magnetic and mechanical moment of an electron in an orbit is given by Eq. (3.43)

$$\frac{\mu_l}{l} = -\frac{e}{2m_e} = -\frac{2\pi\, g_l \beta_e}{h}$$

Besides the orbital angular momentum $\mathbf{l}$, the electron also has a spin angular momentum $\mathbf{s}$. The ratio of magnetic and mechanical moment for the spinning electron is given by Eq. (3.44)

$$\frac{\mu_s}{s} = -2\frac{e}{2m_e} = -\frac{2\pi\, g_s \beta_e}{h}$$

L-S coupling

In case the atom has more than one valence electron, it is assumed that the interaction between the spin angular momentum of the electrons on one hand, the interaction between their orbital motions on the other hand is large compared with the interaction between spin and orbital angular momenta of each electron, that is, LS coupling holds. The magnetic moment μ_L coupled with the total orbital angular momentum $\mathbf{L}$ of all electrons is then

$$\mu_L = -\frac{e}{2m_e}\mathbf{L} \tag{6.1}$$

The magnetic moment μ_S coupled with the total spin angular momentum $\mathbf{S}$ is then

$$\mu_S = -2\frac{e}{2m_e}\mathbf{S} \tag{6.2}$$

In the absence of a magnetic field, the vector $\mathbf{L}$ and $\mathbf{S}$ precess together around their resultant $\mathbf{J}$. When a magnetic field $\mathbf{B}$ is applied, $\mathbf{L}$ and $\mathbf{S}$ couple with it and in the absence of coupling between $\mathbf{L}$ and $\mathbf{S}$, the latter precess independently around $\mathbf{B}$. However, in weak magnetic field, energy corresponding to coupling of $\mathbf{L}$ and $\mathbf{S}$ with $\mathbf{B}$ is smaller than spin-orbit interaction energy. Therefore, under such condition B does not perturb the coupling between L and S. The **L** and **S** precess about their resultant **J**. Because of torque, **J** precesses around **B** at a rate small compared with the precession rate of **L** and **S** about **J** as shown in Fig.6.1.

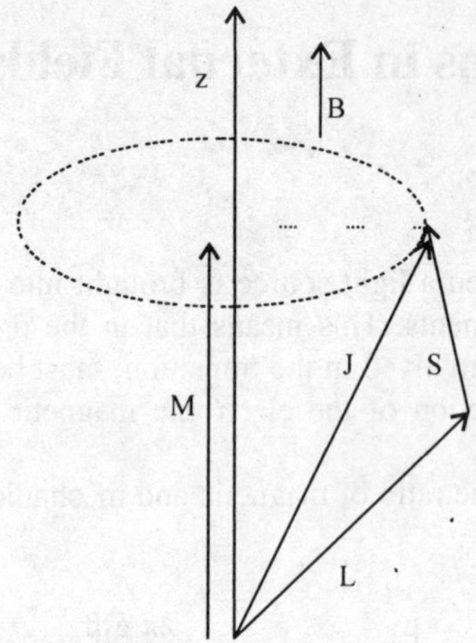

Fig. 6.1 Precession of J about the field direction z in a weak magnetic field B

The total electronic magnetic moment $\mu = \mu_L + \mu_S$ of the atom is not oriented in the same direction as the total angular momentum $J = L + S$ because of different dependence of μ_L and μ_S on **L** and **S**, respectively (Fig. 6.2). Since **L** and **S** precess rapidly about **J**, μ_L and μ_S precess rapidly as well, causing μ to precess about $-J$ at the same rate. Therefore, the component of μ perpendicular to $-J$ averages to zero and the component parallel to $-J$ remains a constant in magnitude μ_J. The component

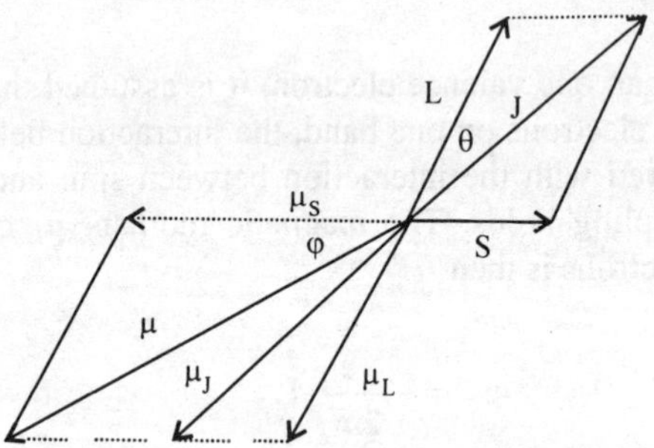

Fig. 6.2 Vectors L, S and J and the associated magnetic moment vectors

of μ along $-J$ axis from Fig.6.2 is

$$\mu_J = \mu_L \cos\theta + \mu_S \cos\varphi \tag{6.3}$$

Using $J = L + S$ or $S = J - L$

$$S^2 = (\mathbf{J} - \mathbf{L}) \cdot (\mathbf{J} - \mathbf{L}) = J^2 + L^2 - 2\mathbf{J.L}$$

$$2JL\cos\theta = J^2 + L^2 - S^2$$

$$\cos\theta = \frac{J^2 + L^2 - S^2}{2JL} \tag{6.4}$$

Similarly,

$$\cos\varphi = \frac{J^2 + S^2 - L^2}{2JS} \tag{6.5}$$

Substituting the values of $\cos\theta$ and $\cos\varphi$ from Eqs. (6.4) and (6.5) , respectively in Eq.(6.3)

$$\mu_J = \mu_L \frac{J^2 + L^2 - S^2}{2JL} + \mu_S \frac{J^2 + S^2 - L^2}{2JS} \tag{6.6}$$

Substituting the values of μ_L and μ_S from Eqs. (6.1) and (6.2), respectively into Eq.(6.6)

$$\mu_J = -\frac{e}{2m_e}\left[\frac{J^2 + L^2 - S^2}{2J} + 2\frac{J^2 + S^2 - L^2}{2J}\right]$$

$$\mu_J = -\frac{e}{2m_e}J\left[1 + \frac{J^2 + S^2 - L^2}{2J^2}\right] = -g\frac{e}{2m_e}J \tag{6.7}$$

where

$$g = 1 + \frac{J^2 + S^2 - L^2}{2J^2} = 1 + \frac{J(J+1) + S(S+1) - L(L+1)}{2J(J+1)} \tag{6.8}$$

g is called Lande's splitting factor.

jj coupling

In the case where interaction between spin and orbital motion of the electron is large compared with the interaction between the spin angular momenta of the electrons on one hand and the interaction between their orbital motions on the other hand, jj coupling holds. In jj coupling, the s and l of each electron are coupled together to form their own resultant j (Fig.6.3). The magnetic moments for the two electrons from Eq. (6.7) are

$$\boldsymbol{\mu}_{j_1} = -g_1 \frac{e}{2m_e}\mathbf{j}_1 \tag{6.9}$$

$$\boldsymbol{\mu}_{j_2} = -g_2 \frac{e}{2m_e} \mathbf{j}_2 \qquad (6.10)$$

where g_1 and g_2 obtained from Eq.(6.8) are

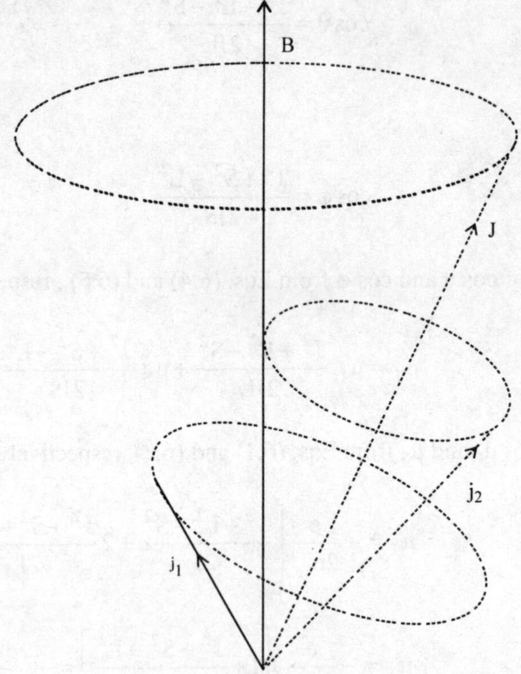

Fig. 6.3 The vector model for jj-coupling in a weak magnetic field

$$g_1 = 1 + \frac{j_1^2 + s_1^2 - l_1^2}{2j_1^2} \qquad (6.11)$$

$$g_2 = 1 + \frac{j_2^2 + s_2^2 - l_2^2}{2j_2^2} \qquad (6.12)$$

$\mathbf{j}_1$ and $\mathbf{j}_2$ couples together to give the resultant $\mathbf{J}$ (Fig.6.3). $\boldsymbol{\mu}_J$ is obtained from the projection of $\mathbf{j}_1$ and $\mathbf{j}_2$ on $\mathbf{J}$.

$$\mu_J = \mu_{j_1} \cos\left(\mathbf{j}_1, \mathbf{J}\right) + \mu_{j_2} \cos\left(\mathbf{j}_2, \mathbf{J}\right) = -[g_1 j_1 \cos\left(\mathbf{j}_1, \mathbf{J}\right) + g_2 j_2 \cos\left(\mathbf{j}_2, \mathbf{J}\right)] \frac{e}{2m_e} \qquad (6.13)$$

But

$$\mu_J = -g \frac{e}{2m_e} J \qquad (6.14)$$

On comparing (6.13) and (6.14)

$$g\,J = g_1 j_1 \cos{(\mathbf{j_1},\mathbf{J})} + g_2 j_2 \cos{(\mathbf{j_2},\mathbf{J})}$$ (6.15)

The angles between $\mathbf{j_1}$, $\mathbf{j_2}$ and $\mathbf{J}$ are constant and $\mathbf{J} = \mathbf{j_1} + \mathbf{j_2}$ or $\mathbf{J} - \mathbf{j_1} = \mathbf{j_2}$. Taking dot product of $\mathbf{j_2}$ with $\mathbf{j_2}$

$$j_2^2 = J^2 + j_1^2 - 2\mathbf{J}\cdot\mathbf{j_1} = J^2 + j_1^2 - 2j_1 J \cos{(\mathbf{j_1},\mathbf{J})}$$

$$\cos{(\mathbf{j_1},\mathbf{J})} = \frac{J^2 + j_1^2 - j_2^2}{2j_1 J}$$ (6.16)

Similarly,

$$\cos{(\mathbf{j_2},\mathbf{J})} = \frac{J^2 + j_2^2 - j_1^2}{2j_2 J}$$ (6.17)

Substituting Eqs. (6.16) and (6.17) in Eq. (6.15)

$$g = g_1 \frac{J^2 + j_1^2 - j_2^2}{2J^2} + g_2 \frac{J^2 + j_2^2 - j_1^2}{2J^2}$$ (6.18)

The magnetic interaction energy resulting from the interaction between the electronic magnetic moment of the atom and an external magnetic field $\mathbf{B}$ (directed along the z axis) is

$$\Delta E = -\boldsymbol{\mu}_J \cdot \mathbf{B}$$ (6.19)

Substituting Eq. (6.7) in Eq. (6.19)

$$\Delta E = g \frac{e}{2m_e} \mathbf{J}\cdot\mathbf{B} = g \frac{e}{2m_e} JB \cos{(\mathbf{J},\mathbf{B})}$$ (6.20)

where g is given by Eq.(6.8) or (6.18) depending whether the coupling is LS or jj. Jcos (JB) is the projection of $\mathbf{J}$ on $\mathbf{B}$ and equal to J_z. The allowed values of J_z are $M\hbar$, where M is magnetic quantum number and can take values from +J to –J, a total of 2J+1 values. The Eq. (6.20) takes the form

$$\Delta E = g \frac{eh}{4\pi\, m_e} MB = g\beta_e MB$$ (6.21)

The magnetic field lifts the degeneracy-giving rise to 2J+1 equidistant sublevels called Zeeman sublevels. The distance between two consecutive sublevels being $g\beta_e B$. If the energy of the atom without magnetic field is represented by E_0, then the energy in the magnetic field is

$$E = E_0 + g\beta_e MB \tag{6.22}$$

The relation between term value T and energy E is $T = -E/hc$, the interaction energy in wavenumber becomes

$$\frac{\Delta E}{hc} = -\Delta T = g\frac{\beta_e}{hc}MB \tag{6.23}$$

$$-\Delta T = gML \tag{6.24}$$

where $L=(\beta_e B/hc)$ is called Lorentz unit and its value is $0.467B$ cm^{-1} when B is measured in Webers per square metre (tesla). The separation between two neighbouring sublevels is gL and is determined by magnetic field B and g-factor belonging to the energy level.

   As an example of Zeeman effect, consider
(a) transition between $^2D_{5/2}$ and $^2P_{3/2}$. According to Eq. (6.8) the g values of $^2D_{5/2}$ and $^2P_{3/2}$ are 4/5 and 4/3, respectively. The splitting of the levels and the allowed transitions are shown in the Fig.6.4. The allowed selection rules for transitions between magnetic sublevels are

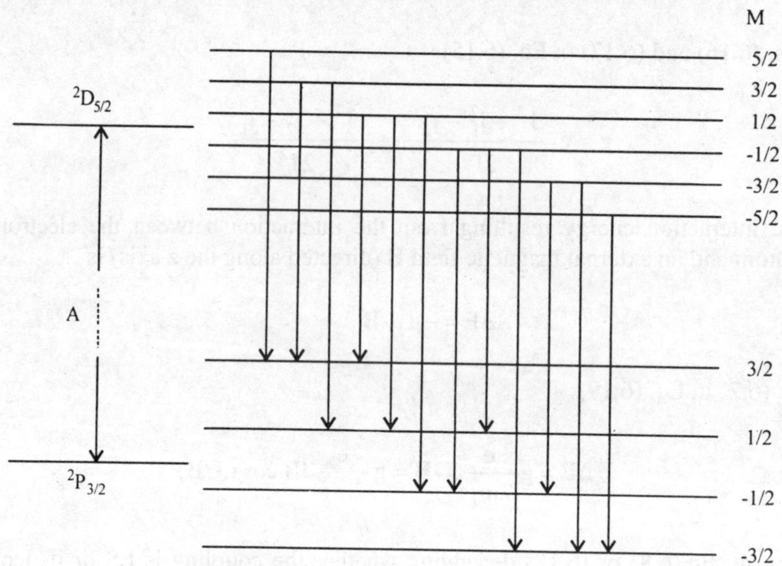

Fig. 6.4 Weak field Zeeman splitting for transition between $^2D_{5/2} \rightarrow {}^2P_{3/2}$

$$\Delta M = 0 \quad (\text{p component or } \pi \text{ component})$$
$$\Delta M = \pm 1 \quad (\text{ s component or } \sigma \text{ component})$$
$$\Delta J = 0, \text{ with } M = 0 \rightarrow M = 0 \text{ is not allowed}$$

   The $^2D_{5/2} \rightarrow {}^2P_{3/2}$ transition splits up into twelve components, four p components and eight s components.
(b) Transition between $^1F_3 - {}^1D_2$. Since $^1F_3$ and $^1D_2$ are singlet, therefore their g values are equal to 1. Fig. 6.5 shows the splitting of the levels in weak magnetic field and allowed transitions between them.

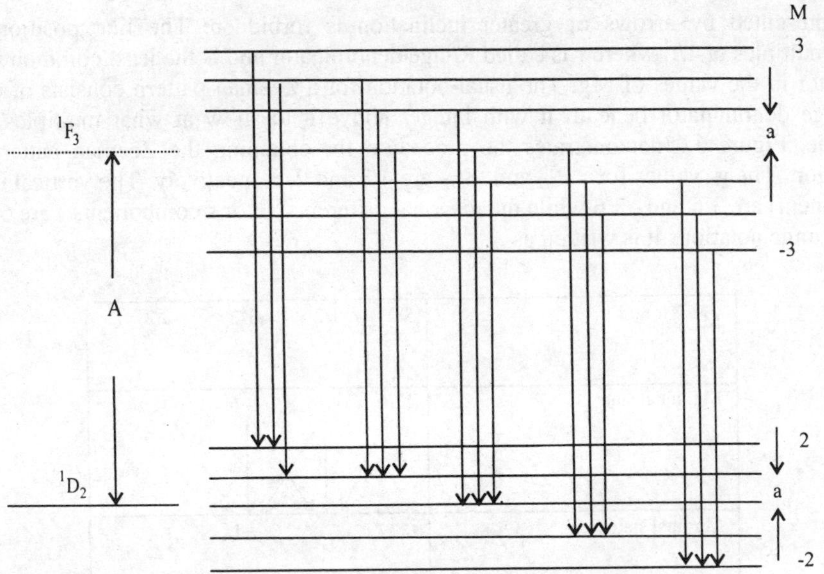

**Fig. 6.5 Zeeman pattern of a fundamental series singlet showing equal
separations in both the initial and final states of the atom**

Table 6.1 lists the allowed transitions from $^1F_3$ to $^1D_2$ in weak magnetic field. From Table 6.1 it is seen

Table 6.1 Allowed transitions from $^1F_3$ - $^1D_2$ and the corresponding energies

M of final state \ M of initial state	3	2	1	0	-1	-2	-3
2	A+a	A	A-a	X	X	X	X
1	X	A+a	A	A-a	X	X	X
0	X	X	A+a	A	A-a	X	X
-1	X	X	X	A+a	A	A-a	X
-2	X	X	X	X	A+a	A	A-a

that there are only three distinct energies at A+a, A and A-a. Thus the spectral line corresponding to $^1F_3$ - $^1D_2$ transition splits into three components in the weak magnetic field, one at the same position and other two are displaced on either side of undisplaced line.

From the above examples it is clear that when levels involved in a transition are singlet we observe only three lines or so called normal Zeeman effect. On the other hand if the levels in a transition involved are different from singlet we observe the so-called anomalous Zeeman effect. Thus when the contribution of spin towards magnetic moment is zero, the normal Zeeman effect is observed and when we take into account the spin contribution we have anomalous Zeeman effect.

The representation of transitions between Zeeman sublevels on a normal energy level diagram often gives rise to confusion. Therefore, the following arrangement is advantageous. Write in a row all the values of M, which will be used as shown in Fig.6.6. Below each value of M put the displacement Mg of the magnetic level of the initial spectral term, which has that M value. Similarly displacement of the levels belonging to the final term is written in the next row. The transitions $\Delta M=\pm1$ (σ or s) are then represented by slanting arrows while $\Delta M= 0$ (π or p) are represented by vertical arrows. Transition that would be represented by arrows of greater inclination is forbidden. The line positions may be expressed as multiples of 1/r, where r is called Runge denominator and is the least common multiple of the denominator in the values of Mg. The usual notation for a Zeeman pattern consists of a long line with the Runge denominator beneath it with integer above it to show at what multiples of 1/r the components lie. Figure 6.6 demonstrates the procedure for obtaining the Zeeman pattern for $^2P_{3/2}$ $\rightarrow {}^2S_{1/2}$ transition. The g values for $^2P_{3/2}$ and $^2S_{1/2}$ are 4/3 and 2, respectively. The vertical differences (π or p component) are 3/6 and –3/6 while the diagonal difference (σ or s components) are 6/6,5/6,-5/6 and –6/6. In Runge notations it is written as

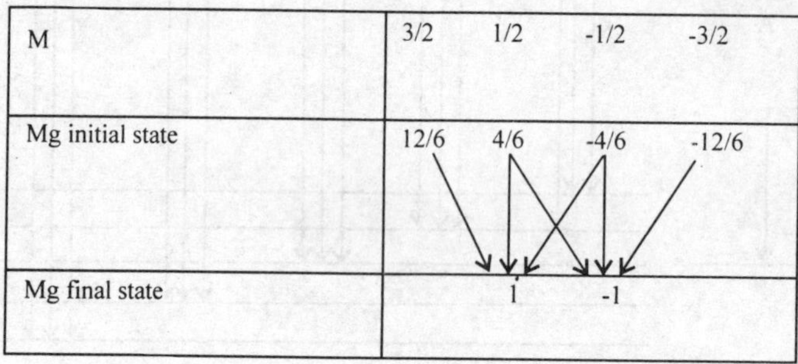

Fig. 6.6 Calculation of Zeeman pattern for $^2P_{3/2} \rightarrow {}^2S_{1/2}$ transition

$$\Delta\bar{v} = \frac{(\pm 3),\pm 5,\pm 6}{6} Lcm^{-1}$$

with two p components being set in parentheses, followed by the four s components.

Intensity rules

(i)   The intensities of the components of one line are symmetrical relative to the position of the original line.

(ii) The sum of the intensities of combinations of a level characterized by the magnetic quantum number M with level M-1, M and M+1 is independent of M.

(iii) Sum of intensities of all p components is equal to the sum of the intensities of all s components

Table 6.2 Intensity rules

	$M \rightarrow M + 1,$	$I = A (J + M +1) (J - M)$
$J \rightarrow J$	$M \rightarrow M - 1,$	$I = A (J- M +1) (J + M)$
	$M \rightarrow M,$	$I = 4AM^2$
	$M \rightarrow M + 1,$	$I = B (J - M) (J - M - 1)$
$J \rightarrow J - 1$	$M \rightarrow M - 1,$	$I = B (J + M) (J + M - 1)$
	$M \rightarrow M,$	$I = 4B (J^2 - M^2)$
	$M \rightarrow M + 1,$	$I = B (J + M +1) (J + M +2)$
$J \rightarrow J + 1$	$M \rightarrow M - 1,$	$I = B (J - M + 1) (J - M + 2)$
	$M \rightarrow M,$	$I = 4B (J + M + 1) (J - M +1)$

When Zeeman effect is observed in a direction perpendicular to the magnetic field, only half of the intensity of the s component is observed. The other half is observed in a direction parallel to the magnetic field. Therefore in studying intensities s components must be multiplied by 2. The intensity formulae are given in Table 6.2. A and B are constants that need not be determined for relative intensities within each Zeeman pattern.

Table 6.2 holds for any coupling scheme. To determine the strongest component the following approximate rule is useful. In case where $J_1 \neq J_2$, the vertical differences in the middle of the scheme and the diagonal differences at the ends give , respectively, the strongest p and s components. If $J_1 = J_2$, the vertical differences at the end of the scheme and the diagonal differences at the centre give, respectively the strongest p and s components with the restriction that M = 0 to M = 0 is forbidden.

Polarisation rules

Polarisation rules are given in Table 6.3

Table 6.3 Polarisation rules

Viewed ⊥ to the field	$\Delta M = \pm 1$; plane polarised ⊥ to B; s or σ component
	$\Delta M = 0$; plane polarised ∥ to B; p or π component
Viewed ∥ to the field	$\Delta M = \pm 1$; circularly polarised; s or σ component
	$\Delta M = 0$; forbidden; p or π component

6.2 Paschen Back Effect

In 1912 Paschen and Back observed that several lines of Li spectrum were normal triplet in a magnetic field. For Li, 670 nm line (corresponding to D_1 and D_2 line of Na, $D_1 - D_2 = 0.013$nm, a normal triplet at 4.5 T). In the Zeeman effect it was assumed that the magnetic field was a weak field. A field is weak when the total spread of the Zeeman pattern of each line is small relative to the spacing of lines themselves. For example, separation of sodium D_1 and D_2 line is 17.18 cm^{-1} while the Zeeman splitting of D_1 is 3.76 cm^{-1} and of D_2 is 4.70 cm^{-1} in a magnetic field of 3T. For D_1 and D_2 , 3 T is therefore, a weak field. However, for Li the separation in zero- field of the two component of the first line of the principal series is 0.3 cm^{-1}. A field of 3 T would produce Zeeman separation of 1.4 cm^{-1}. For Li therefore 3T is a strong field.

In weak field the spin-orbit coupling of the electron is stronger than the coupling between the total angular momentum of the electron and the external magnetic field. As a result of this **J** precesses slowly around **B** in comparison with the precession of **L** and **S** about **J**. As the strength of the magnetic field is increased, the precession frequency of **J** around **B** is also increased. With increasing magnetic field the energy distance between magnetic sublevels also increases from weak to strong. In strong magnetic field the interaction between orbital moment and magnetic field on the one hand, and spin moment and magnetic field, on the other hand are strong in comparison with the interaction between spin and orbital moments. Under these conditions, (i) the vector **L** coupled only to the magnetic field **B**, precesses around the direction of **B** (Fig.6.7) with an angular velocity ω_L. The projection of **L** on **B** takes integer values M_L such that $M_L = +L, L-1,........, -L$ (ii) the vector **S**, also coupled only to the magnetic field **B**, precesses around the direction of this field (Fig. 6.7) with an angular velocity ω_S. The projection of **S** on **B** takes an integer or half integer values M_S such that $M_S = S, S-1, ..., -S$.

Since $\omega_L \neq \omega_S$, the resultant **J** has a length and direction that are continually varying with time, and therefore J is no more a vector nor is J a quantum number. The total energy is therefore determined by the sum of energies of spin and orbital moments in the magnetic field and remaining interaction energy

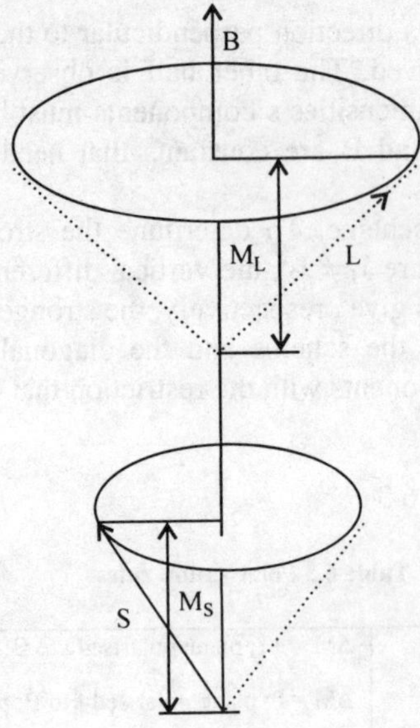

Fig. 6.7 Vector model of the Paschen Back effect. L and S are no longer coupled together

between **L** and **S**. The energy of the orbital moment in the magnetic field is

$$\Delta E_{L,B} = -\mathbf{\mu}_L \cdot \mathbf{B} = -\mu_L B \cos (\mathbf{\mu}_L, \mathbf{B})$$
(6.25)

The energy of the spin moment in the magnetic field is

$$\Delta E_{S,B} = -\mathbf{\mu}_S \cdot \mathbf{B} = -\mu_S B \cos (\mathbf{\mu}_S, \mathbf{B})$$
(6.26)

The energy, using Eqs. (6.1) and (6.2) along with the Fig. 6.7, is

$$E = E_0 + \frac{e\,h}{4\pi\,m_e} B(M_L + 2M_S)$$
(6.27)

The extra energy corresponding to the interaction of S and L can be written from Eqs. (3.15) and (3.64) as

$$\Delta E_{LS} = hc\,A\mathbf{L} \cdot \mathbf{S} = hc\,ALS \cos (\mathbf{L}, \mathbf{S})$$
(6.28)

where A is given by Eq.(3.65).

Since L and S precess independently around B, the angle (**L**, **S**) varies continuously. Therefore in Eq. (6. 28) the average value of cos (**L,S**) is to be used. By means of spherical trigonometry, the average value of this angle is

$$\overline{\cos (\mathbf{L}, \mathbf{S})} = \cos (\mathbf{L}, \mathbf{B}) \cos (\mathbf{S}, \mathbf{B})$$
(6.29)

Substituting Eq. (6.29) in Eq. (6.28)

$$\Delta E_{LS} = hc\ AL\cos(\mathbf{L},\mathbf{B})\ S\cos(\mathbf{S},\mathbf{B}) = hc\ AM_L M_S \qquad (6.30)$$

Thus energy shift from unperturbed energy level is

$$\Delta E = \Delta E_{LB} + \Delta E_{SB} + \Delta E_{LS} = (M_L + 2M_S)\beta_e B + hcAM_L M_S \qquad (6.31)$$

Dividing by hc, the term shift in wavenumber becomes

$$-\Delta T = (M_L + 2M_S)L + AM_L M_S \qquad (6.32)$$

The term value of any strong field level is

$$T = T_0 - [(M_L + 2M_S)L + AM_L M_S] \qquad (6.33)$$

where T_0 is the term value of the hypothetical centre of gravity of the fine structure. The selection rules for Paschen Back effect are $\Delta M_S = 0$ and $\Delta M_L = 0, \pm 1$.

The splitting of $^3S_1$ level in strong field using Eq. (6.33) is shown on the right hand side of the Fig.6.8. At the left of the figure are shown the weak field Zeeman states.

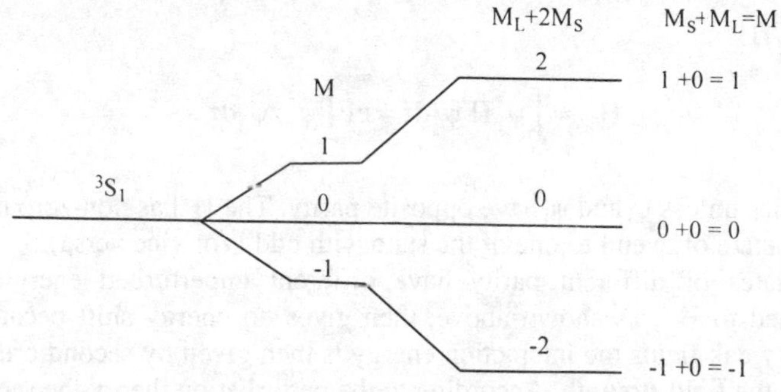

Fig. 6.8 Strong field levels for ^3S

In the intermediate field, where the Zeeman interaction and the spin-orbit interaction are of the same order, J, M_L or M_S are not good quantum numbers. The weak field levels go over to the corresponding strong field level according to the following rules:

(i) Conservation of angular momentum: the sum of the projections of the mechanical moments on the direction of B does not change. M values remain constant.

(ii) No two levels with the same M will cross.

6.3 The Stark effect

When a source of light is placed in an electric field, the spectral lines split up into number of components. Consider an atom in a homogeneous external electric field F in a direction parallel to the z-axis. The Hamiltonian is of the form

$$H = T + V + H' = H_0 + H' \qquad (6.34)$$

where $H_0 = T + V$ is unperturbed Hamiltonian or free field Hamiltonian and perturbed Hamiltonian H' is

$$H' = -\sum_i \mu_i \cdot F = -\sum_i (-er_i) \cdot F = eF \sum_i z_i = eFz \qquad (6.35)$$

H' represents the sum over all electrons i of the interaction energy eFz with the electric field F.

Let ψ_0 be an eigenfunction of the unperturbed Hamiltonian H_0. The expectation value of H' (which is first order correction to the energy) using eigenfunction ψ_0 of H_0 is

$$\overline{H'} = \int \psi_0^* H' \psi_0 d\tau = eF \int \psi_0^* z \psi_0 d\tau \qquad (6.36)$$

The eigenfunction ψ_0 have definite parity, that is, ψ_0 remains unchanged (even parity) or merely changes sign (odd parity) under inversion. Therefore $|\psi_0|^2 = \psi_0^* \psi_0$ is unchanged. However, H' changes sign ($z_i \rightarrow -z_i$) under inversion about the nucleus of the spatial coordinates of all the electrons , that is, $r_i \rightarrow -r_i$. Thus expectation value of H' vanishes. Hence there is no energy shift, which is linear in electric field F. A classical system having an electric dipole moment μ, in electric field **F**, experiences an energy shift of magnitude - $\mu.F$. Since the shift is zero, therefore there is no permanent electric dipole moment in the state ψ_0 .The matrix element of H' for a transition between two unperturbed states ψ_i and ψ_j is

$$H'_{ij} = \int \psi_i^* H' \psi_j d\tau = eF \int \psi_i^* z \psi_j d\tau \qquad (6.37)$$

The above integral vanishes unless ψ_i and ψ_j have opposite parity. The H' has non-zero matrix elements only for a transition from state of even l to one of the state with odd l (or vice versa).

In complex atoms, states of different parity have different unperturbed energies. First order perturbation theory applied to H' , as shown above, then gives no energy shift because expectation value of H' vanishes. For weak fields the interaction energy is then given by second order perturbation theory and is quadratic in the field strength. According to the perturbation theory the second order shift is

$$\Delta E = \sum_{i \neq j} \frac{|H'_{ij}|^2}{E_{0i} - E_{0j}} = e^2 F^2 \sum_{i \neq j} \frac{|z_{ij}|^2}{E_{0i} - E_{0j}} \qquad (6.38)$$

where $z_{ij} = \int \psi_i^* z \psi_j d\tau$ and sum is over all the states except the ground state.

The quadratic dependence of energy shift can also be seen classically. In a homogeneous electric field **F** an atom becomes polarised, that is, centre of electronic cloud no longer coincides with the nucleus. The field induces an electric dipole moment μ which is proportional to the field

$$\mu = \alpha F \qquad (6.39)$$

where α is polarizability which depends upon the orientation of the orbit, that is, of the angular momentum J, to the electric field. The orientation energy of the induced dipole is

$$E = -\mu \cdot F \tag{6.40}$$

since the dipole moment is itself proportional to the field strength, the orientational energy in the Stark effect is proportional to the square of the field strength i.e.

$$E = -\alpha F^2 \tag{6.41}$$

The nature of the forces acting on the electron is purely electrostatic. Therefore the energy of the electron in an orbit of given n and l depend on the inclination of the orbit plane relative to the electric field and is independent of the direction of rotation of motion of the electron in its orbit. The states differing only in sign of M and $-M$ correspond to reversing the direction of motion of all the electrons and do not affect the magnitude of the induced dipole moment and thus have the same energy change due to the applied electric field. For example state $M = 5/2$ has the same energy as that of $M = -5/2$.

As discussed above, the interaction energy is given by second order perturbation theory and is quadratic in the field strength F. However, hydrogen like atoms forms an exception to this rule. The ground state of the hydrogen is non-degenerate and has a definite parity and therefore suffers a quadratic Stark effect. All the states of hydrogen atom with $n > 1$ has some degeneracy in l in nonrelativistic limit. Since H' has non-zero matrix elements for transitions between states of odd and even l, the perturbation H' removes this degeneracy. Let us consider the first excited state of hydrogen, where the principal quantum number $n = 2$, the values of l are 0 and 1. For $l = 1$ the quantum number $m_l = 1, 0, -1$ and for $l = 0$ the value of $m_l = 0$. The eigenfunctions $\psi_{nlml} = R_{nl} Y_{lm}$ are therefore $\psi_1 = \psi_{200}$, $\psi_2 = \psi_{210}$, $\psi_3 = \psi_{211}$ and $\psi_4 = \psi_{21-1}$. The eigenfunctions can be obtained by using Table 2.3 and 2.5. The perturbation H' due to applied electric field is given by Eq. (6.37). The matrix elements $H'_{11} = H'_{22} = H'_{33} = H'_{34} = H'_{43} = H'_{44} = 0$ because in each case the integrand is antisymmetric in z. Further, $H'_{13} = H'_{14} = H'_{23} = H'_{24} = H'_{31} = H'_{32} = H'_{41} = H'_{42} = 0$ because these integrands all contain the factor $\exp(i\varphi)$ and the integration relative to φ is from 0 to 2π. This leaves only the matrix elements H'_{12} and H'_{21} which are given by

$$H'_{12} = H'_{21} = \int \psi^*_{200} H' \psi_{210} d\tau = \int \psi^*_{210} H' \psi_{200} d\tau$$

$$H'_{12} = H'_{21} = \int_0^\infty \int_0^\pi \int_0^{2\pi} R^*_{20} Y^*_{00} R_{21} Y_{10} r^3 \sin\theta \cos\theta \, dr \, d\theta \, d\varphi = -3ea_0 F \tag{6.42}$$

where a_0 is the Bohr radius of hydrogen. The matrix elements can be arranged in the determinant form

$$\begin{vmatrix} -E_1 & -3ea_0F & 0 & 0 \\ -3ea_0F & -E_1 & 0 & 0 \\ 0 & 0 & -E_1 & 0 \\ 0 & 0 & 0 & -E_1 \end{vmatrix} \tag{6.43}$$

on solving it we get

$$E_1 = 3ea_0F; \qquad \chi_1 = \frac{1}{\sqrt{2}}(\psi_1 - \psi_2)$$

$$E_1 = -3ea_0F; \qquad \chi_2 = \frac{1}{\sqrt{2}}(\psi_1 + \psi_2) \qquad\qquad (6.44)$$

$$E_1 = 0; \quad \chi_3 = \psi_3; \quad \chi_4 = \psi_4$$

The Eq.(6.44) gives the first order corrections to the energy and the corresponding zeroth order eigenfunctions. It is seen that out of the four degenerate states, only two states ψ_{200} and ψ_{210} are split but the other two states ψ_{211} and $\psi_{21\text{-}1}$ continue to be degenerate. This is shown in the Fig. 6.9.

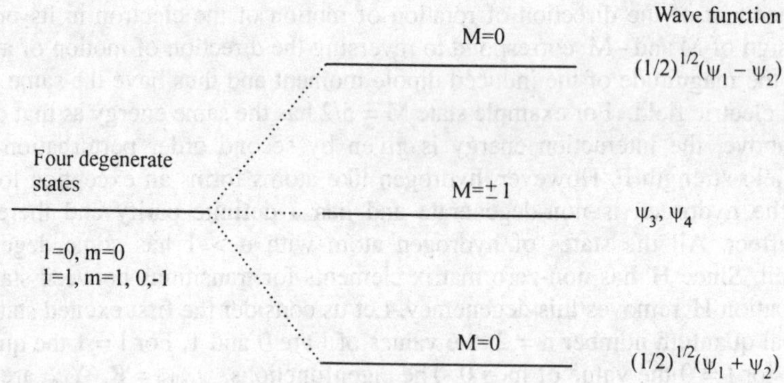

Fig. 6.9 Splitting of the degenerate n = 2 level of atomic hydrogen due to the linear stark effect

Examples

1. For a single electron atom show that $g = \dfrac{[j+(1/2)]}{[l+(1/2)]}$

According to Eq. (6. 8) for single electron atom the g value is

$$g = 1 + \frac{j(j+1) + s(s+1) - l(l+1)}{2j(j+1)}$$

where $j = l \pm s. = l \pm (1/2)$. Putting , $j = l +(1/2)$ and $s = 1/2$ in the above equation

$$g = 1 + \frac{j(j+1) - [1-(1/2)][1+(3/2)]}{2j(j+1)} = 1 + \frac{j(j+1) - (j-1)(j+1)}{2j(j+1)}$$

$$g = 1 + \frac{j-j+1}{2j} = \frac{2j+1}{2j} = \frac{j+(1/2)}{1+(1/2)}$$

If we put $j = l - (1/2)$ in the expression for g

$$g = 1 + \frac{j(j+1) - j(j+2)}{2j(j+1)} = 1 - \frac{1}{2(j+1)} = \frac{j+(1/2)}{j+1} = \frac{j+(1/2)}{1+(1/2)}$$

Thus

$$g = \frac{j+(1/2)}{1+(1/2)} \text{ for } j = l \pm (1/2).$$

2. Show that for a term with orbital quantum number L , $\sum g = (2S + 1)$ if L > S.

From Eq. (6.8)

$$g = \frac{3}{2} + \frac{(S-L)(S+L+1)}{2J(J+1)}$$

and

$$\sum g = \sum_{J=L-S}^{J=L+S} \left[\frac{3}{2} + \frac{(S-L)(S+L+1)}{2J(J+1)} \right]$$

Total numbers of terms are $(2S + 1)$. Therefore, the above expression is

$$\sum g = \frac{3}{2}(2S+1) + \frac{(S-L)(S+L+1)}{2} \sum_{J=L-S}^{L+S} \frac{1}{J(J+1)} \tag{6.45}$$

$$g = \frac{3}{2}(2S+1) + \frac{(S-L)(S+L+1)}{2}\left[\frac{1}{(L-S)(L-S+1)} + \frac{1}{(L-S+1)(L-S+2)} + \ldots + \frac{1}{(L+S)(L+S+1)} \right]$$

Taking sum of first two term of the parenthesis

$$\frac{1}{(L-S+1)}\left[\frac{1}{L-S} + \frac{1}{L-S+2} \right] = \frac{2}{(L-S)(L-S+2)}$$

Taking the sum of third term of parentheses with the above

$$\frac{2}{(L-S)(L-S+2)} + \frac{1}{(L-S+2)(L-S+3)} = \frac{3}{(L-S)(L-S+3)}$$

Proceeding the same way, the sum of n terms is

$$\frac{n}{(L-S)(L-S+n)}$$

Putting n = 2S + 1 in the above sum

$$\frac{2S+1}{(L-S)(L+S+1)}$$

Substituting this in Eq. (6.45)

$$\sum g = \frac{3}{2}(2S+1) - \frac{(S-L)(S+L+1)(2S+1)}{2(S-L)(S+L+1)} = \frac{3(2S+1)}{2} - \frac{2S+1}{2} = 2S+1$$

3. Show that for a term with orbital angular momentum L.

$$\sum g = 2(2L+1) \quad \text{if } S > L$$

We have

$$g = 1 + \frac{J(J+1)+S(S+1)-L(L+1)}{2J(J+1)} = \frac{3}{2} + \frac{S(S+10-L(L+1)}{2J(J+1)}$$

$$\sum g = \sum_{J=S-L}^{J=S+L} \left[\frac{3}{2} + \frac{S(S+1)-L(L+1)}{2J(J+1)} \right]$$

Total number of terms is 2L+1. Therefore, we have

$$\sum_{J=S-L}^{J=S+L} \left[\frac{3}{2} + \frac{S(S+1)-L(L+1)}{2J(J+1)} \right] = \frac{3}{2}(2L+1) + \frac{[S(S+1)-L(L+1)]}{2} \sum_{J=S-L}^{J=S+L} \frac{1}{J(J+1)} \tag{6.46}$$

The second term on the right hand side of the Eq. (6.46) is written as

$$\frac{[S(S+1)-L(L+1)]}{2} \left[\frac{1}{(S-L)(S-L+1)} + \frac{1}{(S-L+1)(S-L+2)} + \frac{1}{(S-L+2)(S-L+3)} + \cdots\cdots + \frac{1}{(S+L)(S+L+1)} \right]$$

Consider sum of first two term in the parentheses

$$\frac{1}{(S-L+1)} \left[\frac{1}{S-L} + \frac{1}{S-L+2} \right] = \frac{2(S-L+1)}{(S-L+1)(S-L)(S-L+2)} = \frac{2}{(S-L)(S-L+2)}$$

Considering the sum of first two term with the third term of the parentheses

$$\frac{2}{(S-L)(S-L+2)}+\frac{1}{(S-L+2)(S-L+3)}=\frac{1}{(S-L+2)}\left[\frac{2}{S-L}+\frac{1}{S-L+3}\right]=\frac{3}{(S-L)(S-L+3)}$$

Similarly the sum of n term can be written as

$$\frac{n}{(S-L)(S-L+n)}$$

Since total numbers of terms are (2L+1), therefore, sum of terms in parentheses is

$$\frac{2L+1}{(S-L)(S+L+1)}$$

The value of $\sum g$ is then given by

$$\sum g = \frac{3}{2}(2L+1)+\frac{(S-L)(S+L+1)}{2}\times\frac{2L+1}{(S-L)(S+L+1)}=2(2L+1)$$

4. Show that the sum of the intensities of combinations of a level characterized by the magnetic quantum number M with the level M-1, M and M+1 is independent of M.

Let us consider J$\rightarrow$J transition. Using Table 6.2 we have

$$I_{M\rightarrow M+1} + I_{M\rightarrow M} + I_{M\rightarrow M-1} = 2A(J+M+1)(J-M) + 4AM^2 + 2A(J-M+1)(J+M)$$

$$= 4AJ(J+1)$$

For J$\rightarrow$J+1 transition

$$I_{M\rightarrow M+1} + I_{M\rightarrow M} + I_{M\rightarrow M-1} = 2B(J+M+1)(J+M+2) +$$
$$+ 4B(J+M+1)(J-M+1) + 2B(J-M+1)(J+M+2)$$

$$= 4B(2J^2 + 5J + 3)$$

For J$\rightarrow$J-1 transition

$$I_{M\rightarrow M+1} + I_{M\rightarrow M} + I_{M\rightarrow M-1} = 2B(J-M)(J-M-1) +$$
$$+ 4B(J^2 - M^2) + 2B(J+M)(J+M-1)$$

$$= 8J(J-1)B$$

5. Calculate the g value for $^2G_{9/2}$ state.

For $^2G_{9/2}$ state, S = 1/2, L= 4 and J = 9/2. From Eq. (6.8)

$$g = 1 + \frac{(9/2)(9/2+1) + (1/2)(1/2+1) - 4(4+1)}{2(9/2)(9/2+1)} = \frac{10}{9}$$

6. Calculate g value for $^3F_2$ state.

For $^3F_2$ state, S = 1, L= 3 and J = 2. From Eq. (6.8)

$$g = 1 + \frac{2(2+1) + 1(1+1) - 3(3+1)}{2 \times 2(2+1)} = \frac{2}{3}$$

7. Calculate g values for sd configuration in jj coupling.

For s electron, l = 0 and s = 1/2 therefore j = 1/2. For d electron l = 2 and s = 1/2 and therefore j values are 3/2 and 5/2. The combinations of j values are $(1/2,3/2)_{2,1}$ and $(1/2,5/2)_{3,2}$. g value can be determined by using the Eq. (6.18)

For s electron ($g = g_1$)

$$g_1 = 1 + \frac{(1/2)(1/2+1) + (1/2)(1/2+1) - 0}{2 \times (1/2)(1/2+1)} = 2$$

For d electron ($g = g_2$) with j = 3/2

$$g_2 = 1 + \frac{(3/2)(3/2+1) + (1/2)(1/2+1) - 2(2+1)}{2(3/2)(3/2+1)} = \frac{4}{5}$$

Similarly $g = g_2$ for d electron with j = 5/2 is 6/5.
Using Eq.(6.18) for J = 2, j_1 = 1/2 and j_2 = 3/2

$$g = 2\frac{6+(3/4)-(15/4)}{12} + \frac{4}{5}\frac{6+(15/4)-3/4}{12} = \frac{11}{10}$$

For J =1, j_1 = 1/2 and j_2 = 3/2

$$g = 2\frac{2+(3/4)-(15/4)}{4} + \frac{4}{5}\frac{2+(15/4)-3/4}{4} = \frac{1}{2}$$

For J = 3, j_1 = 1/2 and j_2 = 5/2

$$g = 2\frac{12+(3/4)-(35/4)}{24} + \frac{6}{5}\frac{12+(35/4)-(3/4)}{24} = \frac{4}{3}$$

For $J = 2$, $j_1 = 1/2$ and $j_2 = 5/2$

$$g = 2\frac{6+(3/4)-(35/4)}{12} + \frac{6}{5}\frac{6+(35/4)-(3/4)}{12} = \frac{16}{15}$$

8. Predict the splitting of $^3P_1$ level when the atom is in an external magnetic field of 2T.

   For $^3P_1$. $L = S = J = 1$, therefore $g = 3/2$. The term split into three components with $M = 1$, 0 and -1. According to Eq. (6.24) the splitting is given by gML cm^{-1}. The splitting is thus $\pm (3/2) L$ cm^{-1} .The value of L is

$$L = \frac{eB}{4\pi\, m_e c} = \frac{1.6022\times10^{-19}\,C\times2T}{4\times3.14\times9.1095\times10^{-31}\,Kg\times2.9979\times10^8\,ms^{-1}} \approx 0.934\ cm^{-1}$$

The splitting is $\pm (3/2) \times 0.934$ cm$^{-1} = \pm1.401$ cm^{-1}. Since 8065 cm^{-1} = 1 eV, therefore the splitting is

$$\pm\frac{3}{2}L\ cm^{-1} = \pm1.401\times\frac{1}{8065}\ eV = \pm1.737\times10^{-4}\ eV$$

9. Calculate the Zeeman pattern and intensities for the transition $^3D_2 \rightarrow {}^3P_2$ in LS coupling.

   For $^3D_2$, $S = 1$, $L = 2$ and $J = 2$. The g value according to Eq. (6.8) is 7/6. For $^3P_2$, $S = L = 1$ and $J = 2$. the g value from Eq.(6.8) is 3/2. The separation factor Mg for both initial and final states are written down in two rows with equal values of M directly below or above each other.

M	2	1	0	-1	-2
Mg(initial)	(14/6)	(7/6)	0	(-7/6)	(-14/6)
Mg(final)	(6/2)	(3/2)	0	-(3/2)	-(6/2)

For $\Delta J = 0$, $M = 0 \rightarrow M = 0$ is not allowed. Hence there are only four p transitions whose difference are given by

$$\pm\left(\frac{14}{6}-\frac{6}{2}\right)L\ cm^{-1} = -(\pm)\frac{2}{3}L\ cm^{-1}, \quad \pm\left(\frac{7}{6}-\frac{3}{2}\right)L\ cm^{-1} = -(\pm)\frac{1}{3}L\ cm^{-1}$$

For s transitions we have

$$\pm\left(\frac{14}{6}-\frac{3}{2}\right)\text{L cm}^{-1}=\pm\frac{5}{6}\text{L cm}^{-1},\ \pm\left(\frac{7}{6}-0\right)\text{L cm}^{-1}=\pm\frac{7}{6}\text{L cm}^{-1},$$

$$\pm\left(0+\frac{3}{2}\right)\text{L cm}^{-1}=\pm\frac{3}{2}\text{L cm}^{-1},\ \pm\left(\frac{7}{6}-\frac{6}{2}\right)\text{L cm}^{-1}=-(\pm)\frac{11}{6}\text{L cm}^{-1}$$

In Runge notations

$$\Delta\bar{v}=\frac{(\pm 2,\pm 4),\pm 5,\pm 7,\pm 9,\pm 11}{6}\text{L cm}^{-1}$$

Intensity of p transitions: We have $\Delta J = 0$ and $M = 0$ to $M = 0$ transition is not allowed. From Table 6.2 $M = 2 \rightarrow M = 2$, $I = 16$ A; $M = 1 \rightarrow M = 1$, $I = 4$A; $M = -1 \rightarrow M = -1$, $I = 4$A and $M = -2 \rightarrow M = -2$, $I = 16$ A. Thus the relative intensity of p components is 4:1:1:4.

Intensity of s components:
For $M \rightarrow M-1$
For $M = 2 \rightarrow M = 1$, $I = 4$A; $M = 1 \rightarrow M = 0$, $I = 6$A; $M=0\rightarrow M = -1$, $I=6$A; $M = -1\rightarrow M = -2$, $I =4$A

For $M \rightarrow M+1$
The intensity for transition $M = -2 \rightarrow M = -1$, $I= 4$A; $M = -1\rightarrow M = 0$, $I = 6$A; $M = 0\rightarrow M = 1$, $I = 6$A; $M = 1 \rightarrow M = 2$, $I = 4$A.
The intensity of s components are in the ratio 2:3:3:2; and 2:3:3:2.

10. The normal Zeeman components of a 500 nm spectral lines are 0.0116 nm apart in a magnetic field of 1.0 T. Find the ratio (e/m) for the electron

The separation of the normal Zeeman components from Eq. (6.22) is $\Delta E = \beta_e B$ and

$$\Delta E = h\,\Delta v = \beta_e B = \frac{eh}{4\pi\,m_e}B$$

$$\Delta v = \Delta\left(\frac{c}{\lambda}\right) = \frac{e}{4\pi\,m_e}B$$

$$\Delta v = \frac{c\,|d\lambda|}{\lambda^2} = \frac{e}{4\pi\,m_e}B$$

$$\frac{e}{m_e} = \frac{4\pi\,c\,|d\lambda|}{B\lambda^2} = \frac{4\times 3.14\times 2.9979\times 10^8\,\text{ms}^{-1}\times 0.0116\times 10^{-9}\,\text{m}}{1.0\,\text{T}\times(500\times 10^{-9}\,\text{m})^2} = 1.748\times 10^{17}\,\text{C/Kg}$$

11. Calculate the separation between the adjacent normal Zeeman components for 4D state of an atom placed in a magnetic field of 1.5 T. Assume that the resultant spin angular momentum is zero.

Since S = 0, the state is singlet and g = 1. Splitting between adjacent Zeeman levels from Eq. (6.24) is

$$\Delta E(cm^{-1}) = L(cm^{-1}) = \frac{\beta_e B}{hc} = \frac{9.27408 \times 10^{-24}\, JT^{-1} \times 1.5\, T}{6.6262 \times 10^{-34}\, Js \times 2.9979 \times 10^8\, ms^{-1}} = 0.70048\ cm^{-1}$$

Since $8065\ cm^{-1} = 1$ eV, thus

$$\Delta E(eV) = \frac{0.70048}{8065}\, eV \approx 0.87 \times 10^{-4}\, eV$$

12. Find the minimum magnetic field needed for the normal Zeeman effect to be observed in a spectral line of 400 nm wavelength when a spectrometer whose resolution is 0.001 nm is used.

The spacing between Zeeman components from Eq. (6.22) is

$$\Delta E = h\, \Delta v = h\, \Delta\left(\frac{c}{\lambda}\right) = h\, \frac{c\,|\, d\lambda\,|}{\lambda^2} = \beta_e B = \frac{e\, h\, B}{4\pi\, m_e\, c}$$

$$B = \frac{4\pi\, m_e c\,|\, d\lambda\,|}{e\lambda^2} = \frac{4 \times 3.14 \times 9.1095 \times 10^{-31}\, Kg \times 2.9979 \times 10^8\, ms^{-1} \times 0.001 \times 10^{-9}\, m}{1.6022 \times 10^{-19}\, C \times (400 \times 10^{-9}\, m)^2} = 1.339\,T$$

13. Calculate the displacement in nm of the two outer Zeeman components from the mercury 184.96 nm line $(6^1S_0 \rightarrow 6^1P_1)$ in a magnetic field of 2 T.

Since the transition is between singlet states therefore g = 1 for both the states. The separation between two Zeeman components is therefore

$$\Delta E = h\, \Delta\left(\frac{c}{\lambda}\right) = hc\, \frac{|\, d\lambda\,|}{\lambda^2} = \beta_e B = \frac{eh}{4\pi\, m_e}\, B$$

$$|\, d\lambda\,| = \frac{e\, B\lambda^2}{4\pi\, m_e c^2} = \frac{1.6022 \times 10^{-19}\, C \times 2.0\,T \times (184.96 \times 10^{-9}\, m)^2}{4 \times 3.14 \times 9.1095 \times 10^{-31}\, Kg \times 2.9979 \times 10^8\, ms^{-1}} = 0.00319\,nm$$

The two outer Zeeman components are at 0.00319 nm on either side of undisplaced component.

14. A certain spectral line is known to result from a transition from a ^3D level to another lower level whose LS designation is unknown. The Zeeman pattern of the line shows 12 components. Find the J value of the upper level and LS designation of the lower level.

It is given that one of the level is ^3D hiving S = 1 and L = 2. Therefore J can be 3, 2 or 1. According to selection rule $\Delta J = 0, \pm 1$ the lower state should also have J an integer. If $\Delta J \neq 0$, then number of p components are half of s components. Thus the number of p components are 4 and accordingly 2J + 1 = 4 or J = 3/2. However, J should be an integer. Therefore, other possibility is that $\Delta J = 0$ and number of components is $2J = 4$ or $J = 2$. Thus lower state has $J = 2$ and the upper state is $^3D_2$. According to selection rule $\Delta L = \pm 1$, $\Delta S = 0$ the lower state is $^3P_2$ as if we take it as $^3F_2$ it has to lie higher for same value of n.

15. Illustrate the Zeeman splitting and the allowed transitions for the sodium D_1 and D_2 line. Calculate the frequency shift for the transition $M = +1/2 \rightarrow M = -1/2$ of sodium D_1 line in a magnetic field of 1 T.

Sodium D_1 line is because of transition from $^2P_{1/2}$ to $^2S_{1/2}$ while sodium D_2 line arises from the $^2P_{3/2}$ to $^2S_{1/2}$ transition. According to Eq. (6.8) the g values of $^2P_{3/2}$, $^2P_{1/2}$ and $^2S_{1/2}$ are 4/3 , 2/3 and 2 respectively. The splitting of the levels and the allowed transitions are shown in the Fig.6.10. The allowed selection rules for transitions between the magnetic sublevels are $\Delta M = 0$, ± 1. The sodium D_1 line split into four components with two s and two p components while sodium D_2 line split into six with two p and four s components.

For sodium D_1 line ($^2P_{1/2} \rightarrow {}^2S_{1/2}$) the shift of $M' = +1/2 \rightarrow M'' = -1/2$ is $\Delta T = M'g'L - M''g''L$ i.e.

$$\Delta T = \left(\frac{1}{2} \times \frac{2}{3} - \frac{-1}{2} \times 2 \right) L\ cm^{-1} = \frac{4}{3} L\ cm^{-1} = \frac{4}{3} \times 0.467 B(T)\ cm^{-1}$$

$$\Delta T = 1.333 \times 0.467 \times 1\ cm^{-1} = 0.6627\ cm^{-1}$$

since $1 cm^{-1} = 2.9979 \times 10^{10}$ Hz

$$\Delta T = 0.6627 \times 2.9979 \times 10^{10}\ Hz = 1.867 \times 10^{10}\ Hz$$

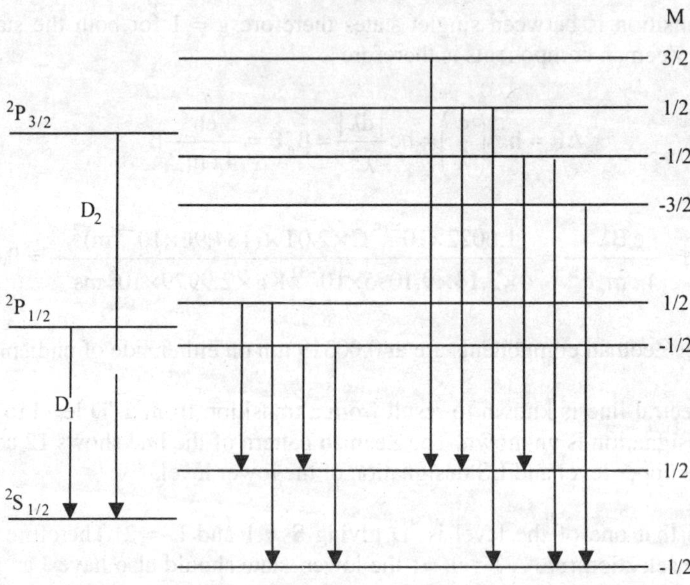

Fig. 6.10 Zeeman splitting for sodium D_1 and D_2 lines

Problems

6.1   Derive an expression for the Zeeman effect splitting of the levels of a singlet.

6.2   Show that for S levels (except $^1S_0$) $g = 2$.

6.3   Show that for all singlet levels (except $^1S_0$) $g = 1$.

6.4   Show that for all levels for which $L = S$ (except levels with $J = 0$) the value of $g = 1.5$.

6.5   Calculate the Lande's g factor for sp configuration in jj coupling.

6.6   Show that for a term with orbital quantum number L

$$g = \frac{L + 2S}{L + S}$$

for the fine structure level with maximum J.

6.7   Show that $g = 3/2$ for levels with $S = L$ (except for levels with $J = 0$).

6.8   Show that for triplet state

$$g = \frac{L + 2}{L + 1} \qquad \text{if } J = L + 1$$

$$= 1 + \frac{1}{L(L + 1)} \qquad \text{if } J = L$$

$$= \frac{L - 1}{L} \qquad \text{if } J = L - 1$$

6.9   Find the total width in wavenumbers of the Zeeman pattern for the doublet transition $^2P_{3/2} \rightarrow$ $^2S_{1/2}$ and $^2P_{1/2} \rightarrow {}^2S_{1/2}$ in an external magnetic field of 3T.

6.10   Calculate g value for $^3F_3$ state.

6.11   Calculate g value for $^2D_{5/2}$ state.

6.12   Calculate the Zeeman pattern for the transition $^2G_{9/2} \rightarrow {}^2F_{7/2}$.

6.13   Calculate the Zeeman pattern for the transition $^3F_4 \rightarrow {}^3D_3$.

6.14   Calculate the Zeeman pattern for the transition $^1D_2 \rightarrow {}^1P_1$.

6.15   How many lines will appear in Zeeman pattern of the transition $^{10}H_{3/2} - {}^{10}G_{1/2}$. By what factor will the total spread of the Zeeman pattern of the transition exceeds the normal Zeeman spread.

6.16   Show the splitting of 2P $\rightarrow$ 2S levels in weak and strong magnetic field. How does each magnetic field level go over to corresponding strong magnetic field level? Show the allowed transitions in weak and strong magnetic field.

6.17   Determine the weak field Zeeman splitting for a line arising from $^3S_1 \rightarrow {}^3P_1$ transition in terms of the normal Zeeman splitting.

6.18   Illustrate the Zeeman splitting and the allowed transitions for rubidium D_2 line.

6.19   Illustrate Paschen-Back effect for the $^3S{-}^3P$ transitions in a strong magnetic field.

7

Hyperfine Structure and Isotope Shift

So far we have considered the nuclei to be positive point charges of infinite mass. However, high precision experiments reveal the splitting of the electronic energy levels, which cannot be explained on the basis that nuclei are point charges of infinite mass. The splitting which is much smaller than the fine structure splitting is of the order of $10^{-3} - 1$ cm^{-1} is called hyperfine splitting or hyperfine structure (hfs). This hyperfine structure can be explained in terms of properties of the nuclei. In 1924, Pauli suggested that nucleus has a total angular momentum **I**, called nuclear spin. The eigenvalue of the operator I^2 is $\hbar^2 I(I+1)$ where I is nuclear spin quantum number . The largest measurable component of the angular momentum **I** is $\hbar I$. I can have integral or half integral values and is different for different states of a nucleus. We shall consider I for the ground state of a given nucleus. The nucleus has a magnetic dipole moment for $I > 0$. If $I \geq 1$, the nucleus also has a quadrupole moment which can interact with electric field gradient (EFG) produced by the atomic electrons.

There are two types of nuclear effects that produce hyperfine structure:

1(a) The interaction between an intrinsic magnetic dipole moment of the nucleus and a magnetic field produced by the motion of the atomic electrons. As nuclear magnetic dipole moment is smaller than the electronic magnetic dipole moment by $\sim 10^{-3}$ the hyperfine splitting is smaller than fine structure splitting by the same order.

(b) The interaction between quadrupole moment of the nucleus, due to a non-spherical symmetrical nuclear charge distribution, and EFG produced by the atomic electrons.

2. The energy levels of isotopes show small displacement relative to each other due to different masses of isotopes and hence produce transitions and magnetic dipole moment that lie at slightly different frequencies.

7.1 Magnetic Hyperfine Structure

The nucleus has a spin angular momentum **I** which is related to nuclear magnetic moment $\boldsymbol{\mu_N}$ by

$$\mu_N = g_I \left(\frac{eh}{4\pi M_p} \right) I = g_I \beta_N I \tag{7.1}$$

where g_I is a dimensionless number called nuclear g factor or nuclear Lande' factor. $\beta_N = e\hbar/2M_p$ is nuclear magneton and has value 5.0509×10^{-27} J T^{-1}. At the site of the nucleus, atomic electrons exhibit an effective magnetic field **B** directed along the atomic axis of rotation **J**, that is,

$$\mathbf{B} = C\mathbf{J} \tag{7.2}$$

where C is a quantity proportional to the internal magnetic field. The interaction of the nuclear magnetic dipole moment in the magnetic field produced by the atomic electrons is given by

$$H_{mhfs} = -\boldsymbol{\mu_N} \cdot \mathbf{B} \tag{7.3}$$

where H_{mhfs} is the Hamiltonian for magnetic hyperfine structure (mhfs). Using Eqs. (7.1) and (7.2) in Eq.(7.3)

$$H_{mhfs} = -C\, g_I \beta_N \mathbf{I} \cdot \mathbf{J} = a\, \mathbf{I} \cdot \mathbf{J} \qquad (7.4)$$

where $a = -C\, g_I\, \beta_N$ is called magnetic dipole interaction constant and contains product of nuclear quantity g_I which is proportional to the moment and electronic quantity C which is proportional to the magnetic field produced by the atomic electrons. The sign of a depends on the sign of g_I which varies for different nuclei. The sign of C can also vary.

When nuclear angular momentum is taken into consideration, F is used as the quantum number for the total angular momentum $\mathbf{F}$ of an atom with quantized values $\hbar\sqrt{F(F+1)}$. According to vector model, with LS coupling between electrons as shown in Fig.7.1, $\mathbf{L}$ and $\mathbf{S}$ precess rapidly around their resultant $\mathbf{J}$. The magnetic field produced by the electrons couples $\mathbf{I}$ and $\mathbf{J}$ and causes these vectors to precess slowly around their resultant $\mathbf{F}$. According to vector addition F can take the following values

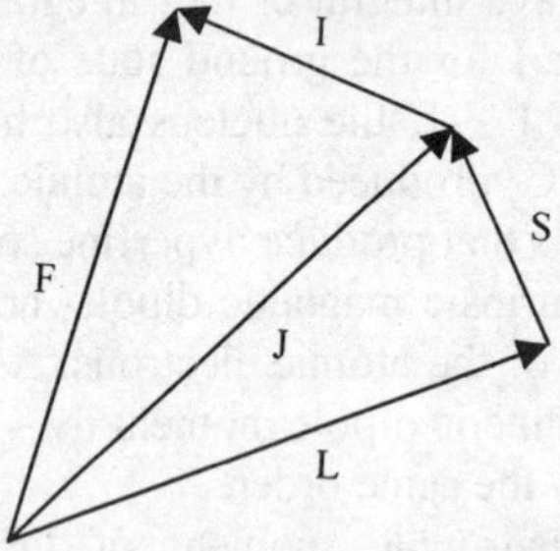

Fig. 7.1 Coupling of angular momentum vectors

$$F = I + J, I + J - 1, \ldots\ldots\ldots\ldots, |I - J| \qquad (7.5)$$

giving $2I+1$ values if $J>I$ and $2J+1$ values if $I>J$. The levels corresponding to F are $2F+1$ fold degenerate and this degeneracy can be removed by the magnetic field. From $\mathbf{F} = \mathbf{J} + \mathbf{I}$, F^2 is

$$F^2 = (\mathbf{J} + \mathbf{I}) \cdot (\mathbf{J} + \mathbf{I}) = J^2 + I^2 + 2\mathbf{J} \cdot \mathbf{I}$$

$$\mathbf{J} \cdot \mathbf{I} = (1/2)[F^2 - J^2 - I^2] = (1/2)[F(F+1) - J(J+1) - I(I+1)] \qquad (7.6)$$

From Eqs. (7.4) and (7.6)

$$E_{mhfs} = (a/2)[F(F+1) - J(J+1) - I(I+1)] \qquad (7.7)$$

The energy difference between two neighbouring hyperfine levels F and F-1 called hyperfine separation from Eq. (7.7)

$$(\Delta E)_{mhfs} = aF \qquad (7.8)$$

is thus proportional to F. This is an example of interval rule. The interaction constant a depends upon the type of orbital of the interacting electrons as well as upon the values of the nuclear magnetic moments. For p, d, f etc. orbitals ($l \neq 0$) which have zero probability density $|\psi(0)|^2$ at the nucleus , the interaction is a dipole-dipole type between the s orbital magnetic moment and the orbital and spin magnetic moment of the electron. For an s electron the dipole-dipole interaction averaged to zero, because of spherical symmetry of the s orbital. However, s orbital has non zero probability density $|\psi(0)|^2$ at the nucleus which gives rise to Fermi contact term.

Depending on the level J considered, the coefficient 'a' could be positive or negative. The nuclear magnetic moment can be in the same direction or in the opposite direction to the vector **I**. The field **B** can be oriented parallel to **J** or in the opposite direction. The field **B**$_L$ created at the nucleus by the orbital motion of an electron is in the opposite direction to the vector **L** due to the negative charge of the electron, whereas the field **B**$_S$ created at the nucleus by the spin magnetic moment   is parallel to **S**. For the assembly of atomic electrons depending on the respective values of **L** and **S** in L-S coupling or different **J** in jj coupling, the projection of the resultant of these fields on **J** representing the mean value of **B** of the field at the nucleus can be either parallel to **J** or in the opposite direction.

Fig.7.2 show the hyperfine structure of $^2P_{1/2}$ - $^2S_{1/2}$ transition of Na. The nuclear spin of Na is 3/2. Therefore, the values of F are 2 and 1 for both $^2P_{1/2}$ and $^2S_{1/2}$ level. According to the selection rules

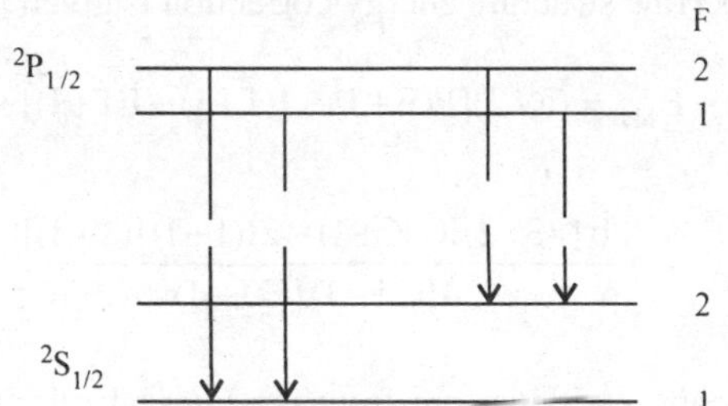

Fig. 7.2 Hyperfine structure of $^2P_{1/2}$ - $^2S_{1/2}$ transition of Na

$\Delta F =0, \pm 1$, transition take place as shown in the Fig.7.2. The transition $^2P_{1/2}$ - $^2S_{1/2}$ split into four. The width of $^2P_{1/2}$ level is about ten times smaller than that of $^2S_{1/2}$ term. For nuclei in which magnetic moment has the same sign as I, the levels of higher F have higher energy and hyperfine structure is normal.

7.2 Electrical Hyperfine Structure

The energy difference between the ground level and the excited levels of an atom is usually 10^{14}Hz. On the other hand the hyperfine splitting between levels of the same J is usually of the order of 10^9 Hz. The effect of hyperfine structure, therefore, appears as very small energy correction. For these corrections to be meaningful, it is necessary to take into account the very small but finite size of the nucleus. The distribution of charge within the nucleus is often not spherically symmetric. Departure from spherically symmetric charge distribution leads to the existence of a quadrupole moment Q of the nucleus. The quadrupole moment Q is positive if the nuclear charge distribution is elongated along the direction of I (prolate) and is negative if the distribution is flattened (oblate). A nucleus whose charge distribution is spherically symmetric has no electric quadrupole moment.

The interaction of electric quadrupole moment with the electric field gradient produced by atomic electrons leads to first order energy shift

$$E_{hfs} = \frac{b\left[\dfrac{3K(K+1)}{2} - 2(I+1)J(J+1)\right]}{4I(2I-1)J(2J-1)} \tag{7.9}$$

with

$$K = F(F+1) - I(I+1) - J(J+1) \tag{7.10}$$

and

$$b = \frac{e^2 q_J Q}{4\pi\varepsilon_0}$$

where q_J is proportional to the electric field gradient. The electric hyperfine structure is of the same order of magnitude but generally somewhat smaller. For $I = 0$ or $1/2$, nuclear quadrupole moment vanishes and therefore the energy shift (Eq. (7.9)) also vanishes. Adding the electric quadrupole correction [Eq. (7.9)], the total hyperfine structure energy correction is given by

$$\Delta E = E_{mhf} + E_{hfs} = (a/2)[F(F+1) - J(J+1) - I(I+1)] + \tag{7.11}$$

$$+ \frac{b\,[\,(3/2)K(K+1) - 2I(I+1)J(J+1)]}{4I(2I-1)J(2J-1)}$$

Electric quadrupole interaction causes a departure from the interval rule, as its dependence on the quantum number F is different from that of the magnetic hyperfine structure. Fig.7.3 shows the magnetic hyperfine structure and electric hyperfine structure for $J = 3/2$.

7.3 Isotope Shift

Isotopes with $I = 0$ have no hyperfine structure. However, in transitions between energy levels in a mixture of $I = 0$ of the same element, a structure may still be observed. This effect is called the isotope shift. Two effects cause the isotope shift: the mass effect and the volume effect:

(a) mass effect

The mass effect can be divided into normal and the specific mass effect. The normal mass effect is due to the motion of the nucleus, as it is not infinitely heavy. For hydrogenic atoms, the reduced mass of the electron

$$\mu = \frac{m_e}{1 + \dfrac{m_e}{M}} \tag{7.12}$$

entering into the Schrödinger equation depends on the mass M of the nucleus and varies therefore with the isotope considered. Energy levels E (M) for an atom whose nucleus has finite mass M is raised above the fictitious level E (∞) for an atom whose nucleus is infinitely heavy

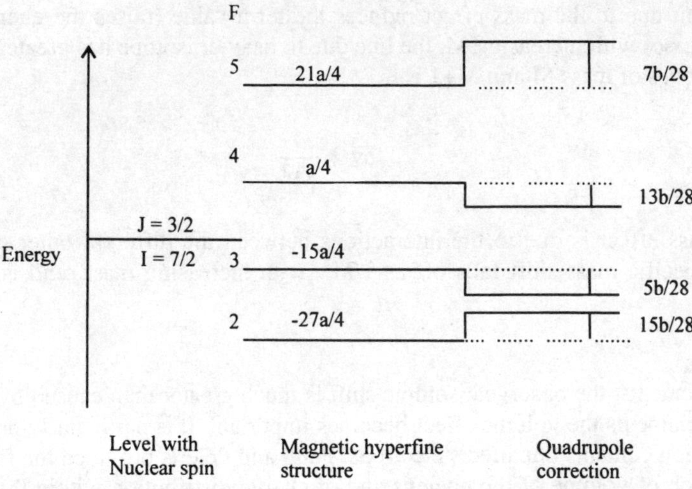

Fig. 7.3 Magnetic hyperfine structure and quadrupole correction for J=3/2

$$E(M) = E(\infty)\frac{M}{M + m_e} \tag{7.13}$$

Using Eq. (7.13)

$$(\Delta E)_M = E(M) - E(\infty) = -E(\infty)\frac{m_e}{m_e + M} \tag{7.14}$$

For another isotope whose mass is M', Eq. (7.14) is of the form

$$(\Delta E)_{M'} = -E(\infty)\frac{m_e}{m_e + M'} \tag{7.15}$$

From Eqs. (7.14) and (7.15)

$$\delta(\Delta E) = (\Delta E)_M - (\Delta E)_{M'} = -E(\infty)\frac{m_e(M' - M)}{(M + m_e)(M' + m_e)}$$

$$\delta(\Delta E) \approx -E(\infty)\frac{m_e\delta M}{MM'} \tag{7.16}$$

as M, M' $\gg m_e$ and $\delta M = M - M'$. The wavenumber shift for the observed spectral line is

$$\delta\bar{\nu} \approx \frac{m_e\delta M}{MM'}\bar{\nu} \tag{7.17}$$

The displacement due to the mass effect reduces the term value (raises the energy) and since this displacement decreases with increasing M, the line due to heavier isotope has greater wavenumber. Eq. (7.16) for two isotopes of mass M and M+1 is

$$\delta\bar{v} \propto \frac{1}{M^2} \tag{7.18}$$

The specific mass effect is due to the interactions between the different outer electrons. Like the normal shift the specific mass shift falls off as $1/M^2$ with increasing mass, and is very small in the heavy elements.

(b) volume effects

For the heavy elements, the observed isotopic shift is much greater than caused by the motion of the nucleus. For such elements the volume effect becomes important. It is particularly prominent when the electron configuration contains unpaired s electrons. Pauli and Peierls provided the first explanation on the basis of variation of volume of the nucleus and of charge distribution within the nucleus between one isotope and another. The nucleus has a charge density over a finite volume. The s electrons have a finite charge probability distribution at the nucleus. The s electrons are then no longer under the influence of pure Coulomb field. The largest relative change in the potential occurs near the origin and therefore the volume effects have great influence for s orbits.

Examples

1. Consider the hyperfine structure of the transition $5d^2 6s \; ^4F_{9/2} \rightarrow 5d^2 6p \; ^4G_{11/2}$. Why might you suspect that the magnetic dipole hyperfine structure of the $^4F_{9/2}$ level is considerably greater than that of the $^4G_{11/2}$ level ?

    The magnetic dipole interaction is largest for configuration with an unpaired s electron. Thus neglecting the contribution of the $5d^2$ electron we assume that the $5d^2 6s$ configuration has a reasonably large structure and $5d^2 6p$ has only a very small structure. The structure of the spectral line therefore, reflects the structure of the level, which is split.

2. Consider the hyperfine structure of the transition $^4F_{9/2} \rightarrow {}^4G_{11/2}$ of La^{139}. Assume that the hyperfine structure of $^4G_{11/2}$ level is negligbly small. Seven well resolved hyperfine components are observed. What is the lower limit to the nuclear spin of La^{139} ?

    The nuclear spin I can be found by counting the number of components of the line. If I > J, for then the components is simply 2J + 1. If I < J, the number of components is 2I + 1. Transitions under consideration have J = 9/2 as splitting of $^4G_{11/2}$ is negligible. Thus if I > J one should expect 10 hyperfine components. The large value of J indicate that J > I and number of components are equal to 2I + 1. Equating 2I + 1 to 7, we have I = 3. As all the hyperfine components are not resolved therefore it can be predicted that I > 5/2.

3. The hyperfine structure of the transition $^4F_{9/2} \rightarrow {}^4G_{11/2}$ show seven well resolved transitions at 0, 111.0, 208.0, 291.6, 361.3, 417.0 and 459.2 mK above the lowest frequency component. Determine the value of nuclear spin from the frequency measurements ignoring the hyperfine structure of $^4G_{11/2}$.

The interval between three components are 111.0, 97.0, 83.6, 69.7, 55.7 and 42.2 mK Let the interval 111.0 mK is due to difference in F and F −1, then the difference between F − 1 and F − 2 is 97 mK. Since the difference follow interval rules therefore

$$111.0 \text{ mK} = aF$$

and

$$97 \text{ mK} = a\,(F-1)$$

Dividing the above two equations

$$\frac{111.0}{97.0} = \frac{F}{F-1}$$

$$F = 8$$

Similarly, using the other intervals F = 8. Maximum value of F = J + I = 8, but J = 9/2. Therefore, I = 8 − 9/2 = 7/2. Nuclear spin = 7/2.

4. Show a schematic diagram of the hyperfine structure splitting of the $^2S_{1/2} \rightarrow {}^2P_{3/2}$ transition for the Na atom (for Na^{23}, I = 3/2).

   For $^2P_{3/2}$, J = I = 3/2 and F = 3, 2, 1, 0. For $^2S_{1/2}$, J = 1/2, I = 3/2 and F = 2, 1. The splitting of the levels is shown in Fig. 7.4. The transitions between components with F = 0, 1, 2, 3 in the $^2P_{3/2}$ term are very small and observed fine structure lines are due to splitting of $^2S_{1/2}$ state into two components with F = 1 and 2, respectively

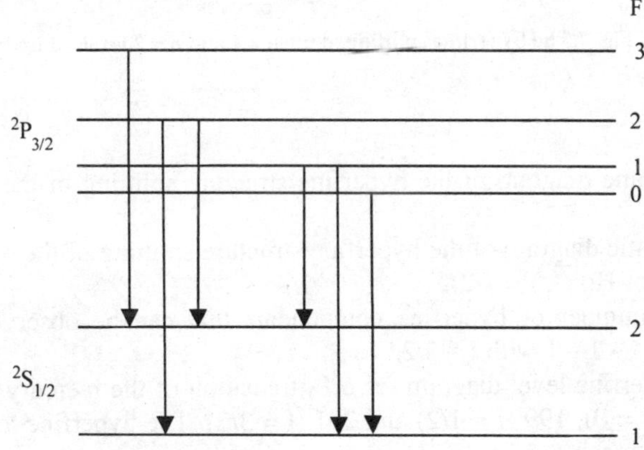

Fig. 7.4 Hyperfine structure splitting of $^2P_{3/2} \rightarrow {}^2S_{1/2}$ transition of Na

5. Obtain the ratio of frequency of lines of hydrogen and deuterium due to isotope shift and also obtain fractional shift.

   From Eq. (7.17)

$$\frac{\delta\overline{v}}{\overline{v}} = \frac{m_e\delta M}{MM'} = \frac{0.511003\ \text{Mev/c}^2}{2\times938.28\ \text{Mev/c}^2} = 2.7\times10^{-4}$$

$$\delta\overline{v} = \overline{v}' - \overline{v}_0 = 2.7\times10^{-4}\,\overline{v}_o$$

$$\overline{v}' = 1.00027\overline{v}_0$$

6.   Show a schematic diagram of the hyperfine splitting of the n = 1 and n = 2 levels of hydrogen (I = 1/2 for hydrogen).

For n = 1, l= 0   we have J = 1/2 and the state is $^2S_{1/2}$. For n = 2, l = 0 and 1 and the corresponding states are $^2S_{1/2}$, $^2P_{1/2}$ and $^2P_{3/2}$. J = 1/2 with I = 1/2 gives F = 1 and 0 while J = 3/2 with I = 1/2 gives F = 2 and 1. This is shown in the Fig. 7.5. As a result of this the level n = 1 split in to 2 levels while n = 2 level splits into four level.

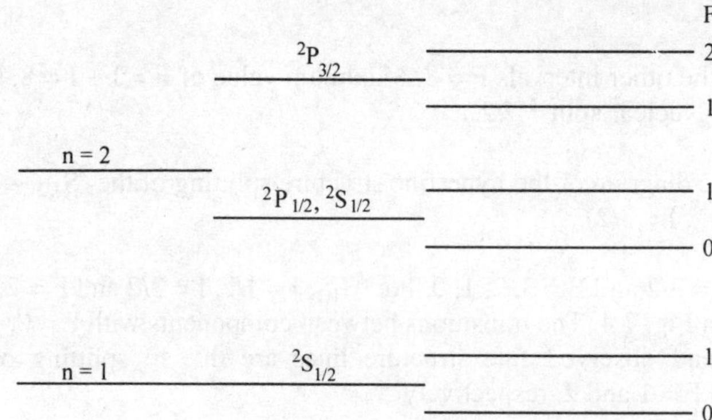

Fig. 7.5 he hyperfine splitting of the n = 1 and n = 2 levels of hydrogen

Problems

7.1   Show a schematic diagram of the hyperfine structure splitting of the n = 1 and n = 2 levels of deuterium

7.2   Show a schematic diagram of the hyperfine structure splitting of the $6\ ^3P_1 \rightarrow 6\ ^1S_0$ transition for the Hg atom (for Hg^{199}, I = 1/2).

7.3   Calculate the number of hyperfine components that can be observed spectroscopically from transition J = 2 → J = 1 with I = 3/2.

7.4   Sketch the hyperfine level diagram for 6^3P_1 transition of the mercury atom for isotopes of mass number 200 (I = 0), 199 (I = 1/2) and 201 (I = 3/2). The hyperfine levels for mass number 201 are inverted.

7.5   Draw the schematic diagram of hyperfine structure levels of λ = 472.2 nm in Bi between $6p^3$ $(1/2,3/2,3/2)_{3/2}$ and $6p^27s(1/2, 1/2, 1/2)_{1/2}$. The nuclear spin is 9/2 and the levels corresponding to $6p^3$ are inverted.

7.6   Sketch the magnetic hyperfine structure and quadrupole correction for J = 1 and I = 3/2.

7.7   Draw the allowed hyperfine transitions for the transition $3d^5 4s^2 (^6S_{5/2}) \rightarrow 3d^5 4s4p\ (^6P_{7/2})$ of Mn^{55} $(I = 5/2)$.

7.8   Show a schematic diagram of the combined magnetic and electric hyperfine structure for $J = 3/2$ and $I = 4$. The hyperfine structure is inverted.

$\boxed{8}$

Width of Spectrum Lines

Purely monochromatic spectral lines do not exist, that is, a transition between two states is not seen as infinitely sharp line but there is a definite line shape, which is independent of any optical system used to observe the spectral lines. The main causes of breadth are:

(i) Natural broadening,
(ii) Doppler broadening and
(iii) Collision broadening.

Consider an isolated atom at rest. The spectral line emitted by this atom shows its natural shape. Let the atom be given thermal energy so as to gain some translational velocity. A velocity component in the direction of the observer will result in a change in frequency of the emitted radiation according to Doppler effect. A Maxwellian distribution of velocities at a given temperature leads to different velocities with various probabilities and thus a certain width called Doppler width will be anticipated at a given temperature. The natural and Doppler broadenings are present even in the absence of neighbours. However, if the atoms are not isolated, there will be an additional broadening effect due to collision among atoms.

Natural broadening and collision broadening are homogeneous while Doppler broadening is inhomogeneous. Homogeneous broadening refers to mechanisms that affect the line shape of every atom or molecule of the sample in the same way. The line shape is Lorentzian. In inhomogeneous broadening atoms or molecules contribute to different parts of the line profile. It is a statistical effect and therefore, broadened line is Gaussian in shape.

8.1 Natural Broadening

According to the laws of classical electrodynamics, an accelerating charge radiates. This radiation carries off energy, which must come at the expense of the particles kinetic energy. Under the influence of a given force, therefore, a charged particle accelerates less than a neutral one of the same mass. The radiation evidently exerts a force back on the charge, a recoil force. Thus a vibrating electric charge is continually damped by the radiation of energy. The energy E of such an oscillator decreases exponentially with time as

$$E = E_0 \exp(-\gamma t) \tag{8.1}$$

where E_0 is initial energy at time $t = 0$. E is the energy at any later time t and $\gamma = \dfrac{2e^2\omega_0^2}{3m_e c^3}$. ω_0 is frequency of emitted radiation.

Classically, spectrum has a Lorentzian shape given by the light intensity I

$$I(\omega) = \frac{\gamma}{\pi}\frac{1}{(\omega_0 - \omega)^2 + \gamma^2} \tag{8.2}$$

where 2γ is full width at half maximum, i.e., the interval between the two points where the intensity drops at half its maximum value (Fig. 8.1).

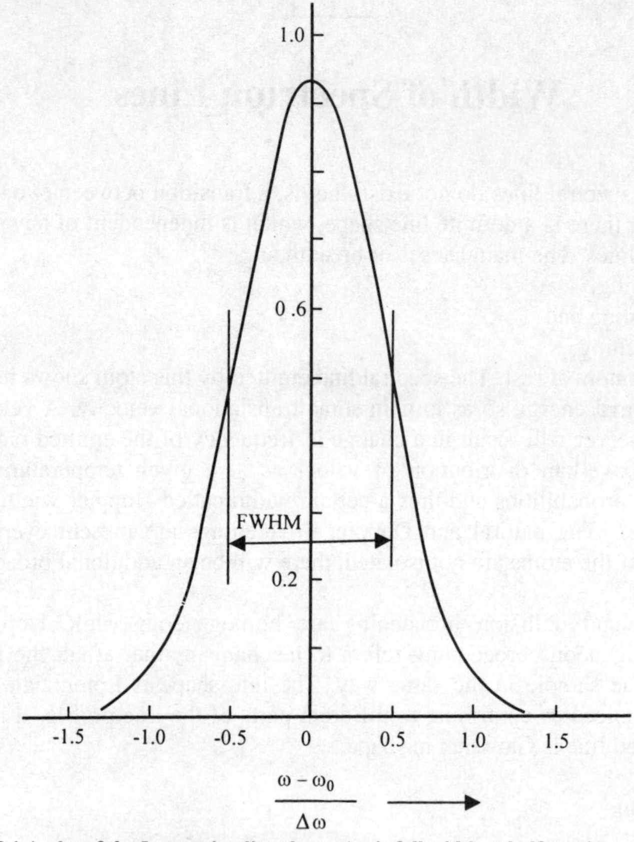

$$\frac{\omega - \omega_0}{\Delta \omega} \longrightarrow$$

Fig. 8.1 A plot of the Lorentzian line shape. $\Delta\omega$ is full width at half maximum (FWHM)

It has already been discussed that even for an isolated atom at rest, a spectral line will be broadened slightly which further implies that there will be indefiniteness in energy, ΔE. Let Δt be the time available for measurement of energy. By Heisenberg's uncertainty principle

$$\Delta E \Delta t \geq \frac{h}{4\pi} \approx \frac{h}{2\pi} \tag{8.3}$$

For a transition from ground state ($\tau = \infty$, $\Delta E = 0$) to a state (τ) we get from Eq.(8.3)

$$\Delta \nu = \frac{1}{2\pi \, \Delta\tau} = \frac{1}{2\pi \, \tau} \tag{8.4}$$

Further,

$$\Delta \nu = \Delta(\frac{c}{\lambda}) = -\frac{c}{\lambda^2} \Delta\lambda \tag{8.5}$$

$$\Delta\lambda \approx \frac{\lambda^2}{2\pi c \tau} \tag{8.6}$$

Now let τ_1 and τ_2 refer to the levels involved in transition then

$$\Delta\lambda = \frac{\lambda^2}{2\pi c}\left(\frac{1}{\tau_1}+\frac{1}{\tau_2}\right) \tag{8.7}$$

8.2 Doppler Broadening

An excited atom drop back to the original state with the excess energy emitted in the form of electromagnetic radiation. For an atom at rest the frequency v_0 of the emitted radiation is related to the corresponding wavelength by

$$v_0 = \frac{c}{\lambda_0} \tag{8.8}$$

where c is the velocity of light. Let us consider that atom while radiating is moving with a velocity v_x towards an observer in x-direction. The frequency of the emitted radiation, according to Doppler effect, is observed to be

$$v = v_0\left(1+\frac{v_x}{c}\right) \tag{8.9}$$

On the other hand, if the atom moves away from the observer, the frequency of emitted radiation is observed to be

$$v = v_0\left(1-\frac{v_x}{c}\right) \tag{8.10}$$

Let us consider an assembly of N atoms moving freely and behaving as the molecules of a gas. The v_x can have all possible values from 0 to ∞. Therefore, according to Eqs. (8.9) and (8.10), all frequencies about v_0 are possible and hence spectral line will have an infinite width. However, since the number of atoms having large velocities is very small, the intensity, which is proportional to number of atoms, falls off rapidly about a central maximum. In thermal equilibrium, the probability $f(v_x)$ that an atom of mass M moving in x direction has velocity between v_x and $v_x + dv_x$ is given·by Maxwell velocity distribution function. The number of atoms in x direction is therefore,

$$dN(v_x) = Nf(v_x)dv_x \tag{8.11}$$

where $f(v_x)$ is the Maxwell velocity distribution function for the x component of the velocity v and is

$$f(v_x) = \left(\frac{M}{2\pi\,k_B T}\right)^{1/2} \exp\left(-\frac{Mv_x^2}{2k_B T}\right) \tag{8.12}$$

From Eq. (8.9)

$$v_x = \frac{c}{v_0}(v - v_0) \tag{8.13}$$

and

$$dv_x = \frac{c}{v_0}\,dv \tag{8.14}$$

From Eqs. (8.13) and (8.14)

$$f(v_x) = \left(\frac{M}{2\pi\,k_B T}\right)^{1/2} \exp\left[-\frac{M}{2k_B T}\frac{c^2}{v_0^2}(v - v_0)^2\right] \tag{8.15}$$

From Eqs. (8.11), (8.14) and (8.15)

$$dN(v_x) = N\frac{c}{v_0}\left(\frac{M}{2\pi\,k_B T}\right)^{1/2} \exp\left[-\frac{M}{2k_B T}\frac{c^2}{v_0^2}(v - v_0)^2\right]dv \tag{8.16}$$

The intensity emitted in the frequency interval v and $v + dv$ is denoted by $I(v)dv$ and is proportional to the number of radiating atoms in the velocity component between v_x and $v_x + dv_x$ irrespective of the value of v_y and v_z. This is because the Doppler shift is based upon the component of velocity of atom moving towards or away from the observer. The $I(v)dv$ is

$$I(v)\,dv = CdN(v_x) = CN\frac{c}{v_0}\left(\frac{M}{2\pi\,k_B T}\right)^{1/2} \exp\left[-\frac{M}{2k_B T}\frac{c^2}{v_0^2}(v - v_0)^2\right]dv$$

$$I(v) = C'\exp\left[-\frac{M}{2k_B T}\frac{c^2}{v_0^2}(v - v_0)^2\right] \tag{8.17}$$

where C and C' are constants of proportionality. The Eq. (8.17) describes a Gaussian as shown in Fig.8.2. The frequency at which intensity drops to half its maximum value is

$$\exp\left[-\frac{M}{2k_B T}\frac{c^2}{v_0^2}(v - v_0)^2\right] = \frac{1}{2} \tag{8.18}$$

Taking natural logarithm of both sides

$$\frac{M}{2k_BT}\frac{c^2}{v_0^2}(v-v_0)^2 = \ln 2$$

$$v - v_0 = v_0\sqrt{\frac{2k_BT\ln 2}{Mc^2}} = 0.35786\times10^{-6}\,v_0\sqrt{\frac{T}{M_N}}$$

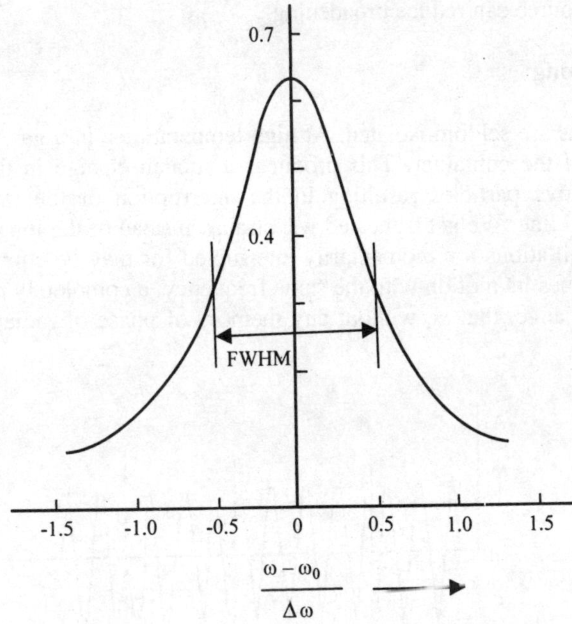

Fig. 8.2 A plot of the Gaussian line shape . $\Delta\omega$ is FWHM

Multiplying the above Equation by 2 to obtain full width at half maximum

$$\Delta v = 7.16\times10^{-7}\,v_0\sqrt{\frac{T}{M_N}} \tag{8.19}$$

where T is temperature in K and M_N is the mass number of the atom. From Eq. (8.8)

$$dv = -\frac{c}{\lambda^2}d\lambda = \left| -\frac{c}{\lambda_0}\right|\frac{d\lambda_0}{\lambda_0}$$

$$\frac{dv}{v_0} = \frac{d\lambda_0}{\lambda_0} \tag{8.20}$$

From Eqs. (8.19) and (8.20)

$$\Delta\lambda = 7.16 \times 10^{-7} \lambda_0 \sqrt{\frac{T}{M_N}}$$

(8.21)

Since Δv is proportional to v_0 , therefore, the Doppler width become less important, the smaller the frequency. In radio frequency spectroscopy the Doppler width is generally quite negligible. The Doppler width is inversely proportional to square root of mass of the atom. Therefore hydrogen lines are diffused whereas Cd and Hg lines are sharp. Further, as Doppler broadening is proportional to $\sqrt{T}$, therefore, cooling the source can reduce broadening.

8.3 Collision Broadening

Atoms in gaseous forms are seldom isolated. At high temperatures, in a gas, atoms collide with other atoms, ions or walls of the container. This produces a sudden change in the atomic radiation. The collisions of the radiative particles resulting in the interruption of the radiative process leads to broadening of a spectral line. We get truncated wavetrains, instead of the long wavetrains, that is, with every collision; the oscillations are momentarily interrupted (or may be completely cutoff). After the collision the atom resumes its motion with the same frequency, a completely random initial phase with a possible amplitude change, that is, without any memory of phase of radiation prior to its collision (Fig. 8.3).

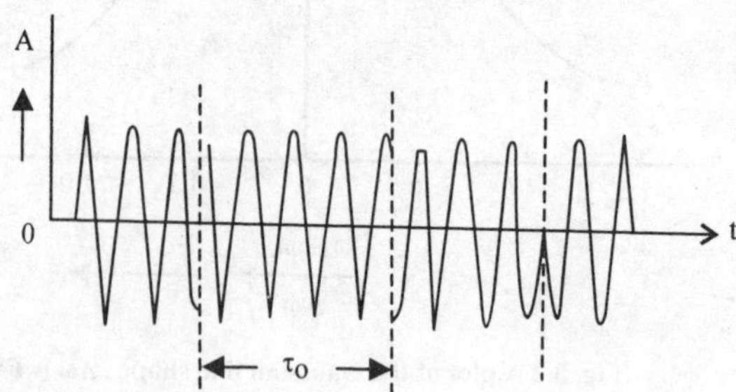

Fig. 8.3 When an atom undergoes random collisions it "sees" a wave like that shown in the figure . At every collision the phase of the wave changes abruptly

In the presence of collisions the linewidth of radiation is greater as compared to that of uninterrupted process (due to inverse proportionality of the spectrum of a wave train to the length of the train) or it can be said that mean life time τ_0 (average) between the collisions is larger than the collision time. The atomic collision frequency depends on the gas pressure therefore; collision broadening is also called pressure broadening. Here the collisions between gas atoms limit their radiative lifetime, which is quite the same as life broadening (without the interruption of the collision in the initial state of the atom).

The relative intensity I as a function of frequency is given by

$$I(\omega) = \text{constant} \frac{1}{(\omega_0 - \omega)^2 + (1/\tau_0)^2}$$

(8.22)

where $\omega_0 = 2\pi v_0$ and v_0 is the frequency with which the atom radiates between the collisions. The intensity drops to half its value at

$$\frac{1}{\tau_0} = \omega_0 - \omega = 2\pi(\nu_0 - \nu)$$

$$\nu_0 - \nu = \frac{1}{2\pi\tau_0}$$

The linewidth at half its maximum intensity

$$\Delta\nu_c = 2(\nu_0 - \nu) = \frac{1}{\pi\tau_0} \qquad (8.23)$$

Examples

1. For Argon ion laser transition 488 nm, calculate the Doppler broadening at 1200 ^{0}C (mass no.of Ar = 40)

From Eq. (8.19)

$$\Delta\nu = 7.16\times10^{-7}\,\frac{2.9979\times10^8\,\text{m/s}}{448\times10^{-9}\,\text{m}}\sqrt{\frac{1473}{40}} \approx 2.67\times10^9\,\text{Hz}$$

2. For sodium D line (5893 Å), calculate the Doppler broadening in Å at 500 K.

From Eq. (8.21)

$$\Delta\lambda = 7.16\times10^{-7}\times5893\times10^{-8}\,\text{cm}\sqrt{\frac{500}{23}}$$

$$\Delta\lambda = 0.0197\times10^{-8}\,\text{cm}$$

3. For potassium D line (7681.94 Å), the Doppler broadening is 770 MHz. Calculate the temperature of potassium vapours.

From Eq. (8.19)

$$T = M_N\left(\frac{\Delta\nu}{7.16\times10^{-7}\times\nu_0}\right)^2 = M_N\left(\frac{\Delta\nu\times\lambda_0}{7.16\times10^{-7}\times c}\right)^2$$

$$T = 39\left(\frac{770\times10^6\,\text{Hz}\times7681.94\times10^{-8}\,\text{cm}}{7.16\times10^{-7}\times2.99792\times10^{10}\,\text{cm s}^{-1}}\right)^2 \approx 296\,\text{K}$$

4. For a He-Ne laser calculate the natural broadening. The spontaneous lifetime $\approx$ 0.1 μs.

From Eq. (8.4)

$$\Delta v = \frac{1}{2\pi\tau} = \frac{1}{2\times 3.14\times 0.1\times 10^{-6}\,\text{s}} \approx 1.6\ \text{MHz}$$

5.   For a He-Ne laser calculate broadening due to collision at a pressure 0.5 Torr at 300 K. The mass of Ne is 20.18 a.u. The atomic radius is 1.12 Å.

Mean free lifetime

$$\tau_0 = \frac{\text{Mean free path}}{\text{Mean velocity}}$$

Mean velocity

$$\sqrt{\frac{8\,k_B T}{\pi\,M}}$$

Mean free path

$$\frac{k_B T}{4\sqrt{2}\,\pi\,P\,a^2}$$

where P is pressure and a is the atomic radius. Thus

$$\tau_0 = \frac{1}{16\,P\,a^2}\sqrt{\frac{M\,k_B T}{\pi}}$$

Now

$$0.5\ \text{Torr} = 1.01325\times 10^5\times (0.5\,/\,760)\ \text{Pa} = 66.66\ \text{Pa}$$

Hence

$$\tau_0 = \frac{1}{16\times 66.66\ \text{Pa}\times (1.12\times 10^{-10}\,\text{m})^2}\sqrt{\frac{20.18\times 1.67\times 10^{-27}\,\text{Kg}\times 1.38\times 10^{-23}\,\text{JK}^{-1}\times 300\text{K}}{3.14}}$$

$$\tau_0 \approx 0.5\ \mu\text{s}$$

$$\Delta v = \frac{1}{\pi\,\tau_0} = \frac{1}{3.14\times 0.5\times 10^{-6}\,\text{s}} \approx 0.64\ \text{MHz}$$

6. For a laser transition 510.5 nm, the Doppler linewidth is 2.3 x 10^9 Hz at 1500 ^{0}C.Calculate the mass number of the lasing element.

From Eq. (8.19)

$$\frac{T}{M_N} = \left(\frac{\Delta v}{v_0}\right)^2 \frac{10^{14}}{(7.16)^2}$$

$$M_N = \frac{Tc^2 \, (7.16)^2}{(\Delta v)^2 \, \lambda_0^2 \times 10^{14}}$$

$$M_N = \frac{1773 \, K \times (2.99792 \times 10^{10} \, cm \, s^{-1})^2 \times (7.16 \times 10^{-7})^2}{(2.3 \times 10^9 \, Hz)^2 \times (5105 \times 10^{-8} \, cm)^2}$$

$$M_N = 59.3 \approx 59$$

The mass number is 59.

7. An atom emits a spectral line of wavelength 400 nm and its width is found to be 10^{-19} cm. Assuming that this spectral line arises from the transition of atom from an excited energy level to the ground state, estimate the life time of the excited energy level.

From Eq. (8.6)

$$\tau \approx \frac{\lambda^2}{2 \pi c \, \Delta \lambda} = \frac{(400 \times 10^{-9} \, m)^2}{2 \times 3.14 \times 2.9979 \times 10^8 \, ms^{-1} \times 10^{-21} \, m}$$

$$\tau \approx 8.5 \times 10^{-2} \, s.$$

Problems

8.1   For He-Ne laser transition 632.8 nm, calculate Doppler broadening at 400 K (mass number of Ne = 20).

8.2   For the vibrational transition of CO_2 leading to 10.6 μm radiation at 300 K, obtain the Doppler linewidth in Å.

8.3   Consider two level system with $E_1 = -13.6$ eV and $E_2 = -3.4$ eV. Assume spontaneous emission coefficient $A_{21} \approx 6 \times 10^8$ s^{-1}. What is the natural broadening.

8.4   For sodium D line (5893 Å), calculate the Doppler width in cm^{-1} at 298 K.

8.5   Obtain the collision broadening if the mean collision time is $\approx 10^{-6}$ s.

8.6   For potassium D line (7681.94 Å), calculate the Doppler broadening in Hz at 2000 K.

8.7   The spontaneous life time of the sodium levels corresponding to D_1 line is 16 ns. Calculate the natural width.

8.8   The Doppler linewidth of a CO_2 laser for a transition at 10.6 μm is 6.1 x 10^7 Hz. Calculate the gas temperature.

8.9   In which region will Doppler broadening be more prominent in radio frequency region or in visible region ?

9

Electron Spin Resonance

9.1 Electron Spin Resonance

Electron spin resonance (ESR) and electron paramagnetic resonance (EPR) are synonymous terms. Zavoisky first observed ESR absorption in 1945 at Kazan and shortly later by Cummerow and Halliday in USA. This leads to an important method for obtaining information about paramagnetic substances. Essentially it forms a branch of high resolution spectroscopy using frequencies in the microwave region ($\nu \sim 10^9$ - 10^{11} Hz). ESR has been defined as the form of spectroscopy concerned with microwave induced transitions between magnetic energy levels of electron having a net angular momentum. ESR differs from simple microwave spectroscopy in being concerned with paramagnetic materials.

9.2 Substances, Which Can Be Investigated by ESR

Since ESR requires the presence of the unpaired electrons in the sample being studied; its range of applications is restricted to paramagnetic substances and to substances that can be converted to a paramagnetic form with sufficient stability for a spectrum to be observed. Paramagnetism occurs in :

1. Atom and ions: all configurations with an odd number of electrons must possess angular momentum and therefore must be paramagnetic.
2. Molecules and molecular ions: molecules such as NO and NO_2 have odd number of electrons and are therefore paramagnetic. The molecules such as O_2 although having an even number of electrons, have a ground state with a partially filled molecular shell and is thus paramagnetic.
3. Transition group impurities: these are atoms or ions with incomplete 3d, 4d, 5d, 4f or 5f shell. However, not all the valence states of these transition metal ions are paramagnetic. The most commonly observed paramagnetic ions are V^{4+}, VO^{2+}, Ti^{3+}, $Cr^{3+,}$ Mn^{2+}, Fe^{3+}, Fe^{2+}, Co^{2+}, Ni^{2+}, Cu^{2+}, Pd^{2+}, Ru^{2+}, Os^{4+}, Gd^{3+}, Eu^{2+}, Mo^{5+}, Ir^{4+} etc.
4. Donors and acceptors in semiconductors such as phosphorous donor impurities in silicon.
5. Colour centres (e.g. V_k , F centres).
6. Organic and inorganic radicals.
7. Activators and coactivators in phosphors, such as self activated ZnS.
8. Radiation damage centres.
9. Conduction electrons.

9.3 Resonance Condition

For a free electron the only magnetic moment is that which is associated with the spin. Classically the energy of interaction between the magnetic moment $\mathbf{\mu_e}$ and magnetic field $\mathbf{B}$ is

$$E = -\mathbf{\mu_e} \cdot \mathbf{B}$$

(9.1)

The magnetic moment is a vector, which is collinear with spin angular momentum vector **S**. The operators for the vectors are related by

$$\boldsymbol{\mu}_e = -\gamma \mathbf{S} \qquad (9.2)$$

where γ is the magnetogyric ratio, or the ratio between magnetic and mechanical moments. The value of γ is given by g_e ($e/2m_e$), g_e is positive dimensionless factor (2), however, quantum electrodynamical calculations show it to be 2.0023. $\boldsymbol{\mu}_e$ is antiparallel to **S** because of the negative charge of the electron. Eq. (9.2) can be written as

$$\boldsymbol{\mu}_e = -g_e \beta_e \mathbf{S} \qquad (9.3)$$

Where S is now dimensionless. To obtain the quantum mechanical Hamiltonian, we replace $\boldsymbol{\mu}_e$ by the appropriate operator in Eq. (9.1)

$$H = g_e \beta_e \mathbf{S} \cdot \mathbf{B} \qquad (9.4)$$

If the magnetic field is assumed to be in z direction, $B_x = B_y = 0$ and $B_z = B$ and scaler product becomes

$$H = g_e \beta_e S_z B \qquad (9.5)$$

where β_e is Bohr magneton given by e $\hbar$ /2m$_e$ The eigenvalues of Eq. (9.5) are just the multiples of the eigenvalues of S$_z$ and are given by

$$E = g_e \beta_e BM \qquad (9.6)$$

where M = 1/2 and –1/2 .
Substituting the values of M in Eq. (9.6)

$$E = \pm \frac{1}{2} g_e \beta_e B \qquad (9.7)$$

The lowest state has the negative sign and corresponding to magnetic moment aligned parallel to the magnetic field and hence spin antiparallel to it. The difference in energy between the two states is given by

$$\Delta E = h\nu = E_2 - E_1 = g_e \beta_e B \qquad (9.8)$$

The separation between two levels increases linearly with the magnetic field (Fig.9.1). Transition between the two levels can be induced by a microwave magnetic field B$_1$ at right angle to B. The transitions, which give rise to ESR spectra, are magnetic dipole in origin and parity does not change. The selection rule for the magnetic quantum number M is Δ M = ±1. This is in contrast to the intrinsically far more strongly allowed electric dipole transitions observed in electronic spectra where the parity of the states must differ.

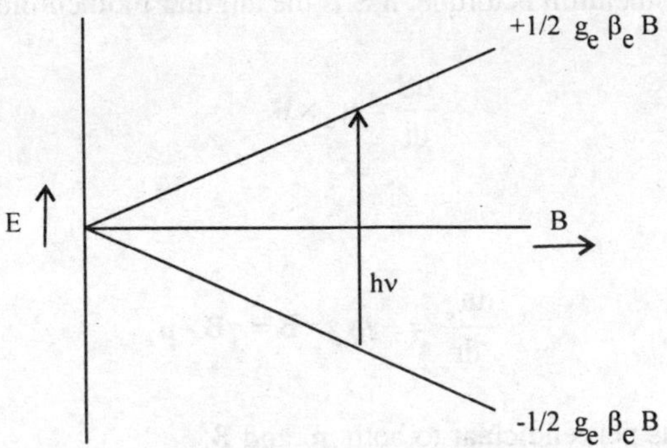

Fig. 9.1 The energy level of an electron in a magnetic field. Application of microwave radiation of energy hν causes resonance at a field given by hν /g_e β

Substituting the values of constants in Eq. (9.8) and $g_e = 2$ we have

$$B(mT) = 35.724\nu \ (GHz) \tag{9.9}$$

The conventional ESR spectrometer operates at different frequency bands are listed in Table 9.1

Table 9.1. Representation frequency and magnetic field for resonance at g = 2 for various wave band used in the ESR spectroscopy

Band	λ (cm)	ν(GHz)	B(T)
S	9.0	3	0.11
X	3.0	9	0.33
K	1.2	24	0.85
Q	0.8	35	1.25
E	0.4	70	2.50

Eq. (9.8) establishes a triple dependence

1. Magnetic field B bring forth the appearance of spin sublevel and determine the energy difference between them.
2. The microwave frequency energy quantum   hν causes transitions from M= -1/2 state to M= +1/2 state, attended by absorption of energy, and produces an absorption signal.
3. The g-value defines the change in the position of the absorption line in the spectrum under given hν and B depending on the features particular to the state of paramagnetic electron in the study of the specimen, thus presenting the characteristic of the substances in condition of resonance.

9.4 Description of ESR by Precession

When a magnetic dipole is place in a magnetic field the torque **N** acting on it is

$$\mathbf{N} = \boldsymbol{\mu}_e \times \mathbf{B} \tag{9.10}$$

Rate of change of angular momentum is torque, if $\mathbf{S}$ is the angular momentum

$$\frac{d\mathbf{S}}{dt} = \boldsymbol{\mu}_e \times \mathbf{B} \qquad (9.11)$$

From Eqs. (9.2) and (9.11)

$$\frac{d\boldsymbol{\mu}_e}{dt} = -\gamma \boldsymbol{\mu}_e \times \mathbf{B} = \gamma \mathbf{B} \times \boldsymbol{\mu}_e \qquad (9.12)$$

the change in $\boldsymbol{\mu}_e$ at any time is perpendicular to both $\boldsymbol{\mu}_e$ and $\mathbf{B}$.

Taking magnetic field in z-direction i.e. $\mathbf{B} = B\mathbf{k}$ and $\boldsymbol{\mu}_e = (\mu_x\mathbf{i} + \mu_y\mathbf{j} + \mu_z\mathbf{k})$ we have

$$\frac{d\boldsymbol{\mu}_e}{dt} = \gamma(B\mathbf{k}) \times (\mu_x\mathbf{i} + \mu_y\mathbf{j} + \mu_z\mathbf{k}) = \gamma(\mu_x B\mathbf{j} - \mu_y B\mathbf{i}) \qquad (9.13)$$

since Eq. (9.13) relates two vectors, the components on each side must be identical. Using Eq. (9.13) we have

$$\frac{d\mu_x}{dt} = -\gamma\mu_y B$$

$$\frac{d\mu_y}{dt} = \gamma\mu_x B \qquad (9.14)$$

$$\frac{d\mu_z}{dt} = 0$$

These equations are satisfied by

$$\mu_z = \text{constant}$$

$$\mu_x = \cos\omega_L t \qquad (9.15)$$

$$\mu_y = \sin\omega_L t$$

where

$$\omega_L = \gamma B \qquad (9.16)$$

This $\boldsymbol{\mu}$ precess about $\mathbf{B}$ with a constant frequency ω_L making a fixed angle with the direction of static magnetic field B (Fig.9.2). The frequency ω_L is known as Larmor frequency.

Let us consider a second coordinate system (x',y',z') rotating about z axis at an angular velocity ω (Fig.9.3). The connection between the laboratory and rotating frames is given by

$$\left(\frac{d\mathbf{S}}{dt}\right)_{lab} = \left(\frac{d'\mathbf{S}}{dt}\right)_{rot} + \boldsymbol{\omega} \times \mathbf{S} \qquad (9.17)$$

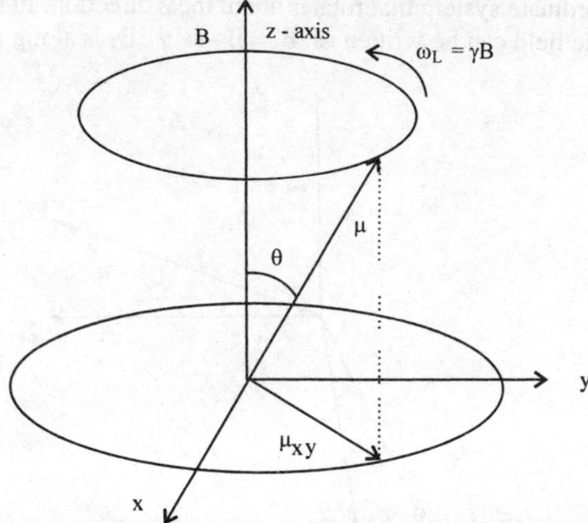

Fig. 9.2 Precession of magnetic moment μ around a constant magnetic field B

$$\left(\frac{d'S}{dt}\right)_{rot} = \left(\frac{dS}{dt}\right)_{lab} + S \times \omega = \mu \times (B - \frac{\omega}{\gamma}) \tag{9.18}$$

$$(d'\mu/dt)_{rot} = \gamma B' \times \mu \tag{9.19}$$

with

$$B' = B - \frac{\omega}{\gamma} \tag{9.20}$$

The time rate of change of μ, measured in the rotating system can be calculated, therefore, as if the rotating system was a stationary one having an effective magnetic field B' along z-axis. If B' is constant and $\omega = \gamma B$, the effective field B' vanishes. The magnetic moment μ would see no torque in the rotating frame and would remain constant with respect to it.

ESR experiments do not use a constant B but rather employ time dependent components, which are usually small, compared to the constant component. Let total magnetic field have a z component which is of constant magnitude B and it also have an oscillating component along x axis and with a peak to peak amplitude of $4B_1$ such that $B_1 << B$. The time dependent field can be written as

$$B_1 = i2B_1\cos\omega\,t = B_1(i\cos\omega\,t + j\sin\omega\,t) + B_1(i\cos\omega\,t - j\sin\omega\,t) \tag{9.21}$$

The first term is a field of constant amplitude B_1 rotating about z-axis at frequency ω in the same sense as electronic precession. The second term is a similar field but with the opposite sense of the electronic precession. The component rotating opposite to the sense of the electronic precession does not cause resonance. Neglecting this component the total field B_e can be written as

$$B_e = iB_1\cos\omega\,t + jB_1\sin\omega\,t + kB \tag{9.22}$$

B_1 is constant in a coordinate system that rotates about the z direction. In this rotating frame it is shown by Eq. (9.19) that static field can be written as $\mathbf{B'} = \mathbf{B} - \omega/\gamma$. B_1 is along x' axis of rotating coordinate

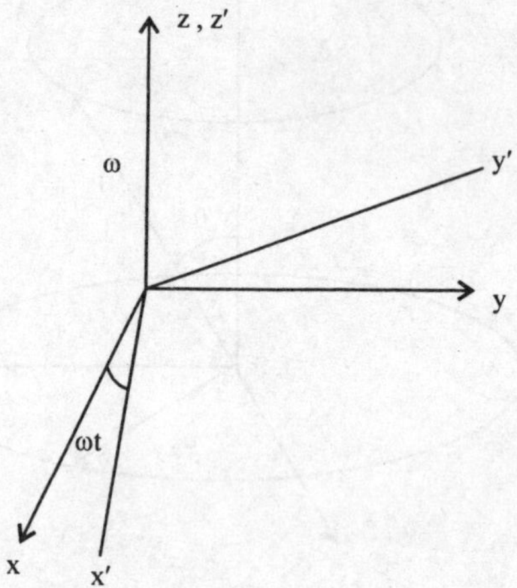

Fig. 9.3 Fixed laboratory coordinates x , y , z and rotating coordinates x' , y' , z'
for the angular velocity ω along z-axis. The primed and unprimed axes
being the same at t = 0

system. B' is along z or z' axis. When $\mathbf{B} = \omega/\gamma$ the condition for magnetic resonance occurs and $B_e = B_1$. For away from magnetic resonance B' is either nearly parallel to B or antiparallel to B as $B_1 \ll B$. Thus the motion of magnetic moment is then of a precession about B_1 with angular velocity γB_1 and every half cycle of this motion it changes from being parallel to antiparallel and back again. In the rotating system this motion take place in the plane normal to B. Since $B_1 \ll B$, the precession about B_1 occurs at a much lower velocity than that at which B_1 rotates in the laboratory system. Thus in the latter system the motion of the angular momentum and magnetic moment consists of a rapid motion about B at an angle to B which varies from 0 to π and back again. Thus when $\mathbf{B} = \omega/\gamma$ the magnetic moment assume to be initially parallel to B can be completely reversed by an application of rotating field B_1, and it will occur no matter how small is the value of B_1.

9.5 Quantum Mechanical Description of ESR

When an unpaired electron is placed in a constant magnetic field B (which we take parallel to the z - axis), the interaction with its magnetic moment is described by the Hamiltonian given by Eq. (9.4). The only non-vanishing matrix element of S_z is $\langle M |S_z| M \rangle = M$ and the energy eigenvalue of the Eq. (9.4) is a stationary state is given by Eq. (9.6).

Let a linearly polarised oscillating magnetic field of amplitude $2B_1$ ($B_1 \ll B$) is applied parallel to x-axis in the xy plane. The oscillating field is given by $2B_1\cos wt$. The total field is therefore,

$$\mathbf{B}_0 = \mathbf{i}2B_1\cos\omega\, t + \mathbf{k}B \tag{9.23}$$

The Hamiltonian (9.4) becomes

$$H = g_e\beta_e S_z B + 2g_e\beta_e B_1 S_x \cos\omega\, t = H_0 + H_1 \tag{9.24}$$

The effect of oscillatory perturbation is evaluated with the use of time dependent perturbation. The wavefunction can be written as summation over all states $|u_n>$

$$\Psi = \sum_n C_n |u_n> \exp\left(-\frac{i2\pi E_n t}{h}\right)$$

(9.25)

where C_n is amplitude and E_n the energy of the nth state. For $S = 1/2$, the states are $|1/2>$ and $|-1/2>$. Denoting $|1/2>$ by $|M'>$ and $|-1/2>$ by $|M>$, the wave function can be written as

$$\Psi = C_{M'}|M'> \exp\left(-\frac{i2\pi E_{M'}t}{h}\right) + C_M |M> \exp\left(-\frac{i2\pi E_M t}{h}\right)$$

(9.26)

$$\Psi = C_{M'}\Psi_{M'} + C_M \Psi_M$$

(9.27)

where

$$\Psi_{M'} = |M'> \exp\left(-\frac{i2\pi E_{M'}t}{h}\right)$$

(9.28)

$$\Psi_M = |M> \exp\left(-\frac{i2\pi E_M t}{h}\right)$$

(9.29)

The Schrödinger wave equation is

$$H\Psi = \frac{ih}{2\pi}\frac{\partial \Psi}{\partial t}$$

(9.30)

On substituting the value of Ψ in Eq.(9.30) from Eq.(9.26) we obtain

$$H_1\Psi = \frac{ih}{2\pi}\left\{|M'> \exp\left(-\frac{i2\pi E_{M'}t}{h}\right)\frac{dC_{M'}}{dt} + |M> \exp\left(-\frac{i2\pi E_M t}{h}\right)\frac{dC_M}{dt}\right\}$$

(9.31)

Multiplying Eq. (9.31) from left by $\Psi_M{}^*$

$$<M|H_1|M'> C_{M'}\exp\left[-\frac{i2\pi (E_{M'}-E_M)t}{h}\right] + <M|H_1|M> C_M =$$

$$\frac{ih}{2\pi}<M|M'> \exp\left[-\frac{i2\pi (E_{M'}-E_M)t}{h}\right]\frac{dC_{M'}}{dt} + \frac{ih}{2\pi}<M|M>\frac{dC_M}{dt}$$

(9.32)

Since H_1 has no matrix element between the eigenstates of S_z the Eq. (9.32) reduces to

$$\frac{dC_M}{dt} = \frac{2\pi C_{M'}}{ih} < M | H_1 | M' > \exp\left[-\frac{i2\pi (E_{M'} - E_M)t}{h} \right] \tag{9.33}$$

Similarly,

$$\frac{dC_{M'}}{dt} = \frac{2\pi C_M}{ih} < M' | H_1 | M > \exp\left[-\frac{i2\pi (E_M - E_{M'})t}{h} \right] \tag{9.34}$$

Now

$$< M | H_1 | M' > = < M | 2g_e\beta_e B_1 S_x \cos\omega t | M' > = g_e\beta_e B_1 < M | S_x | M' > [\exp(i\omega t) + \exp(-i\omega t)] \tag{9.35}$$

Eqs. (9.33) and (9.34) can be written, by using $\dfrac{2\pi(E_{M'} - E_M)}{h} = \omega_{M'M}$ in the form

$$\frac{dC_M}{dt} = \frac{2\pi g_e\beta_e B_1 C_{M'}}{ih} < M | S_x | M' > \{\exp[i(\omega - \omega_{M'M})t] + \exp[-i(\omega + \omega_{M'M})t]\} \tag{9.36}$$

$$\frac{dC_{M'}}{dt} = \frac{2\pi g_e\beta_e B_1 C_M}{ih} < M' | S_x | M > \{\exp[i(\omega + \omega_{M'M})t] + \exp[-i(\omega - \omega_{M'M})t]\} \tag{9.37}$$

We assume that at t = 0, the system is in state Ψ_M so that $C_M = 1$. therefore, Eq.(9.37) gives

$$\int_0^{C_{M'}} dC_{M'} = \frac{2\pi g_e\beta_e B_1}{ih} < M' | S_x | M > \int_0^t \{\exp[i(\omega + \omega_{M'M})t] + \exp[-i(\omega - \omega_{M'M})t]\} dt$$

$$C_{M'} = \frac{2\pi g_e\beta_e B_1}{ih} < M' | S_x | M > \left\{ \frac{\exp[i(\omega + \omega_{M'M})t] - 1}{i(\omega + \omega_{M'M})} + \frac{\exp[-i(\omega - \omega_{M'M})t] - 1}{i(\omega - \omega_{M'M})} \right\} \tag{9.38}$$

Near resonance the denominator of one of the two terms in the Eq. (9.38) will be small and that term will dominate the expression for $C_{M'}$. The Eq. (9.38) is therefore,

$$C_{M'} = \frac{2\pi g_e\beta_e B_1}{ih} < M' | S_x | M > \frac{\exp[-i(\omega - \omega_{M'M})t] - 1}{i(\omega - \omega_{M'M})} \tag{9.39}$$

Multiplying Eq. (9.39) by $C_{M'}^*$, and simplifying

$$C_{M'}C_{M'}^* = \frac{16\pi^2 g_e^2\beta_e^2 B_1^2}{h^2} |< M' | S_x | M >|^2 \frac{\sin^2[(\omega - \omega_{M'M})t/2]}{(\omega - \omega_{M'M})^2}$$

$$= \frac{4\pi^2 g_e^2\beta_e^2 B_1^2}{h^2} |< M' | S_x | M >|^2 \eta[\omega - \omega_{M'M}, t] \tag{9.40}$$

where

$$\eta[\omega - \omega_{M'M}, t] = \frac{4\sin^2[(\omega - \omega_{M'M})t/2]}{(\omega - \omega_{M'M})^2} \qquad (9.41)$$

Eq. (9.40) gives the probability that at a time t, the system will be in a stationary state $|M'>$ if it originally was in a state $|M>$, when inducing radiation has a particular frequency. The excited states of quantum states do not have perfectly defined energies, but are usually broadened. Further, perturbation is not a pure oscillating at a single frequency but rather a mixture of frequencies in a band of greater or lesser width centred at ν. As a result of these all transitions are broadened in some way. Let us take this to be a distribution of $\nu_{M'M}$ where the relative density of ν values in the range from $\nu_{M'M}$ to $\nu_{M'M} + d\nu_{M'M}$ is given by $g(\nu_{M'M})d\nu_{M'M}$ where

$$\int_0^\infty g(\nu_{M'M})d\nu_{M'M} = 1 \qquad (9.42)$$

The transition probability per unit time $W_{M'M}$ is given by the time derivative $\dfrac{d(C_{M'}C_{M'}^*)}{dt}$ integrated over a distribution of frequencies

$$W_{M'M} = \frac{4\pi^2 g_e^2 \beta_e^2 B_1^2}{h^2 t} \int |< M'|S_x|M >|^2 \, \eta[2\pi\nu - 2\pi\nu_{M'M}, t]g(\nu_{M'M})d\nu_{M'M} \qquad (9.43)$$

If t is long enough so that η is sharper function of $2\pi\nu_{M'M}$ than is g, for then $\eta[2\pi\nu - 2\pi\nu_{M'M}, t]$ may be replaced by $t\delta(\nu - \nu_{M'M})$ and Eq.(9.43) reduces to

$$W_{M'M} = \frac{4\pi^2 g_e^2 \beta_e^2 B_1^2}{h^2 t} g(\nu)t|< M'|S_x|M >|^2 = \gamma^2 B_1^2 g(\nu)|< M'|S_x|M >|^2 \qquad (9.44)$$

$<M'|S_x|M>$ is the quantum mechanical matrix element of the x component of the electron spin operator and is zero unless $M = M' \pm 1$ i.e. $\Delta M = \pm 1$. The Eq. (9.44) furnishes the selection rule $\Delta M = \pm 1$ so that transitions are permitted only between adjacent energy levels. For an unpaired electron $M' = 1/2$ and $M = -1/2$ and there is only one transition , so Eq. (9.44) becomes

$$W = \frac{1}{4}\gamma^2 B_1^2 g(\nu) \qquad (9.45)$$

9.6 Relaxation Mechanisms

Under thermal equilibrium conditions, the populations of the two levels with $M = -1/2$ and $M = 1/2$ are determined by the Boltzmann distribution

$$\frac{N_1}{N_2} = \exp\left(\frac{h\nu}{k_B T}\right) \qquad (9.46)$$

where hv is the energy difference between the two levels. N_1 and N_2 are the population of M= -1/2 and M = 1/2 levels, respectively. However, N_1-N_2 is very small. The phenomenon of ESR absorption depends on this difference. The *a priori* probability of transition from lower level to upper level equals that from transition from upper level to lower level. Because of N_1>N_2 results in an excess of upward transitions and a net absorption of energy from microwave field. However, this leads to an increase of N_2, which will continue until N_1= N_2 and net absorption of energy will tend to zero. That this situation does not occur it means that there must therefore be other mechanisms by means of which energy absorbed and stored in the upper level can be dissipated in such a manner as to allow return to lower level to maintain the population difference. Such mechanisms are called relaxation processes.

(a) spin-lattice relaxation

In the absence of magnetic field the electron spins are randomly oriented in space and their resultant moment is zero. When a static magnetic field **B** is applied, the spins become aligned parallel or antiparallel to **B**, with slightly less spins having the parallel alignment to give the population difference. As a result of this magnetization M_z is observed. However, a certain time interval is needed to obtain equilibrium value, which is same as that needed for the spins to become again randomly oriented when the magnetic field is suddenly switched off. This reorientation time is measured by T_1. Thus the spin-lattice relaxation time T_1 measures the characteristic time for recovery of the magnetization of the paramagnetic system along the static field direction after the equilibrium has been disturbed.

The interpretation of the spin-lattice relaxation time is based on the interaction between the spins and lattice vibrations (phonons). The spin-magnetic moment is not influenced directly by vibrations of the lattice. The coupling of the lattice vibrations with magnetic spin states occurs indirectly through residual spin-orbit coupling. The crystalline electric field in ionic compounds or chemical bonds in molecular free radical raise the degeneracy of the orbital states, usually leaving an orbital singlet as ground state. The orbital singlet behaves like an atomic S state. However, orbital magnetic field is not completely quenched because of a second –order admixture of the singlet orbital ground state with higher orbital states. Therefore, there is slight orbital magnetic field acting on the spin moment. The orbital moments are strongly coupled to the lattice through strong crystalline electric fields. If the lattice is vibrating, interatomic distances vary and therefore crystalline field also vary. As a result of this the residual orbital magnetic field acting on the electron spin is effectively modulated by all vibrational modes (we are concerned with the phonons having frequency 3-30GHz, the region in which ESR work is usually carried out). The probability that a magnetic spin transition will be induced by such modulated at resonance frequency is proportional to the square of the residual orbital field component that are transverse to **B**. In general one expect large value of T_1 with decrease of temperature because of the freezing of lattice vibrations. Three processes are proposed for spin-lattice relaxation.

(i) Direct Process

It involves phonons of the same frequency as the ESR resonance quantum hv (Fig.9.4). Only a

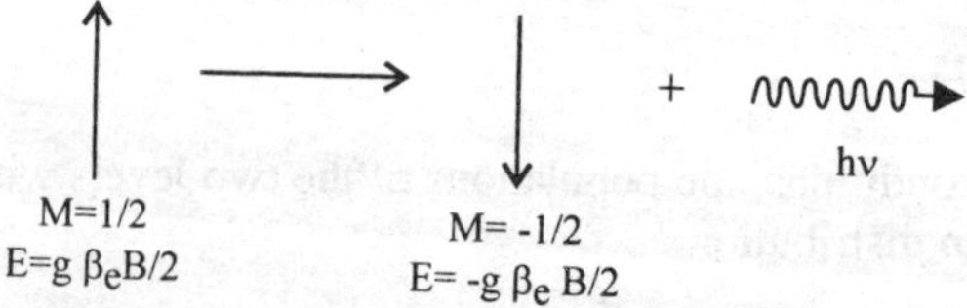

**Fig. 9.4 Schematic of the Direct process. A spin makes a transition to
the lower state emitting a phonon at the resonant frequency v**

small fraction of the normal distribution of the thermal energy is concentrated in vibrational frequencies as low as the ESR frequencies. Therefore, phonons at the resonance frequencies are normally scarce. The process is prominent at low temperatures.

(ii) Raman Process

This is a two-phonon process (Fig. 9.5). In this process lattice phonons are scattered by the spins, with the ESR frequency add to or subtracted from the frequency of scattered phonons. For this method to be effective, there must be lattice modes having difference

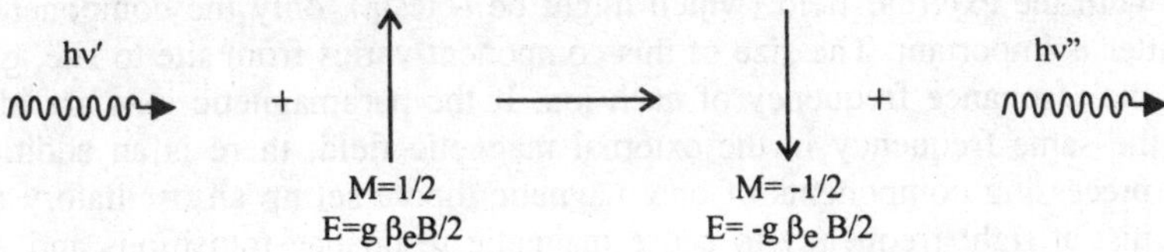

Fig. 9.5 Relaxation by the Raman process where phonon hv' is absorbed and phonon hv" is emitted accompanied by a down transition of the spin

frequencies equal to the ESR frequency. Such vibrations, however, can be in the higher frequency region, more densely populated at room temperature. In this process, many phonons pair can participate since the only requirement is that their frequency difference be equal to ESR frequency. Effectiveness of this process decreases as the temperature is decreased since the vibration at higher frequencies would be frozen out at low temperatures.

(iii) Orbach Process

This involve absorption of a phonon by direct process to excite the spin system from the upper level to a much higher level at an energy δ above the ground doublet, then dropping back into lower Zeeman level of the ground state by emitting another phonon of slightly different energy (Fig.9.6). Thus the paramagnetic ion is indirectly transferred from upper Zeeman level to the lower Zeeman level of the ground doublet. It is more restricted than the Raman process because two specific phonons are involved.

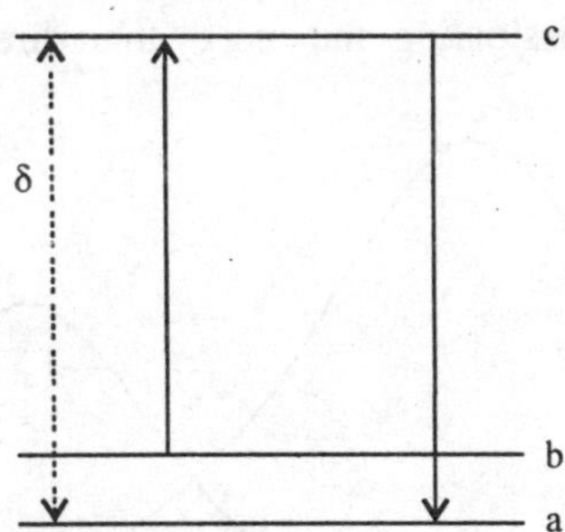

Fig. 9.6 Energy level diagram to illustrate the Orbach process. Level b and a relax by way of two direct transitions involving a third level c

At liquid He temperature only the Direct Process is usually significant whereas Raman and Orbach process become dominant at higher temperatures.

(b) spin-spin relaxation

It contains all the mechanisms where by the spins can exchange energy amongst themselves, rather than giving it back to the lattice, or molecular system as a whole.

(i) Dipole interaction

It arises from the influence of the magnetic field of one of the paramagnetic ion on the dipole moments of neighbouring ions. The actual local field at any given site will depend on the arrangement of the neighbours and the direction of their dipole moments. If the external magnetic field acts on the paramagnetic compound, the local field at each ion must be added vectorially to it. If the local field is small compared with the external field (which might be ~ tesla), only the component of the former parallel to the latter is important. The size of this component varies from site to site, giving a random displacement to the resonance frequency of each ion. If the paramagnetic ions are identical, so that they precess at the same frequency in the external magnetic field, there is an additional resonance interaction. The precessing components of one magnetic dipole set up an oscillatory field at another dipole which is just at right frequency to cause magnetic resonance transitions and vice versa. The mutual interaction produces resonance transitions that are equivalent to the exchange of quanta between neighbouring ions. Thus the quanta are exchanged among neighbouring ions by mutual spin flip and spins are in thermal equilibrium is disturbed, it is re-established exponentially with time constant T_2 which is called spin-spin relaxation time. $T_1 \sim 10^{-6}$ sec and $T_2 \sim 10^{-10}$ sec. T_2 is independent of temperature.

(ii) Exchange Coupling

This is important only in undiluted crystals where the paramagnetic sites are so close together that the orbital of the unpaired electrons overlap. Here the spins interact electrostatically through a short-range interaction, which is called exchange interaction. This results in a change in width of absorption lines in the ESR spectrum.

(c) cross relaxation

Let us consider a sample having two spin systems with different resonance frequencies. If the tails of the two absorption curves corresponding to the spin systems overlap as shown in (Fig 9. 7), a flip-flop spin exchange can take place in the region of the overlap. This process is known as cross relaxation. Since spin-spin relaxation times are often much shorter than spin lattice relaxation time, this may be more effective process for dissipating the energy than direct transfer to the lattice.

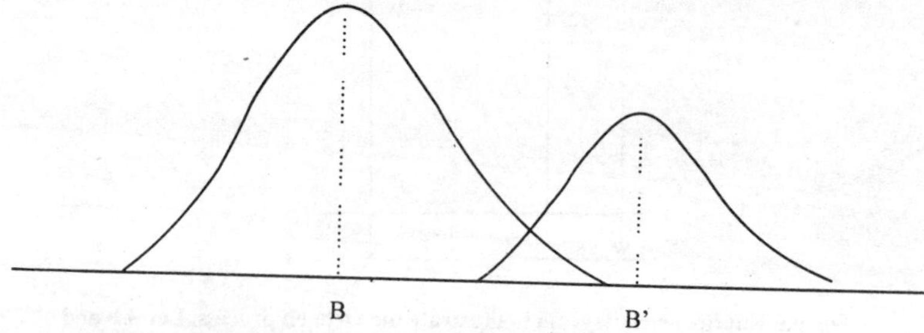

Fig. 9.7 Two overlapping lines centred on fields B and B'

Let us assume for example T_1^X (for species X) is very long and that T_1^Y (for species Y) is very short. If a quantum of energy absorbed by a spin of type X is given to a spin Y through a spin exchange, the quantum will have a high probability of being transferred to the lattice by the Y spin before it is transferred back to the X spin through a reverse exchange (because of short T_1^Y). The spin-lattice relaxation time of the X spin system is thus effectively reduced.

9.7 Paramagnetic Ion in a Crystalline Field

When the paramagnetic ion in incorporated in a solid there are interaction between it and its surrounding. The ions (ligands) surrounding the paramagnetic ion produce a strong electrostatic field (crystalline field) at the ion site. This electrostatic field reflects the symmetry of the site of the paramagnetic ion. The electrons localized on the ions and moving in this field experiences a Stark splitting in their orbital levels, and their degeneracies are partially lifted depending on the symmetry of the site. In the case of local cubic symmetry group 1,2,3 or 4 -fold degenerate levels are found, but in case of lower symmetry, levels very often are either singlet or degenerate in pairs only. An important theorem concerning the residual degeneracy is due to Kramer's : In a system containing an odd number of electrons, at least two-fold degeneracy must remain in the absence of a magnetic field.

Let us consider the effect of crystalline field on the d orbitals (Sec.2.5) of an ion. The five d orbitals are divided into two groups (i) three four-lobe orbitals d_{xy}, d_{yz}, d_{zx} pointing between the cartesian axes e.g. d_{xy} between x and y axes in the xy plane (ii) two orbitals of the same type but pointing towards the ligands and disposed along the cartesian axes: $d_{x^2-y^2}$ along x and y axes while d_{z^2} along z axis. The d orbitals are five-fold degenerate in a free atom.

In an octahedral coordination, the electrons in $d_{x^2-y^2}$, d_{z^2} orbitals are subjected to greater repulsion by the negatively charged ligands than the d_{xy}, d_{yz}, d_{zx} orbitals as the lobes of later are pointing out along the bisector of the cartesian axes. Therefore $d_{x^2-y^2}$, d_{z^2} orbitals are higher in energy than the d_{xy}, d_{yz}, d_{zx} orbitals. Thus in an octahedral environment the five-fold degeneracy is partly removed and the energy level is split into two levels: the lower triply degenerate level and the higher doubly degenerate level (Fig. 9. 8). These two sets of orbitals are labelled by different designation, which can be compared as follows:

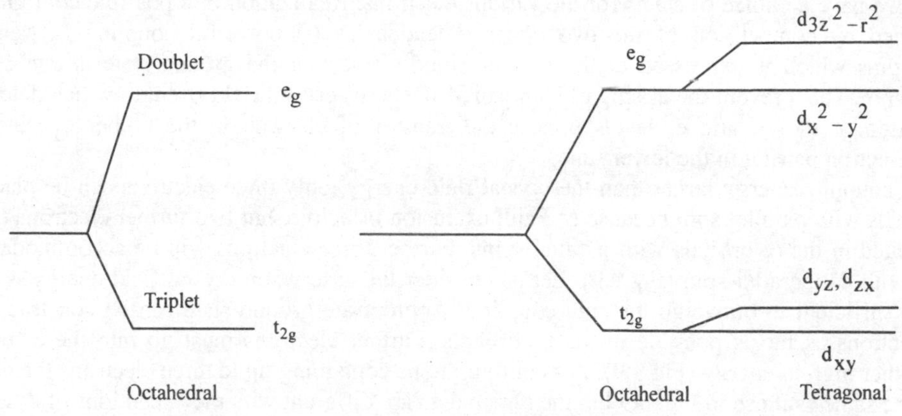

Fig. 9.8 Energy level of a single d electron in crystal fields of (a) octahedral symmetry and (b) tetragonal symmetry

$$d_{x^2-y^2}, d_{z^2} \quad : e_g \text{ or } d\gamma \text{ or } \Gamma_3$$

$$d_{xy}, d_{yz}, d_{xz} : t_{2g} \text{ or } d\varepsilon \text{ or } \Gamma_5$$

Distortion of the regular octahedron leads to further splitting of the d orbitals. Let the neighbouring ions along the z axis are moved closer, to produce a field of tetragonal symmetry. This leads to lesser energy of the $d_{x^2-y^2}$ orbital than the d_{z^2} orbital and to lesser energy of d_{xy} orbital than the d_{xz}, d_{yz} orbitals (Fig.9.8). Elongation of octahedral along the z axis leads to the similar splitting but with reversed position of the levels.

The transition metal complexes can be divided into three broad classes according to the strength of the crystalline field.

(a) Weak Crystal Field (rare earths and actinides group i.e. 4f and 5f)

$$H_{ee} > H_{LS} > H_{CF}$$

H_{ee} is interelectron interaction, H_{LS} is spin-orbit interaction and H_{CF} is crystal field interaction.

(b) Intermediate Crystal Field (Iron group i.e. 3d)

$$H_{ee} > H_{CF} > H_{LS}$$

(c) Strong Crystal Field (Pd and Pt group i.e. 4d and 5d ions and also in 3d series if there is considerable amount of covalent bonding between paramagnetic ion and the ligands e.g. the ligands are CN^-).

$$H_{CF} > H_{ee} > H_{LS}$$

9.8 High Spin and Low Spin States

Distribution of d electrons between t_{2g} and e_g states arising in the octahedral crystal field can be different in their dependence on the crystal field strength. For configuration containing more than one d electron we have a choice of states for the various electrons. Realization of a possible configuration is determined by competition of the two opposite tendencies (i) powerful coupling between the electrons spins which is expressed in the first of Hund's rule that the ground state is one of the maximum spin. This prevent the arising of spin paired states (ii) crystal field strength which determine the separation of the t_{2g} and e_g levels oppose the transfer of electron in the higher e_g state thus favouring electron pairing in the lower state.

For spin coupling energy larger than the crystal field energy, only three electrons can be placed in the t_{2g} orbitals with parallel spin because of Pauli exclusion principle, but two further electrons can be accommodated in the e_g orbitals with parallel spins. Further three electrons will be accommodated in t_{2g} orbital with antiparallel spin (Fig.9.9). Let us consider the case when crystal field energy is much larger, and sufficient to outweigh the spin coupling. Appropriate ground state can be constructed by placing electrons as far as possible in the t_{2g} orbitals. Further electrons must go into the e_g orbital, which is rather high in energy (Fig 9.9). For configurations containing up to three electrons the ground states are the same as those in Fig. 9.9 but the remainders are different with the exceptions of d^8 and d^9. The maximum spin multiplicity (S=3/2) is obtained for d^3 and then decreases as far as d^6.

In the case of intermediate crystalline field the high spin state arise i.e. states with maximum number of unpaired electrons in accordance with Hund's rule. In the case of strong crystalline field the low spin states arise with the spin S equal to 0 or 1 due to pairing of electrons. The low spin states, correspond to the case of strong crystalline field, are possible in the d^4 to d^7 configuration only and occur mainly in ions of 4d and 5d groups, in the 3d group they occur only in compounds with ligands such as CN^-.

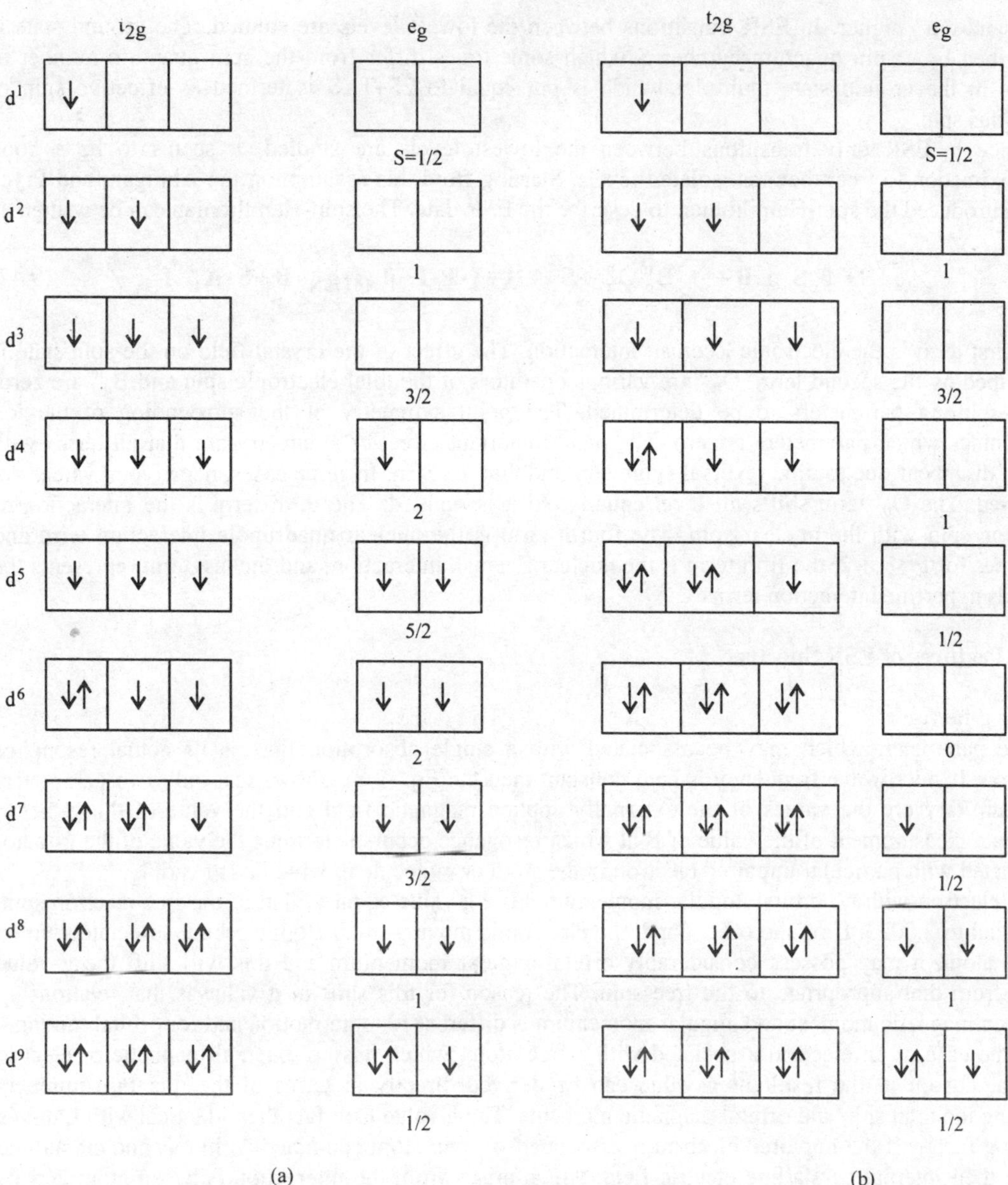

Fig. 9.9 Construction of the ground states of the d-configurations in an octahedral field using the strong crystal field approach and assuming (a) the spin coupling to be stronger than the crystal field energy, (b) crystal field energy to be stronger than the spin coupling energy

9.9 Effective Spin and Spin-Hamiltonian

The results of ESR measurements are summarized in a so-called spin-Hamiltonian. The formalism is adopted to the conditions of paramagnetic ions inserted in crystals. The ground state of these ions is degenerate or is split not more than some cm^{-1}. The next higher state being several hundred or thousand cm^{-1} higher. In ESR transitions between the lowest levels are studied. The ground state is described by a spin quantum number S which some times differ from the spin quantum number of levels in the ground state multiplet which is put equal to 2S+1. S is termed as effective spin or fictitious spin.

Since in ESR only transitions between the lowest levels are studied, it seems to be a good approximation to treat them as isolated levels. Starting from this assumption the Abragam and Pryce first introduced the spin-Hamiltonian to describe the ESR data. The spin-Hamiltonian can be written as

$$H = \beta_e S \cdot g \cdot B + \sum B_n^m O_n^m + S \cdot A \cdot I + I \cdot P \cdot I - \beta_N I \cdot g_N \cdot B + S \cdot A_L \cdot I_L \qquad (9.47)$$

The first term is the electronic Zeeman interaction. The effect of the crystal field on the spin state is described by the second term. O_n^m are various operators of the total electronic spin and B_n^m are zero-field splitting parameters to be determined. The point symmetry of the surrounding of the ion determines which parameters is zero. The most important effect of symmetry are that all terms with odd n disappear due to time reversal symmetry and that $m \leq n$. In some cases, negative m values are required. The O_0^0 term shifts all level equally, so it is omitted. The third term is the interaction of electron spin with the nuclear spin. The fourth term is the nuclear quadrupole interaction term and vanishes for $I \leq 1/2$, the fifth term is the nuclear Zeeman interaction, and the last term represents the ligands hyperfine interaction terms.

9.10 Features of ESR Spectra

(i) the g-factor

The parameter, which may be associated with a single absorption line, is its actual resonance position. If microwave frequency is held constant then the Eq. (9.8) shows that only two parameters that can vary are the values of the externally applied magnetic field and the value of the g-factor. Hence, a measurement of the value of B at which resonance occurs determine the value of the g-factor associated with particular unpaired electron in the atom or molecule in which it is residing.

An electron with no orbital angular momentum has a g value equal to that of the free electron spin, i.e. equal to 2.0023. If on the other hand, the electron is moving in an atomic orbital associated with a single atom, it may possess considerably orbital angular momentum and this will shift the g value away from that appropriate to the free spin. The reason for this shift in g value is that relationship between magnetic moment and angular momentum is different for spin motion and for orbital motion.

In the case of an electron associated with a free atom, which has no external magnetic or electric field acting on it, the resultant g value can be derived directly in terms of the quantum numbers defining the total spin and orbital magnetic moments. The g value is in fact then identical with Lande's splitting factor. If the unpaired electron is associated with an atom contained within a solid crystalline lattice then internal crystalline electric field, which arises from the interaction between atom and its surrounding, will also be acting on it. These electric fields act on the orbital states of the atom and can vary radically alter their energies, and, as a result, the simple Lande's factor can no longer be applied. Here the position of the resonance for a fixed frequency is determined by the competitive effects of the environment, which electric field tend to 'quench' (the electronic orbital motion interacts strongly with the crystalline fields and become decoupled from the spin, a process called 'quenching') the orbital angular momentum. The more complete the quenching, the closer the g factor approaches the free electron value. For example, g = 2.0036 in the free radical α, α'-diphenyl-β-picryl hydrazyl (DPPH) which is very close to the free electron value of 2.0023. It equals 1.98 in many chromium compounds, and it sometimes exceeds 6 for Co^{2+}. Thus the amount of quenching varies with the spin system.

In the general case therefore the quantitative field frequency relation is not known *a priori* and we write the generalized resonance condition $h\nu = g\,\beta_e B$, where g is the g-factor, and is determined from the ESR spectrum. The above relation can be regarded as giving the separation between eigenvalues of a spin-Hamiltonian $H = \beta_e\,\mathbf{S.g.B}$, so that g is a measure of the effective magnetic moment associated with an angular momentum $\mathbf{S}$ such that there are (2S+1) energy levels in an applied magnetic field. Since the magnitude of the orbital part of the magnetic moment depends on the crystal field its magnitude is usually different for different direction of applied magnetic field $\mathbf{B}$, and shows an angular variation, which follows the symmetry of the crystal field. The g value may then be anisotropic by an amount, which depends on the magnitude of the orbital contribution to the moment, and on the asymmetry in the crystalline field. The two general procedures for determining g are (i) the direct method where frequency and magnetic field are measured independently at the resonance abosrption peak and g is calculated from Eq. (9.8), (ii) the comparison method where an unknown specimen is measured along with a system of known g value (e.g. DPPH).

(ii) fine structure

The spectra of oriented systems of total spin greater than one half often show (2S) features, indicating unequal separation between the 2S+1 Zeeman levels. This structure is called fine structure. The fine structure reflects a splitting of the 2S+1 levels even in the absence of a magnetic field, a phenomenon called zero-field splitting. If trivalent chromium ions existed in free state and were not subjected to the electrostatic fields inside the crystal or molecule, then all possible orientations of S = 3/2 total spin vector would have same energy, and hence all would be degenerate on the energy level diagram, as shown in Fig. 9 10. The application of an external magnetic field would then cause the

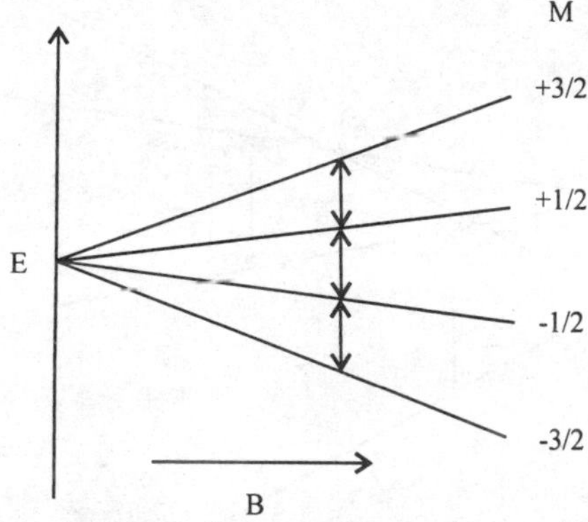

Fig. 9.10 Energy levels for Cr^{3+} ion (S=3/2). At a given frequency
all transitions take place at the same value of B

spins to take up different quantized orientations with respect to this field. The possible orientations varying from those with component of M= +3/2 in the direction of the field through those with orientation corresponding to M = +1/2, -1/2 to the other extreme of −3/2. The energies of these different orientations will diverge with the application of external magnetic field, by amounts proportional to the resolved spin quantum number. Transition from one state to another take place according to selection rule $\Delta M = \pm 1$. It is evident that for a given frequency all the possible transitions will take place at the same value of the externally applied magnetic field, and hence a single absorption line will be produced from the overlapping transitions (Fig.9.10).

If the Cr^{3+} ion is present in a crystal, it will be acted on by the internal electrostatic field and the effect of the latter on the energy levels of the ion is profound. First of all they split the ground state into a number of components. The number of such components and the extent of the splitting depends on the symmetry of crystalline electric field and on its strength but electric field itself cannot remove the two-fold degeneracy associated with the states of the same resolve quantum numbers. The action of the internal electric field will therefore be as indicated on the left hand side of (Fig. 9.11), and zero-field electronic splitting will be produced between ±1/2 and ±3/2 levels as indicated. Application of external magnetic field will now remove the two-fold degeneracy of these spin levels and these will diverge in the way shown in Fig. 9.11. At high enough values of B they will be arranged in an energy order +3/2 to –3/2 as indicated. The allowed transitions between them no longer occur at the same value of B and three separate transitions will now be obtained. The separation between these is a direct reflection of the separation produced in zero magnetic field by the electrostatic internal fields and can therefore be used as a measure of this quantity. In the spin-Hamiltonian, Eq. (9.47), this term is represented by $\sum B_n^m O_n^m$. The intensity of fine structure lines is proportional to [S(S+1)- M(M-1)] where S is the spin of the ion, and M is greater value of M for level between which the resonance transition is taking place. For determination of g value, to a first approximation the average of fine structure lines position gives the value of B to be used in Eq. (9.8).

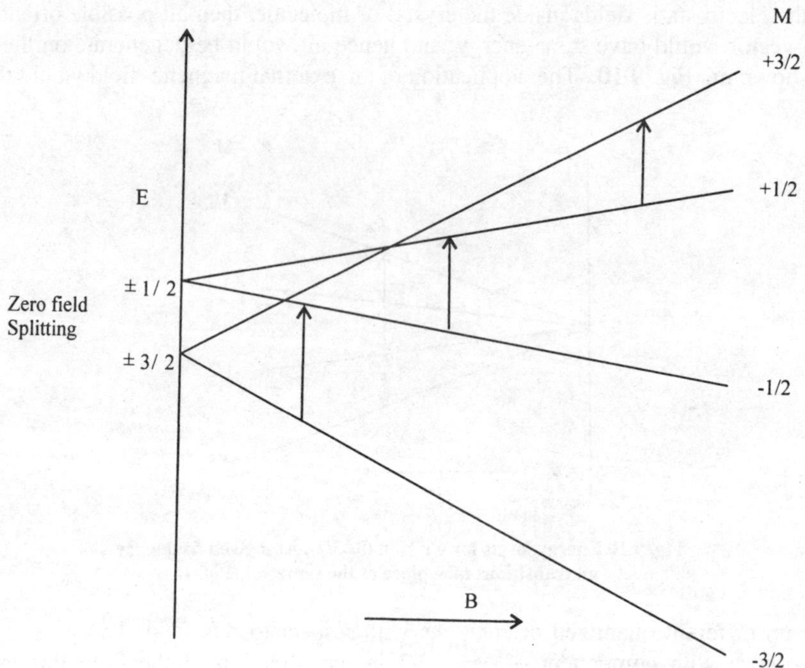

Fig. 9.11 Energy levels of Cr^{3+} in a crystal. At a given frequency, the three allowed transitions take place at different values of B

(iii) hyperfine structure

If the nucleus of the paramagnetic ion has a magnetic moment, it will interact with the electronic moment, resulting in hyperfine structure in the ESR spectrum. The nuclear moment produce a magnetic field B_N at the magnetic electron. Thus an unpaired electron experiences the magnetic field, which is applied externally as well as the magnetic field due to the nuclear magnetic moment. The resonance condition (Eq.9.8) is modified to

$$h\nu = g\beta_e \, | \mathbf{B} + \mathbf{B}_N |$$

(9.48)

The nuclear moment is space quantized in the direction of the internal magnetic field (which is due largely to the electrons) and can take up 2I+1 positions, $\mathbf{I}$ being the nuclear spin. There are, therefore (2I+1) possible values of $\mathbf{B}_N$ and the same number of total magnetic field which satisfy resonance condition [Eq. (9.8)] when the frequency ν is constant. The nuclear moments are smaller than the electron moments by a factor of 10^3. Therefore, according to Eq. (9.46) in a given magnetic field the nuclear spin states are more nearly equally populated than are the electron spin states. Therefore, at room temperature there will be an equal number of nuclei in all of the (2I+1) orientations. Hence there will be an equal number of unpaired electrons experiencing the (2I+1) different values of the total magnetic field resulting hyperfine lines of equal intensity in each electronic transition. To a first approximation the separation between hyperfine lines are equal.

Therefore, to a first approximation hyperfine coupling constant can be determined by taking the average of the separations between hyperfine lines. Further spacing between the components are independent of field. The centre of a hyperfine group defines the position of fine structure line. Thus one can distinguish a hyperfine structure from fine structure where the lines have unequal intensities (Fig.9.12). The selection rule for hyperfine transition is $\Delta m = 0$ i.e. in electronic transition the orientation of nuclear spins remain unchanged. In the spin-Hamiltonian (9.47) this term is represented by $\mathbf{S.A.I}$ where A is called hyperfine coupling constant.

The magnitude of the hyperfine coupling depends on whether the interacting electron is in an s or a p or d orbital. Since s orbital has high electron density at the nucleus, the hyperfine coupling constant will be large. Because of spherical symmetry of s orbital, A will be independent of direction. This interaction is called the isotropic hyperfine coupling (A_{iso}) or the Fermi contact interaction. The p and d orbital, where there is no electron density at the nucleus, the electron is to be found at some distance away from the nucleus. The interaction between it and then nucleus will be of two magnetic dipoles and consequently it will be small and depend on the direction of the orbital relative to the applied magnetic field as well as their separation. This interaction is called anisotropic hyperfine coupling. In liquids anisotropic coupling is averaged to zero and only isotropic coupling A_{iso} is observed.

The isotropic interaction is determined by the expression

$$A_{iso} = \frac{8\pi}{3} g_N \beta_N \, | \psi(0) |^2 \text{ gauss}$$

(9.49)

$|\psi(0)|^2$ is the electron density at the nucleus. A_{iso} will be more intensive, the greater the magnetic moment of the nucleus and the higher the density of the s electrons on the nucleus. A_{iso} will be a direct measure of the s character of the unpaired electron since only s orbitals have a non zero values of the wave function at the nucleus. The anisotropic hyperfine coupling is given by

$$A_{anis} = \frac{g_e \beta_e g_N \beta_N}{2r^3} < 3\cos^2\alpha - 1 > < 3\cos^2\theta - 1 >$$

(9.50)

where r represents the unpaired electron - nucleus distance, α is the angle between r and one of the principal axes of the anistropic tensor A; θ is the angle between this same axis and the magnetic field. The sign < > denote the mean value.

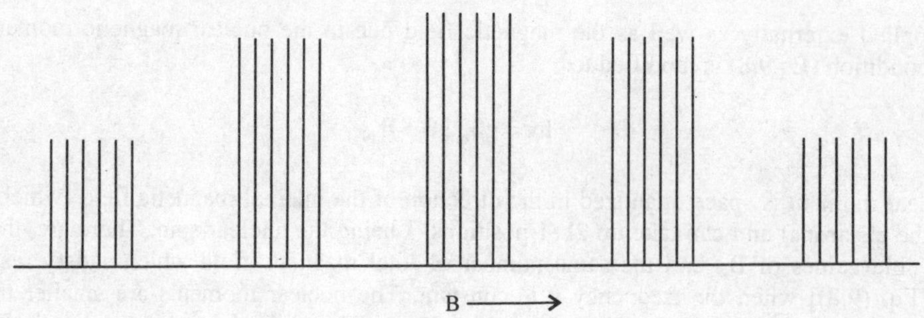

Fig. 9.12 A stick diagram showing the ESR spectrum of Mn^{2+} in a single crystal. S = I = 5/ 2

Nuclei with $I \geq 1$ have electric quadrupole moment Q because their charge distribution deviate from spherical symmetry. The quadrupole moment can interact with electric field gradient in an atom, molecule or crystal. The electric field gradient is set up by the anisotropic distribution of electric charges on the paramagnetic ion and its immediate neighbours. The effect of quadrupole interaction on the resonance spectrum is usually complicated because it is accompanied by a large magnetic hyperfine structure interaction. Therefore, there is a competition between the electric field gradient, and the magnetic field from the unpaired electrons, in fixing the orientations of the nucleus. The result is that, for a general orientation of B, the intensities and spacing of the usual 2I+1 hyperfine lines corresponding to $\Delta m = 0$ are modified by Q and there are also additional lines corresponding to the normally forbidden transitions with $\Delta m = \pm 1, \pm 2$ etc. The effect of quadrupole interaction is represented by **I.Q.I** in the spin-Hamiltonian (9.47)

The nuclear magnetic moment β_N interact directly with the applied magnetic field B with energy ~ $\beta_N B$. The effect of this interaction is very small on the ESR spectra and this is represented in the spin-Hamiltonian (9.47) by $\beta_N \mathbf{B}. g_N .\mathbf{I}.$

(iv) ligand hyperfine structure (superhyperfine structure)

The unpaired electrons can also interact with the nuclei of the near neighbouring atoms, if their nuclei are magnetic resulting in the observation of a structure called superhyperfine or ligand hyperfine structure. The is represented by the last term of the spin-Hamiltonian (9.47). Under high field approximation, the isotropic spectrum may be obtained to first order by splitting the unperturbed electronic levels in to $2I_i+1$ for nucleus i, then each of these levels is further split into $2I_j+1$ levels by nucleus j and so on. Thus the total number of energy levels for a given value of M is $(2I_i+1)(2I_j+1)(2I_k+1)$ .... or can be written as

$$\prod_i (2I_i + 1) \qquad (9.51)$$

Since transitions between hyperfine levels occur with no change of nuclear spin quantum number i.e., $\Delta m = 0$, this is also the maximum number of observable ESR lines. Since each interacting nuclei of spin I causes (2I+1) number of hyperfine energy levels for a given electronic level. For n nuclei of spin I, the maximum number of levels is

$$N = (2I+1)^n \qquad (9.52)$$

when the different nucleus have equivalent hyperfine coupling constant, the number of hyperfine components will be

$$N = 1 + 2\sum_i I_i = 1 + 2T \qquad (9.53)$$

where $T = \sum I_i$. If n equally coupling nuclei have the same spin I the number of hyperfine components are

$$N = 2nI + 1 \qquad (9.54)$$

When unpaired electron has more than one coupling nucleus the relative intensities of the ESR hyperfine components are not necessarily equal because of the possibility that the components arising from the different nuclei may fall at the same frequency (Fig.9.13). If an unpaired electron is coupled to several equivalent protons, the relative intensities of such a multiplet can be calculated from probability theory. For $I = 1/2$, each nucleus must point up ($m = 1/2$) or down ($m = -1/2$) in the static magnetic field B. The outside components of the multiplet are given by the combination of the n nuclei pointing either all up or all down. The outer most components are assigned unit intensity. The component next to these will have all of the nuclei except one pointing in the same direction. Thus $k = n - 1$ must point in the same direction. This can happen n-1 time. For the next component all of the nuclei except two will be pointing out in the same direction and this can happen $n - 2$ time and so on. Generally the number of combination of n thing taken k at a time is

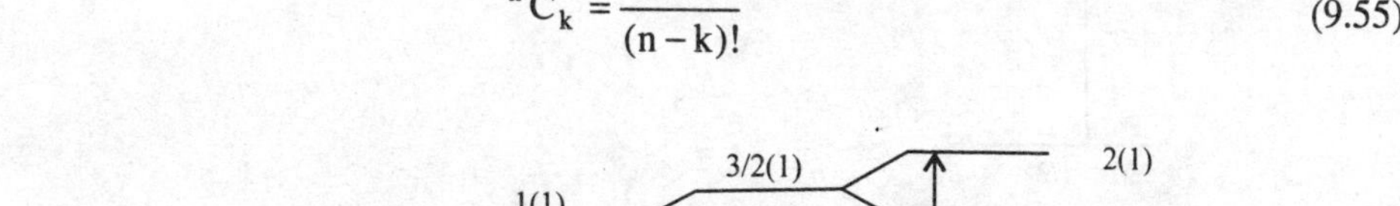

$$^n C_k = \frac{n!}{(n-k)!} \qquad (9.55)$$

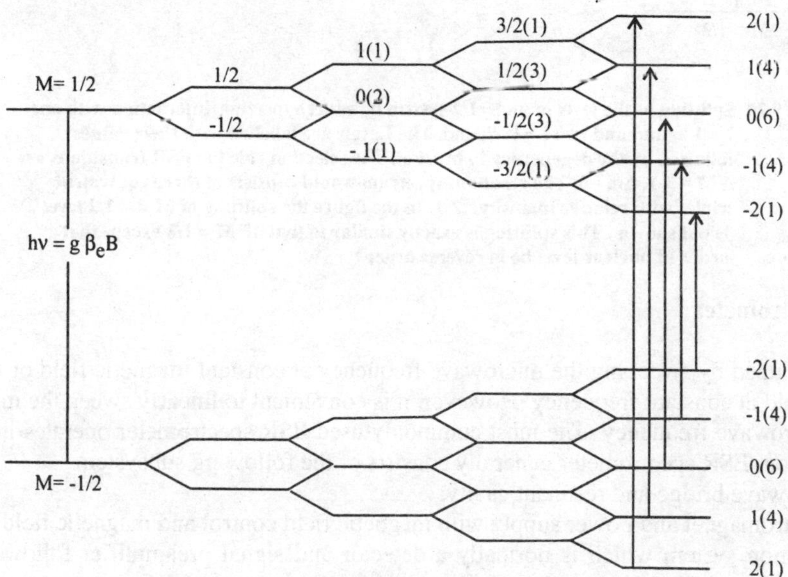

Fig. 9.13 Splitting of the levels of an S =1/2 system by small ligand hyperfine interaction with 1,2,3 and equivalent I = 1/2 ligand nuclei. The levels are labelled with their m values followed by the degeneracy in brackets. The selection rules for ESR transitions are $\Delta M = \pm 1, \Delta m = 0$. The resultant spectrum consists of quintet of relative intensity 1 : 4 : 6 : 4 : 1

For an unpaired electron, coupling with several nuclei of different hyperfine coupling constant or spin can be obtained by making diagrams of the relative displacements by different nuclei and counting the component that coincide (Fig.9.14).

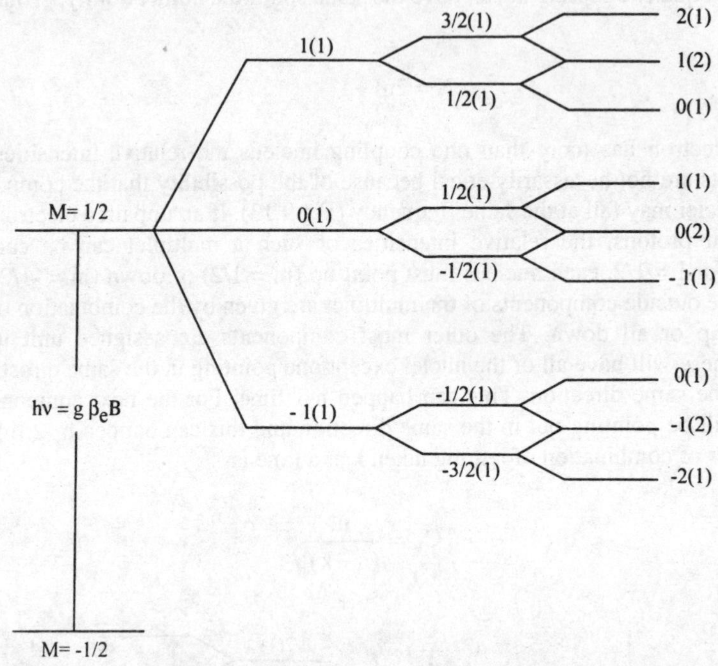

Fig. 9.14 Splitting of the level of an S=1/2 system by superhyperfine interaction with one I = 1 ligand and two I = ½ ligand. The Levels are labelled with their values followed by the degeneracy in bracket. The selection rule for ESR transitions are Δ M = ± 1, Δm =0. The resultant spectrum would consists of three equivalent triplet with relative intensity1:2:1. In the figure the splitting of M = - 1/2 level is not shown . This splitting is exactly similar to that of M = 1/2 except that order of nuclear level be in reverse order

9.11 ESR Spectrometer

ESR can be obtained by sweeping the microwave frequency at constant magnetic field or by sweeping the magnetic field at constant frequency. However, it is convenient to linearly sweep the magnetic field at constant microwave frequency. The most commonly used ESR spectrometer operates in the X band microwave region. ESR spectrometer generally consists of the following subsystem

(a) A microwave bridge and resonant cavity,

(b) An electromagnet and power supply with magnetic field control and magnetic field sweep, and

(c) A detection system which is normally a detector and signal preamplifier followed by phase sensitive detection at the magnetic field modulation frequency, a CRO and recorder display.

A block diagram of an ESR spectrometer is shown in Fig.9.15. It consists of a microwave source may be either Klystron (vacuum tube) or a Gunn oscillator (a solid state device) either of which provides a monochromatic coherent source of electromagnetic radiation. The frequency of microwave can be swept on either side of the centre frequency by about ±100MHz for the purpose of tuning the microwave circuit and cavity. It is critical to maintain the microwave frequency at constant value so

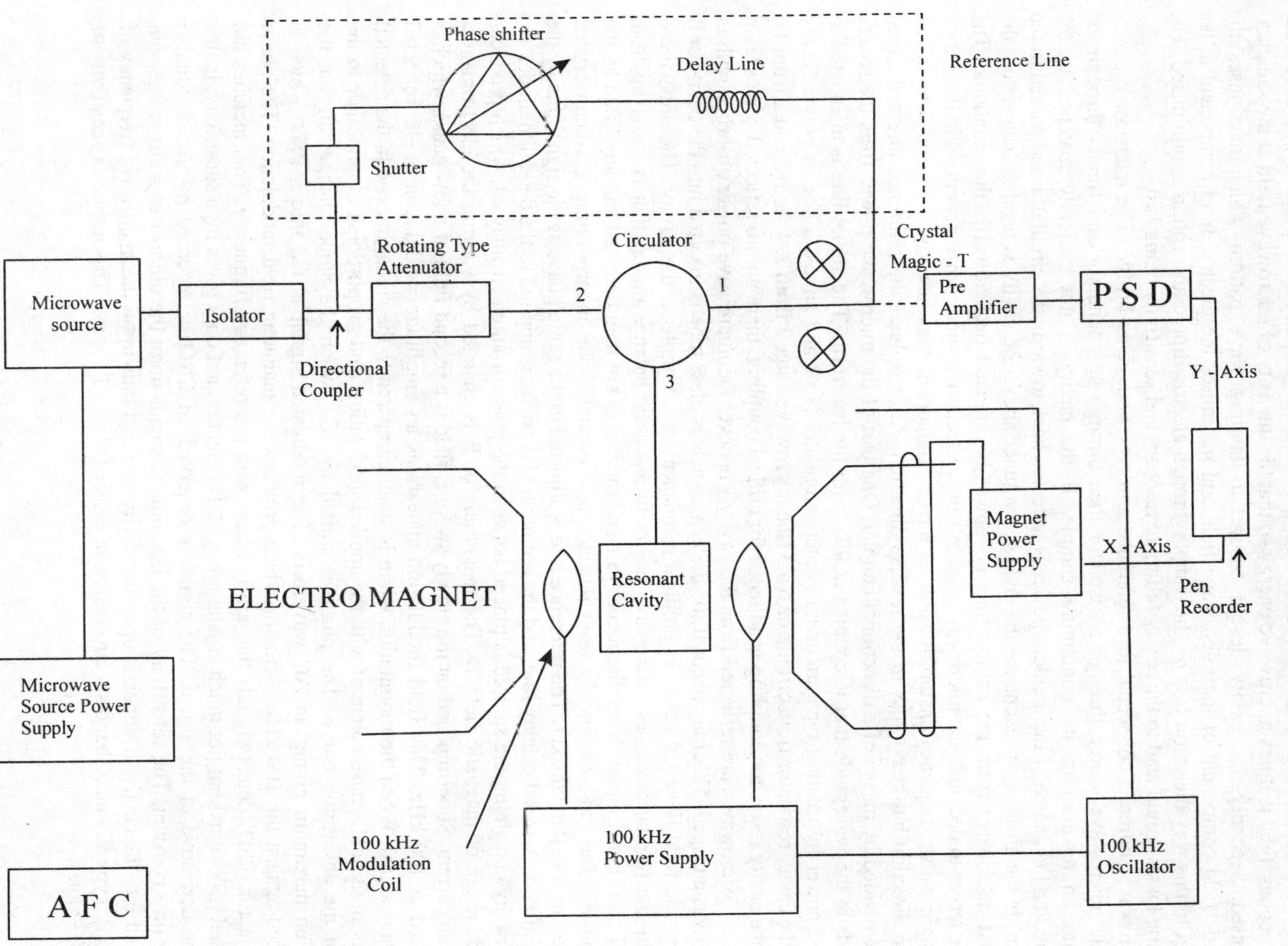

Fig. 9.15 Block diagram of a typical X-band ESR spectrometer

that the signal will not drift on a constant magnetic field. Therefore an automatic frequency control (AFC) circuit is used. The sample is placed inside a resonant cavity. This constitutes a hollow conducting box with dimensions designed to support a standing microwave pattern. The cavity has two functions. First it stores microwave energy so that the intensity of microwave field at the sample is maximised. Secondly, the cavity design ensures a standing wave pattern which maximises the magnetic field component of the microwave field and minimises the electric field component at the sample. In this way dielectric loss by the sample through electric dipole absorption is minimised. The dielectric losses would lead to the cavity failing to resonant and sensitivity being lost.

Microwave source is connected to a reference arm type homodyne bridge. The microwave passes from the microwave source through an isolator then through an attenuator and finally through the circulator to resonant cavity containing sample. If the cavity is not critically matched to the transmission line, some of the incident power is reflected back towards the circulator and passed to the detector. In the absence of resonance the slide screw tuner above the cavity is used to balance out the reflected microwave from the cavity, thus no power is reflected back towards the circulator. The isolator serves to decouple the microwave source from reflection or other noise generated in the bridge that might cause frequency instabilities. It is ferrite component that has low insertion loss in the forward direction but heavy loss in the reverse direction. After isolator the reference line and signal line are divided by means of a directional coupler. One part of the microwave power from the source proceeds to the waveguide that is coupled at 90^0 with the baseform. The signal line is attenuated to required power by rotating type attenuator, enters circulator No.2 and almost all of its power finally enters the cavity resonator from circulator No.3 (the microwave enter 3 from 2, 1 from 3 and 2 from 1). When the cavity resonator coupling is adjusted for critical coupling, there are no reflected waves from the cavity. Microwaves are reflected from the cavity whenever the microwave power absorbed with in the cavity is changed. This may occur if the dielectric such as glass tube is inserted into the cavity or if the magnetic field is varied so as to produce resonance for a sample in the cavity. The unbalanced cause by the resonance causes microwave power to enter the balance mixer, that is made up from magic T and crystal mount from circulator No.1 and the waves are then detected and amplified by the preamplifier. The most commonly used detector is a crystal diode incorporating a silicon crystal attached to a tungsten whisker. The reference line is adjusted to the same phase as the signal line by the phase shifter, it passes the delay line, and the optimum bias power is applied to the detection diode.

The cavity containing the sample is placed between the poles of an electromagnet, which produces the wide range of magnetic field B. The magnitude of B is changed by varying the electromagnet excitation current. Superimposed on the slowly swept B field is a second field of a few gauss, which is modulated at 100 kHz. This field modulation appears as an amplitude modulation of the reflected microwave signal. When the modulation width is small compared to the signal linewidth, the detected output is sinusoidal, phase coherent with the applied modulation, and proportional in amplitude to the slope of the absorption curve. The phase is shifted by 180^0 when the slope changes sign at the absorption maximum giving an AC waveform. The modulated signal is fed into a PSD, which is referenced against the 100 kHz oscillator which provides the magnetic field modulation. The phase shift detector (PSD) detect signals that are in phase with the reference frequency. This includes the modulated ESR signal but not much crystal noise. PSD detection not only gives high sensitivity, it also takes the derivative of the signal. The signal is observed on CRO or recorded by use of suitable filtering time constant. The sensitivity of the technique depends upon the number of spins, which can be packed into the cavity. It depends upon the cavity size and thus upon the microwave frequency of the source. Typically an X-band spectrometer can detect $10^{15} - 10^{16}$ spins. This implies a sensitivity of $10^{-7} - 10^{-8}$ moles.

Examples

1. Calculate the ESR resonance frequency of an unpaired electron (g = 2) in a magnetic field of 0.335T

   From Eq. (9.8)

$$\nu = \frac{g\beta_e B}{h} = \frac{2\times 9.27\times 10^{-24}\times 0.335}{6.626\times 10^{-34}} = 9.377\,\text{GHz}$$

2. Calculate the g value if the methyl radical show ESR at 0.329 T in an ESR spectrometer operating at 9.230 GHz.

   From Eq. (9.8)

$$g = \frac{h\nu}{\beta_e B} = \frac{6.626\times 10^{-34}\times 9.23\times 10^{9}}{9.27\times 10^{-24}\times 0.329} = 2.0044$$

3. The ESR spectrum of an unknown sample shows a single line at 0.335 T at a frequency. The DPPH with g = 2.0036 show a line at 0.3375 T at the same frequency. Determine the g value of ESR line of unknown sample.

   The resonance condition (Eq.9.8) for unknown sample becomes

$$h\nu = g_{sample}\beta_e \times 0.335\,\text{T}$$

   For DPPH the resonance condition is

$$h\nu = 2.0036\times\beta_e \times 0.3375\,\text{T}$$

   Equating above two equations

$$g_{sample} = \frac{2.0036\times 0.3375\,\text{T}}{0.335\,\text{T}} = 2.0185$$

4. Compute the population difference of two states of an electron spin in a magnetic field of 0.3 T at 300 K.

   From Eqs. (9.8) and (9.46)

$$\frac{N_{-1/2}}{N_{1/2}} = \exp\,(h\nu/k_B T) = \exp\,(g\beta_e B/k_B T)$$

   For g = 2, the above equation is

$$\frac{N_{-1/2}}{N_{1/2}} = \exp\left(\frac{2\times9.2741\times10^{-24}\ JT^{-1}\times0.3T}{1.3807\times10^{-23}\ JK^{-1}\times300K}\right) \approx 1.001344$$

Since

$$N_{-1/2} + N_{1/2} = 1$$

$$N_{-1/2} - 1.001344N_{1/2} = 0$$

Solving these two equations

$$N_{-1/2} = 0.500357$$
$$N_{1/2} = 0.4996642$$

and

$$N_{-1/2} - N_{1/2} = 0.0006928$$

5.   The ESR spectrum of Gd^{3+} in $Sm_2Mg_3(NO_3)_{12}.24H_2O$ single crystals show seven fine structure lines corresponding to allowed selection rule $\Delta M = \pm 1$. Determine the effective spin of Gd^{3+} in this single crystal.

According to $\Delta M = \pm 1$ transitions can take place between adjacent levels. Therefore, there must be eight levels between which seven fine structure lines corresponding to allowed transitions are observed. Effective spin can be determined by equating the number of levels to $2S+1$. Thus $2S+1 = 8$ or effective spin $S = 7/2$.

6.   Determine the relative intensity of the five fine structure transitions of Fe^{3+} in a single crystal. The effective spin is 5/2.

The intensity of a fine structure transition is proportional to $S(S+1) - M(M-1)$. For $S = 5/2$, the values of M are 5/2,3/2,1/2, -1/2, -3/3, -5/2. The intensity of $5/2 \rightarrow 3/2$ transition is $[S(S+1) - M(M-1)] = 35/4 - 15/4 = 5$

Similarly

The intensity of $3/2 \rightarrow 1/2$ transition is 8

The intensity of $1/2 \rightarrow -1/2$ transition is 9

The intensity of $-1/2 \rightarrow -3/2$ transition is 8

The intensity of $-3/2 \rightarrow -5/2$ transition is 5

Therefore the relative intensities of five fine structure transitions are 5:8:9:8:5.

7. The ESR spectrum of a radical with two equivalent nuclei of a particular kind is split into five lines of intensity ratio 1:2:3:2:1. What is the spin of the nuclei?

    According to Eq. (9.54), the total number of lines for equivalent nuclei is $2nI + 1$. Therefore equating this to five gives $2nI = 4$. Since there are two nuclei hence $4I = 4$ or $I = 1$. The spin of the nuclei is therefore 1.

8   Predict the intensity distribution in the hyperfine lines of the ESR spectrum of $.CH_2CH_3$

    In $.CH_2CH_3$, there are two sets of inequivalent protons. One set contain three equivalent protons while the other set contains two equivalent protons. Two protons of second set split the energy levels in three of statistical weight 1:2:1. Now each of these energy levels further split up because of three equivalent protons of first set in to four with statistical weight 1:3:3:1. Therefore one observes a triplet (1:2:1) of quartet (1:3:3:1).

9. The separation of two lines (splitting) in a free radical spectrum is given by 75.0 MHz and g = 2.0050. Express the splitting in gauss and in cm^{-1}.

Using Eqs. (9.8) and (9.9) we have

$$B(mT) = \frac{71.448 \; v \; (GHz)}{g}$$

Substituting the values of g and v in the above we have

$$B(mT) = \frac{71.448 \times 75 \times 10^{-3}}{2.005} = 2.672 \, mT$$

To convert frequency in to wavenumber, divide the frequency by the velocity of light i.e. splitting (in cm^{-1}) is

$$\frac{v}{c} = \frac{75 \times 10^6 \, Hz}{2.9979 \times 10^{10} \, cm \, s^{-1}} \approx 25.0175 \times 10^{-4} \, cm^{-1}$$

10. A radical containing two equivalent protons shows a three line spectrum with intensity distribution 1:2:1. The lines occur at 330.2 mT, 332.5 mT and 334.8 mT. What is the hyperfine coupling constant for each proton? What is the g value of the radical given that the spectrometer is operating at 9.319 GHz?

    Since the protons are equivalent, therefore the hyperfine splitting constant for both the protons are identical. The separation between the adjacent lines gives the hyperfine coupling constant. From the given field positions of the lines, the difference between two adjacent lines is 2.3 mT. Hence the hyperfine coupling constant is 2.3 mT. The value of g can be determined from the position of ESR line if there is no hyperfine splitting. This position is obtained by taking the average of all the three line positions i.e. (330.2 + 332.5 + 334.8)/3 = 332.5 mT. From Eq. (9.8) we have

$$g = \frac{6.6262 \times 10^{-34}\,\text{Js} \times 9.319 \times 10^{9}\,\text{Hz}}{9.2741 \times 10^{-24}\,\text{JT}^{-1} \times 0.3325\,\text{T}} = 2.0025$$

11. Predict the intensity distribution in the hyperfine lines of the ESR spectra of $.CD_3$.

$.CD_3$ contains three deuteron nuclei of spin I =1. Therefore each of the two electronic levels will split into 2nI +1 level i.e. 7 levels. Because of selection rule $\Delta m = 0$ seven lines will be observed with intensity 1:3:6:7:6:3:1 as shown in (Fig.9.16)

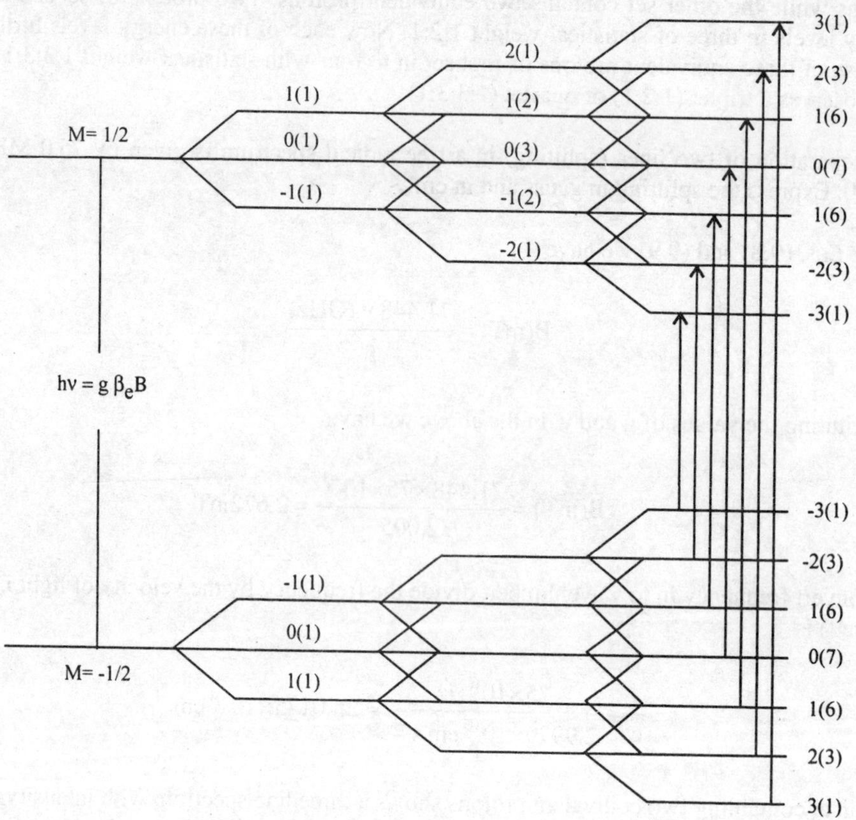

Fig. 9.16 **Splitting of the levels of $.CD_3$ system by the ligand hyperfine interaction with three equivalent deuteron (I = 1) ligand nuclei. The levels are labelled with their m values followed by the degeneracy in brackets . The selection rules for ESR transitions are $\Delta M = \pm 1, \Delta m = 0$. The resultant spectrum consists of septet of relative intensity 1: 3 : 6 :7: 6 : 3 : 1**

12. The ESR spectrum of Cu^{2+} in $Cs_2C_2O_4.H_2O$ shows four lines of equal intensity at 268.9 mT, 282.3 mT, 295.5 mT and 307.2 mT.   Determine the effective spin and nuclear spin of Cu^{2+}. Also determine the value of hyperfine coupling constant and g value of Cu^{2+} in this sample given that DPPH line (with g = 2.0036) occurs at 337.5 mT.

Since all the lines of the spectrum are of equal intensity therefore it is hyperfine structure and number of lines are equal to 2I+1. Hence I = 3/2. Further there is only one set of four hyperfine lines which are centred at an electronic transition which is between two levels and hence S= 1/2. To a first approximation the hyperfine splitting is determined by taking the average of the separation between the lines which is (13.4+13.2+11.7)/3= 12.76mT. The position where the hyperfine structure is centred is obtained to a first approximation by taking the average of positions of all the four lines which is 288.47 mT. Hence g value can be obtained from the expression

$$h\nu = g_{DPPH}\beta_e B_{DPPH} = g_{sample}\beta_e B_{sample}$$

$$g_{sample} = \frac{g_{DPPH}\ B_{DPPH}}{B_{sample}} = \frac{2.0036\times 337.5\ mT}{288.47\ mT} = 2.344$$

13. The experimental hyperfine coupling constant A for the hydrogen atom is 1420.4 MHz, g = 2.002256. Compare the value of A with that calculated using Fermi Interaction term.

For hydrogen atom the electron wavefunction is

$$\psi = \left(\frac{1}{\pi a_0}\right)^{3/2} \exp\left(-\frac{r}{a_0}\right)$$

where a_0 is radius of Bohr's first orbit ($a_0 = 0.52918 \times 10^{-8}$ cm).

By putting r = 0,we have

$$A_{iso} = \frac{8\pi}{3}g_N\beta_N \mid \psi(0)\mid^2 = \frac{8\times 5.5854\times 5.05082\times 10^{-24}\ erg\ gauss^{-1}}{(0.52918\times 10^{-8}\ cm)^3} \approx 507.66\ Gauss$$

From Eq. (9.9) we have

$$B(mT) = \frac{71.448\ \nu\ (GHz)}{g}$$

From the above

$$\nu\ (GHz) = \frac{gB(mT)}{71.448} = \frac{2.002256\times 50.766}{71.448}\ GHz = 1422.66\ MHz$$

14. Predict the intensity distribution and number of lines in the hyperfine structure of the ESR spectra of cyclooctatetraene anion .

   In cyclooctatetraene anion radical unpaired electron interact with eight equivalent protons. According to Eq. (9.54) each M = ±1/2 level will split up into N = (2 x 8 (1/2) + 1) = 9 hyperfine levels and according to selection rule $\Delta M = 1$, $\Delta m = 0$, nine ESR lines will be observed. The relative intensity of these transitions are given by Eq. (9.51) and intensities are found to be in ratio 1:8:28:56:70:56:28:8:1.

Problems

9.1   Predict the intensity distribution in the hyperfine lines of $.CD_2CD_3$.

9.2   A radical containing two equivalent proton with hyperfine constant 2.0 mT and 2.6 mT gives a spectrum centred on 332.5 mT. At what fields do the hyperfine lines occur and what are their relative intensity.

9.3   Predict the intensity distribution in the hyperfine lines of the ESR spectra of $.CH_3$.

9.4   The ESR spectrum of a radical with a single magnetic nucleus is split into four lines of equal intensity. What is the spin of the nuclei ?

9.5   A radical containing three equivalent protons shows a four line spectrum with an intensity distribution 1:3:3:1. The lines occur at 331.4 mT, 333.6 mT, 335.8 mT and 338.0 mT. What is the hyperfine coupling constant for each proton? What is the g value of the radical given that spectrometer is operating at 9.332 GHz ?

9.6   Predict the form of the ESR spectrum of radical containing three equivalent N^{14} nuclei (I = 1).

9.7   The ESR spectrum of VO^{2+} in $(NH_4)_2SeO_4$ show eight lines of equal intensity at 276.2mT, 296.8 mT, 317.7 mT ,339.6mT, 359.6 mT, 380.4 mT, 401.8 mT and 422.8 mT. Determine the effective spin and nuclear spin of VO^{2+}. Also determine the value of hyperfine coupling constant and g value of VO^{2+} in this sample given that DPPH line (with g = 2.0036) occurs at 335.0 mT.

9.8   Calculate the hyperfine splitting constant of the deuterium atom (I=1).

9.9   Compute the population difference of two states of an electron spin K band (30GHz) at 300 K.

9.10   ESR spectrum of Cr^{3+} in $AlCl_3.6H_2O$ shows three fine structure lines. Obtain effective spin and relative intensity of various fine structure transitions.

10

Nuclear Magnetic Resonance

10.1 NMR

Nuclear magnetic resonance (NMR) was discovered in 1945 independently by Bloch and co-workers and by Purcell and co workers. NMR has been defined as the form of spectroscopy concerned with radio frequency induced transitions between magnetic energy levels of the nucleus having a nuclear angular momentum. This is the same type of magnetic resonance phenomenon that occurs in ESR as discussed in Ch.9, except that it is the magnetic dipole moment of the nucleus that is involved. As a result of mass difference the nuclear magnetic moment is smaller than that of electron. The resonance are correspondingly lower in frequency and fall in the radio frequency region. Another important difference from ESR is that there is no longer a restriction to molecules with unpaired electrons. Any molecule with a magnetic nucleus will give an NMR spectrum. The NMR and ESR are similar in the sense that while NMR deals with nuclear ground state; ESR deals with electronic ground state.

10.2 The Principle of the Phenomenon

Consider a nucleus having a magnetic moment μ_N. Classically the energy of interaction between the nuclear magnetic moment and magnetic field **B** is

$$E = - \mu_N \cdot B \tag{10.1}$$

The magnetic moment is a vector, which is collinear with angular momentum **J**. The operator for the vectors are related by

$$\mu_N = \gamma\, J \tag{10.2}$$

where $\gamma = g_N \dfrac{e}{2M_p}$ is called gyromagnetic ratio, g_N is nuclear g factor and M_p is mass of the proton. γ varies with the state it can have positive or negative value and is characteristic of the nuclei.

We define a dimensionless angular momentum operator **I** by

$$J = (h/2\pi)\, I \tag{10.3}$$

I^2 has eigenvalues I (I+1) where I is an integer or half integer. The component I_z has eigenvalue m where m can take values I, I-1, I-2,..... , -I. From Eqs. (10.2) and (10.3)

$$\mu_N = \gamma\, (h/2\pi)I \tag{10.4}$$

The quantum mechanical Hamiltonian from Eqs. (10.1) and (10.4)

$$H = -\gamma\,(h/2\pi)\,I \cdot B \tag{10.5}$$

If the magnetic field is assumed to be in the z direction, i.e. $B_x = B_y = 0$ and $B_z = B_0$, the Hamiltonian is

$$H = -\gamma\,(h/2\pi)\,I_z B_0 \tag{10.6}$$

The eigenvalues of Hamiltonian are just the eigenvalues of I_z, that is

$$E_m = -\gamma\,(h/2\pi)B_0 m \tag{10.7}$$

For $I = 3/2$, $m = 3/2, 1/2, -1/2, -3/2$ and the energies from Eq. (10.7) are

$$E_{3/2} = -\frac{3\gamma h}{4\pi}B_0$$

$$E_{1/2} = -\frac{\gamma h}{4\pi}B_0$$

$$E_{-1/2} = \frac{\gamma h}{4\pi}B_0 \tag{10.8}$$

$$E_{-3/2} = \frac{3\gamma h}{4\pi}B_0$$

The energy levels for $I = 3/2$ in a constant magnetic field are shown in Fig.10.1

Fig. 10.1 Energy levels of Eq. (10.8)

The separation between levels of Eq. (10.7) is

$$\Delta E = \gamma(h/2\pi)B_0 \tag{10.9}$$

The separation between the levels increases linearly with B_0. Since energy is related to frequency, hence Eq. (10.9) is

$$\Delta E = h\nu = \gamma(h/2\pi)B_0$$

$$\nu = \frac{\gamma B_0}{2\pi} \tag{10.10}$$

$$\omega = \gamma B_0 \tag{10.11}$$

ω is called Larmor frequency. For H^1 in normal magnetic field (2.35 – 18.6 T) the frequency is in the range 100 – 800 MHz. Putting the value of γ in Eq. (10.10)

$$\nu = g_N \frac{eB_0}{4\pi M_p}$$

$$h\nu = g_N \frac{eh}{4\pi M_p} B_0 = g_N \beta_N B_0 \tag{10.12}$$

where $\beta_N = eh/4\pi M_p$ is called nuclear magneton. Eq. (10.10), (10.11) or (10.12) is called resonance condition. The resonance for NMR may be achieved by varying either B_0 or by varying driving frequency. Fig. 8.2 shows the energy levels for a proton $(I = 1/2)$.

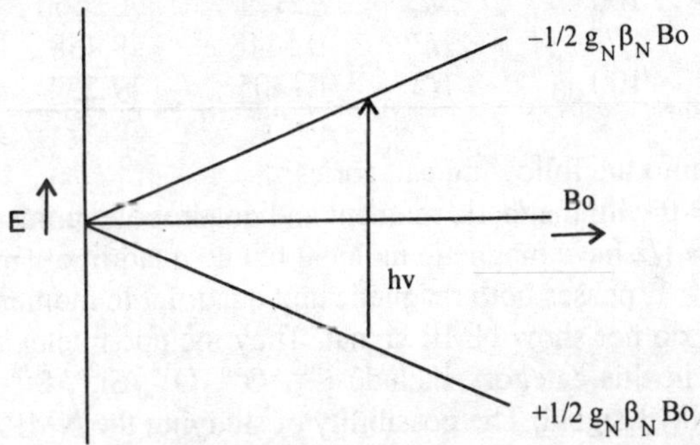

Fig. 10.2 The energy levels of proton in a magnetic field. Application of radio-frequency of energy hv causes resonance at a field given by Eq. (10.12)

From Eq. (10.7)

$$E_{\pm 1/2} = \pm(-1)\frac{h}{4\pi}\gamma B_0 = \pm(-1)\frac{1}{2}g_N \beta_N B_0 \tag{10.13}$$

The separation between the two levels increases linearly with the magnetic field and transition between them can be induced by magnetic field component B_1 of radio-frequency, which is at right angle to B_0. The transitions, which give rise to NMR spectra, are magnetic dipole in origin. The selection rule for the magnetic quantum number m is $\Delta m = \pm 1$.

The phenomenon of NMR can also be understood on the basis of precession as explained for ESR in

Sec.9.4, but with the difference that electronic magnetic moment is replaced by nuclear magnetic dipole moment. Similarly, quantum mechanical description of NMR phenomenon is on the line of ESR as described in Sec. 9.4 with nuclear moments replacing electronic moments.

10.3 Types of Nuclei Viewed from the Standpoint of NMR

NMR requires the presence of nuclear magnetic moment associated with non-zero nuclear spin. The occurrence of non-zero nuclear spin is common in the periodic table, and thus NMR can be observed in isotopes of most elements. Table 10.1 lists some of the nuclei used in NMR studies.

Table 10.1 NMR properties of selected nuclei

Element	Z	A	Natural Abundance %	Spin	Magnetic Moment (β_N)	$\gamma/2\pi$ $(10^6 s^{-1} T^{-1})$	Resonance frequency (MHz) at 2.348 T
H	1	1	99.985	1/2	2.79255	42.573	100
Li	3	7	92.58	3/2	3.2559	16.545	38.87
Be	4	9	100	3/2	-1.1774	-5.983	15.06
B	5	11	80.42	3/2	2.6886	13.662	32.08
C	6	13	1.108	1/2	0.702	10.702	25.15
N	7	14	99.63	1	0.404	3.079	7.23
F	9	19	100	1/2	2.6283	40.068	94.09
Na	11	23	100	3/2	2.2171	11.2666	26.43
Si	14	29	4.79	1/2	-0.5548	-8.458	19.87
P	15	31	100	1/2	1.1305	17.234	40.48

The nuclei can be divided into the following categories:
(a) All the nuclei with I = 0 with magnetic moment and quadrupole moment are zero.
(b) All the nuclei with I = 1/2 have magnetic moment but no quadrupole moment.
(c) All the nuclei with I ≥ 1, posses both magnetic and quadrupole moment.

The nuclei of category (a) do not show NMR signal. They include nuclei with an even Z and even mass number A. The nuclei in this category include C^{12}, O^{16}, O^{18}, Si^{28}, S^{32} and Ca^{40}. The nuclei of categories (b) and (c) show NMR signal. The possibility of studying the NMR spectra of nuclei having spin I different from zero is conditioned by the magnetic moment and natural abundance.

The splitting of the ground level of the nucleus in the presence of magnetic field is small because of low values of nuclear magnetic moment. Hence the population of these sublevels are very close to each other. For example, at normal temperature population of lower sublevel of proton with m = 1/2 is only 1.0000066 times greater than that of higher sublevel with m = -1/2. Because of this difference, the NMR transitions with absorption occur. The intensity of such transitions is low, therefore, the specimen to be studied should contain large number of nuclei or is of considerable bulk (~ 0.5 cm^3).

10.4 High Resolution and Broad Line NMR

Two basic branches exist in NMR spectroscopy
(a) High resolution NMR is usually used to study molecules with spin I nuclei dissolved in low viscosity solvents.
(b) Broad line NMR is observed in solids

The most obvious difference between the NMR spectra of liquids and solids lies in the linewidth of the observed resonance. In solids, the width of the lines lies within the range 1-100 kHz whereas in liquids it is in the range 1-100 Hz. The difference is due to rapid molecular motion in liquids or solutions greatly reducing the magnetic coupling of the nuclear dipole. However, this coupling is preserved in the solids. The NMR spectra of liquids are very sharp and as a result the very small but important electron-coupled spin-spin interaction (Sec. 10.7) and chemical shifts (Sec. 10.6) can be observed. On the other hand, the anisotropic interactions are lost. The broad line NMR studies these interactions. The interaction between the spin of the two nuclei is the dipole-dipole coupling between their magnetic moments. The effect of dipolar field upon a nucleus will depend upon where this nucleus is located within the dipolar field. This shift in NMR frequency due to dipolar field is

$$\Delta\omega_{dip.} = \text{constant } \frac{1 - 3\cos^2\theta}{r^3} \qquad (10.14)$$

where θ is the angle between the line connecting the interacting nuclei and external magnetic field and r is the internuclear distance. The constant of proportionality is the magnitude of the magnetic moment of the interacting nuclei. Thus anisotropic shift in NMR resonance frequency, which is sensitive to internuclear distance, provides a measure of nuclear geometries. The main application of broad line NMR is to the determination of internuclear distance and other crystal parameters.

10.5 Relaxation Mechanisms

Under thermal equilibrium conditions, population of the two levels with $m = -1/2$ and $m = 1/2$ are determined by the Boltzmann distribution

$$\frac{N_1}{N_2} = \exp\left(\frac{h\nu}{k_B T}\right) \qquad (10.15)$$

where $h\nu$ is the energy difference between the two levels. N_1 and N_2 are the population of $m = 1/2$ and $m = -1/2$ levels, respectively. However, $N_1 - N_2$ is very small. The phenomenon of NMR absorption depends on this difference. The *a priori* probability of transition from lower level to upper level equals that from transition from upper level to lower level. Because $N_1 > N_2$, there in an excess of upward transitions and a net absorption of energy from radio frequency field. However, this leads to an increase of N_2, which will continue until $N_1 = N_2$ and net absorption of energy will tend to zero. That this situation does not occur it means there must be other mechanisms by means of which energy absorbed and stored in the upper level can be dissipated in a manner so as to allow return to lower level to maintain the population difference. Such mechanisms are called relaxation processes.

spin lattice relaxation

Each nuclear spin in the system is not entirely isolated from rest of the assembly of the molecules commonly referred to as the lattice. Each nucleus sees a number of nearby nuclei both in the same molecule and the other molecules. The neighbouring nuclei are in motion relative to the observed nucleus. The electronic and nuclear motion of the molecule produces fluctuating magnetic fields. The observed nuclear magnetic moment, which is precessing about the applied magnetic field, also experiences the fluctuating magnetic fields of its neighbours. Since the motion of each molecule and its neighbours are random, therefore there will be a broad range of frequencies describing them. The fluctuating magnetic fields will have components of right frequency in a direction perpendicular to the static field. This oscillating field causes transition from upper level to lower level. The lattice absorbs the energy exchanged in this process and the population difference between the levels is maintained.

The relaxation time in NMR is defined as the time required for the excited state population to fall $1/e$ of its original population. Spin-lattice relaxation time is also known as longitudinal relaxation time and is denoted by the symbol T_1. $T_1 \approx 10^{-2}$ to 10^4 s for solids and from 10^{-2} to 10^2 s for liquids.

spin-spin relaxation

In this mechanism, spins can exchange energy amongst themselves, rather than giving it back to the lattice. The spin exchange process is brought about when a precessing nucleus generates a localized magnetic field at the site of the neighbouring nucleus with a different nuclear spin orientation. The magnetic field produced by the precessing nucleus can be resolved into a static component along the direction of static field and into an oscillating component. This oscillating component has a frequency corresponding to resonance frequency as the nuclear dipole was precessing with Larmor frequency about the static field. This will induce a transition in the neighbouring nucleus. If this spin exchange or flip-flop occurs, the nuclei will exchange magnetic energy states with no overall change in the energy of the system but with a shortening of the lifetime of each. This will have a pronounced effect on the uncertainities in the energy level separation. These spin-spin relaxation times are known as transverse relaxation times and are denoted by the symbol T_2. For solids $T_2 \approx 10^{-4}$ s and for liquids $T_2 \approx T_1$.

From uncertainity relation

$$\Delta E \Delta \tau \sim (h/2\pi)$$

$$\Delta v \sim \frac{1}{2\pi\Delta\tau} \tag{10.16}$$

where $\Delta\tau$ is relaxation time. $\Delta\tau \approx 10^{-4}$s, $\Delta v \approx 0.11 \times 10^4$ Hz. When T_2 is more important relaxation time, it can be calculated from the shape of the absorption line. The relaxation time T_2 is

$$\frac{1}{T_2} = \pi \Delta v_{1/2} \, Hz \tag{10.17}$$

where $\Delta v_{1/2}$ is the half-width of the line at half the absorption peak height.

10.6 Chemical Shift

The principal effect underlying the usefulness of NMR technique is chemical shift. This refers to the fact that the field at the nucleus is not external field B_0 but one which is modified by the chemical environment. The origin of the chemical shift is that a nucleus in a molecule is surrounded by an electric charge cloud that is a reflection of a chemical bonding about the nucleus. The shape of this charge cloud is usually not spherical but is of complicated form. External magnetic field induces motion in the charge cloud such that an additional magnetic field called local field is developed. This field is proportional to B_0 and algebraically added to it. Thus the molecular electron cloud effectively shields the nucleus from the external field. This shielding interaction causes the magnetic field seen by the nucleus in the molecule to be different from B_0. The resonance condition given by Eq. (10.11) takes the form

$$v = \frac{\gamma B_{eff}}{2\pi} = \frac{\gamma}{2\pi}(B_0 - B_{local}) = \frac{\gamma}{2\pi}B_0(1-\sigma_i) \qquad (10.18)$$

B_{eff} is the algebraic sum of B_0 and local field. σ_i is the shielding constant for nuclei i. The σ is less than 10^{-5} for proton and less than about 10^{-3} for most other nuclei. Since the magnitude of shielding depends on the orientation of the molecule relative to B_0, σ is a second rank tensor. However, for gases and liquids orientational dependence is averaged out and σ may be treated as a scalar. It is convenient to regard σ as the sum of positive diamagnetic contribution σ_d and a negative paramagnetic contribution σ_p and hence write $\sigma = \sigma_d + \sigma_p$. σ is positive if diamagnetic contribution dominates and is negative if the paramagnetic contribution dominates.

The effect of screening is to decrease the spacing of the nuclear magnetic energy levels of H^1 as the effective field seen by the nucleus is decreased. At constant frequency of radio frequency field, the value of B_0 will have to be increased to achieve the resonance condition given by Eq. (10.12). Thus if the resonance absorption positions are expressed on a scale with B_0 increasing from left to right, the absorption peaks for more shielded nucleus will be on the right hand side of the spectrum. This is because more screened nucleus requires higher field to satisfy resonance condition.

The frequency for resonance for a nucleus in a reference compound in analogy with Eq. (10.18) is

$$v_r = \frac{\gamma B_0}{2\pi}(1-\sigma_r) \qquad (10.19)$$

Difference of frequency for the resonating nucleus in a sample and in the reference compound (having the same nucleus) is obtained from Eqs. (10.18) and (10.19)

$$v_i - v_r = \frac{\gamma B_0}{2\pi}(\sigma_r - \sigma_i) \qquad (10.20)$$

From Eqs. (10.19) and (10.20)

$$\frac{v_i - v_r}{v_r} = \frac{\sigma_r - \sigma_i}{1-\sigma_r} \qquad (10.21)$$

Since $1 \gg \sigma_r$,

$$\frac{v_i - v_r}{v_r} = \sigma_r - \sigma_i = \delta \qquad (10.22)$$

δ is called chemical shift. It is the frequency difference in Hz divided by the operating frequency in MHz. The value of chemical shift is expressed in parts per million (ppm). The use of dimensionless scale unit has the great advantage as the chemical shift values so expressed are independent of the value of B_0 (or v) of any particular NMR spectrometer. However, the spacing of the chemical shift position between two given nuclei depends on B_0 (or v).

The chemical shift scale for a given nucleus is established by choosing a substance as a standard and defining its chemical shift as zero. For proton, C^{13} and Si^{29}, tetramethylsilane $(CH_3)_4Si$ (TMS) is used as a reference material because (i) it is miscible with most of the organic solvents, (ii) it is inert and (iii) highly volatile (~300 K) and therefore can easily be removed after the measurement have been made. The structure of TMS is shown in Fig. 10.3(a). In TMS, protons are highly shielded and give a narrow resonance line. For water soluble substances DSS (2,2-dimethyl-2 silapentane-5-sulphonate) [Fig. 10.3(b)] may be used as a reference. The proton in the methyl group of DSS gives a strong resonance. The compound C_6F_6, $CFCl_3$ and trifluroacetic acids are used as a reference for F^{19} while Cl^- (aq.) and H_2O are used as reference for Cl^{35} and O^{17}, respectively.

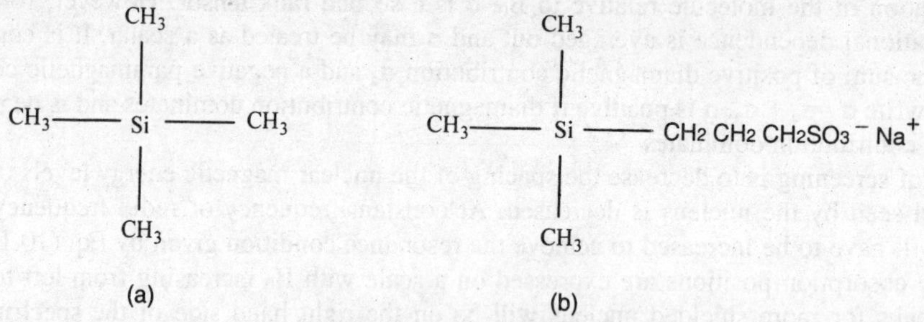

Fig. 10.3 Structure of (a) TMS (b) DSS

In the H^1 NMR spectrum, the less screened region is on the left hand side of the spectrum with TMS on the right. If B_0 is kept constant, then the effective magnetic field seen by the nucleus goes on increasing as one moves away from TMS signal position on the left. Thus to observe resonance, the frequency according to resonance condition, should also increase. Thus frequency increases to the left of TMS. On the right of TMS signal, the nucleus is more screened and hence effective field seen by the nucleus is also less. Therefore, for this low magnetic field, the resonance frequency is also low. The frequency therefore decreases on the right of TMS signal. Alternatively, if the frequency is kept constant and B_0 is varied, then highly screened TMS will show signal at low magnetic field. Thus magnetic field increases from left to right. The left hand side region of TMS is termed as high frequency and low magnetic field region while right hand side of TMS is low frequency, high field region. In H^1 NMR spectroscopy, all those protons on high field of TMS have negative chemical shift value. This scale is called δ scale. An alternate system, which is generally used for defining the position of resonance relative to the reference, is tau (τ) scale. On this scale the reference is assigned the arbitrary position of 10 and the values of other resonance are given by

$$\tau = 10 - \delta$$

Table 10.2 Typical proton chemical shifts

Compound	δ
TMS	0
CH_4	0.2
CH_3 in C_2H_5OH	1.17
$(CH_3)_2CO$	2.1
$(CH_3)_2O$	3.2
CH_2 in C_2H_5OH	3.59
C_2H_4	5.5
C_6H_6	7.2
$CHCl_3$	7.2
CH_3OCHO	8.03
CH_3CHO	9.716

The proton resonance shifts are restricted to a very small range (Table 10.2). The H^1 chemical shifts in organic compounds can be correlated with the electronegativity of neighbouring groups, type of carbon bonding and hydrogen bonding. Protons attached to or near electronegative groups or atoms such as O, OH, halogens, CO_2H, NH_3^+ and NO_2 experience a lower density of shielding electrons and are resonant at lower value of B_0. Protons removed from such groups appear at higher value of B_0. For example, in low-resolution spectrum, that is, showing only chemical shifts, of ethyl alcohol furnishes an illustration of these effects. Since oxygen is more electronegative than C, the electron density on H in O-H bond will be less than on H in the C-H bond. The OH proton of CH_3CH_2OH is least shielded as it is attached to the more electronegative oxygen atom. Out of CH_2 and CH_3, proton of CH_2 is expected to be less shielded as it is directly attached to the oxygen atom. The observed lines are shown on a stick diagram in Fig. 10.4. The number of protons in OH, CH_2 and CH_3 are 1,2 and 3, respectively. Therefore, the intensity of lines should also be in the same ratio, which is observed experimentally.

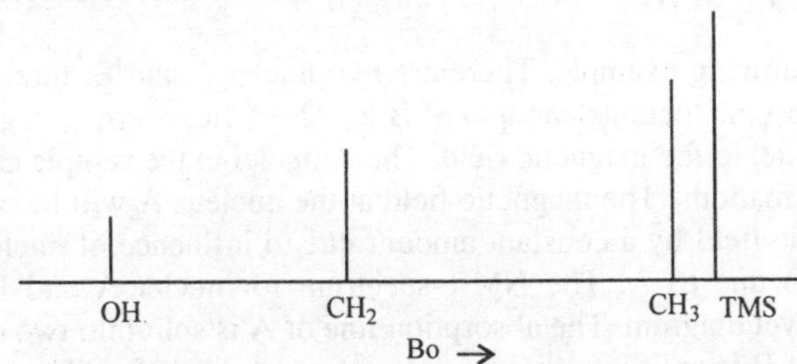

Fig. 10.4 A stick diagram showing the low resolution H^1 NMR spectrum of ethyl alcohol

10.7 Spin-Spin Coupling

If a resonance line is examined more closely using high-resolution techniques, it is often observed that it is composed of several finely spaced lines. The interaction between the spins of the neighbouring nuclei in a molecule may cause the splitting of the lines in the NMR spectrum. This interaction occurs via the bonding electrons. The coupling between the nuclei is a two-stage process, in which the first nucleus A creates a magnetic field affecting its electron shell, which in turn influence the electron shell around the other nucleus B. This changes the magnetic field at the nucleus B and hence changes the nuclear frequency. The effect is mutual. The magnitude of the effect for a particular pair of nuclei depends on the following factors:

(a) The coupling of two nuclei via polarization of the intervening electron spins depends upon the electron density at each nucleus. As s electrons have appreciable density at the nuclei, therefore spin-spin interaction depends on s character in the bonds.

(b) Coupling of two nuclei A and B occur via the orbital motion of the valence electrons. The nuclear magnetic moment of a nucleus A will induce currents in its electron cloud and the resulting magnetic field will be felt by the other nucleus B and vice versa. This coupling may be unimportant for proton- proton interaction.

(c) Coupling between equivalent nuclei or group of equivalent nuclei is not directly observable. A group of nuclei is chemically equivalent if nuclei in the group are related by a symmetry operation of the molecule and have same chemical shift. The nuclei experiencing different environment or having different chemical shifts are called non-equivalent.

(d) The magnetic moment of the nuclei and is directly proportional to $\gamma_A\gamma_B$ where γ_A and γ_B are the magnetogyric ratios of the interacting nuclei A and B.

(e) The coupling generally decreases rapidly as the number of interacting bonds increases and is not usually observed over more than four or five bonds.

The magnitude of the coupling interaction is measured in Hz. It is called coupling constant and is denoted by the symbol J. The coupling constant can be either positive or negative. The magnitude of the multiplet separation resulting from spin-spin interaction is independent of the strength of the applied magnetic field. Although the absorption positions of a group (in Hz) changes with applied magnetic field strength, the spacing of spin- spin coupled multiplet do not. The spacing is determined by the spin-spin coupling interaction.

For describing systems of nuclear spins within the molecule, the following notation is usually adopted. Non equivalent nuclei of the same species which have chemical shift difference comparable to the mutual coupling are denoted by A, B, C,....etc. The another group of nuclei separated from this group (A, B, C,...) by large chemical shifts are denoted by X, Y, Z, ..etc. The nuclei of second group have chemical shift difference among themselves comparable to mutual coupling. Equivalent nuclei are denoted by the same symbol and the number of equivalent nuclei as subscript, e.g., PF_3 would be referred to as AX_3.

Let us consider the following example. There are two nuclei A and B, three bonds away from one another in a molecule. Assume that nuclear spin of B is 1/2. In the external magnetic field, spin of B is either parallel or antiparallel to the magnetic field. The A nuclei in the sample can each experience one of the two different perturbations. The magnetic field at the nucleus A will be either greater or smaller than the external magnetic field by a constant amount due to influence of nucleus B. A similar effect would be observed by B due to A. The NMR spectrum for nuclei A and B reflects the splitting observed in the energy level diagram. The absorption line of A is split into two components centred on δ_A and absorption line of B is split into two components centred on δ_B . The separation between two split absorption lines is called the J coupling constant or the spin -spin splitting constant and is a measure of the magnetic interaction between two nuclei.

Now consider a molecule with three spin 1/2 nuclei. One of them is of type A while the other two are of type B. Type B nuclei are three bonds away from nucleus A. In some molecules, the two B nuclear spins will oppose the magnetic field, in other both spins lie with the magnetic field while in remainder they will be oriented in the opposite direction. This is shown by an arrow diagram in Fig.10.5. Since

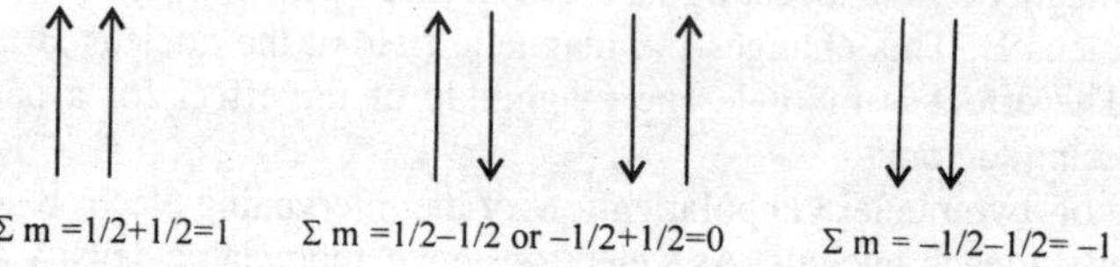

Fig. 10.5 An arrow diagram showing the possible configurations of two spin 1/2 nuclei

the B spins can be paired in opposition in two different ways, there will be twice as many molecules with them in this state as there are with them in each of the other two possible configurations. Thus the magnetic field seen by the spin A has three possible values due to four possible spin configurations of the two B nuclei. The levels with $\Sigma m = 0$ is two-fold degenerate as there are two components of spins giving the same value of $\Sigma m = 0$. Therefore, we have three lines with intensity in the ratio 1: 2:1. When $\Sigma m = 0$ there is no change from the external magnetic field seen by A and hence no change in position corresponding to this configuration. The resultant NMR spectra have three lines. The centre absorption line of those centred at δ_A is twice as high as the either of the outer two as shown by stick diagram (Fig. 10.6). This is because there were twice as many transitions in the energy levels. The peaks at δ_B are taller because there are twice as many B type nuclear spins than A type spins.

Let us now consider a molecule with four spin 1/2 nuclei; one type A and three of type B. Let the

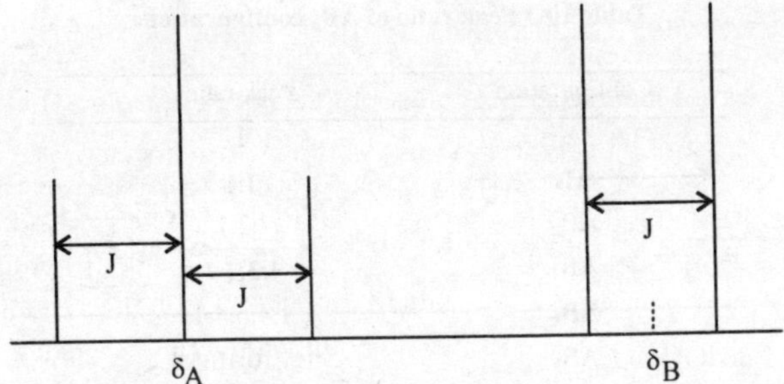

Fig. 10.6 A stick diagram showing the splitting due to two spin 1/2 nuclei

type B nuclei be three bonds away from type A nucleus. In some molecules all the three B spins are either oriented parallel or antiparallel to the magnetic field. In some molecules, one spin is oriented parallel to the field and other two are antiparallel to the field. In the remainder molecules, two spins are oriented parallel to the field and one antiparallel to the field. The resulting eight possible configurations are shown in the arrow diagram (Fig.10.7).

$$\uparrow\uparrow\uparrow \qquad\qquad \Sigma m = \frac{1}{2}+\frac{1}{2}+\frac{1}{2}=\frac{3}{2}$$

$$\uparrow\uparrow\downarrow \text{ or } \downarrow\uparrow\uparrow \text{ or } \uparrow\downarrow\uparrow \qquad \Sigma m = \frac{1}{2}+\frac{1}{2}-\frac{1}{2}=-\frac{1}{2}+\frac{1}{2}+\frac{1}{2}=\frac{1}{2}-\frac{1}{2}+\frac{1}{2}=\frac{1}{2}$$

$$\uparrow\downarrow\downarrow \text{ or } \downarrow\uparrow\downarrow \text{ or } \downarrow\downarrow\uparrow \qquad \Sigma m = \frac{1}{2}-\frac{1}{2}-\frac{1}{2}=-\frac{1}{2}+\frac{1}{2}-\frac{1}{2}=-\frac{1}{2}-\frac{1}{2}+\frac{1}{2}=-\frac{1}{2}$$

$$\downarrow\downarrow\downarrow \qquad\qquad \Sigma m = -\frac{1}{2}-\frac{1}{2}-\frac{1}{2}=-\frac{3}{2}$$

Fig. 10.7 An arrow diagram showing the possible configurations for three spin 1/2 nuclei

Since the levels with $\Sigma m = 1/2$ and $-1/2$ are three fold degenerate, hence there would be three times as many degenerate transitions in the energy level diagram for these transitions. Thus the magnetic field seen by the nuclei has four possible values all different from the external field due to four possible spin configurations of three B nuclei. Therefore NMR spectra would consist of four lines centred at δ_A with intensity ratio 1:3:3:1. This is shown in the stick diagram (Fig .10.8). The peaks at δ_B are taller because there is thrice as many B type spins as there are A type spins.

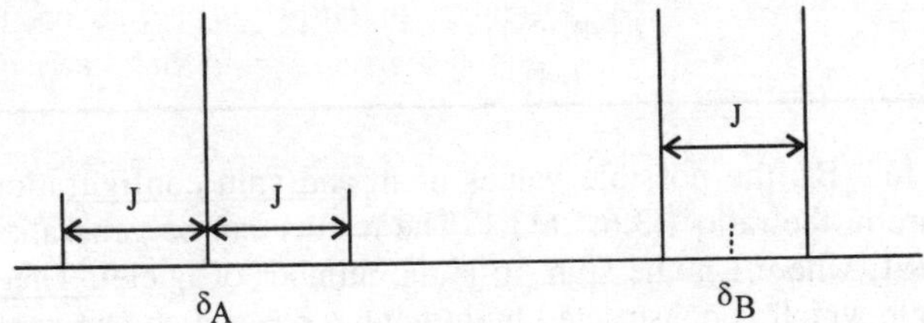

Fig. 10.8 A stick diagram showing the splitting due to three spin 1/2 nuclei

The lines of B are split into components because of spin $I = 1/2$ of nucleus A which can have two m values 1/2 and $-1/2$. These results can be generalised and are given in Table 10.3

Table 10.3 Peak ratio of AB_n configuration.

Configuration	Peak ratio
A	1
AB	1:1
AB_2	!:2:1
AB_3	1:3:3:1
AB_4	1:4:6:4:1
AB_5	1:5:10:10:5:1
AB_6	1: 6:15: 20:15: 6: 1

The series is called Pascal's triangle and can be obtained from the coefficients of the expansion of the $(x + 1)^n$ where n is the number of B nuclei in the above table, for example, if $n = 2$

$$(x+1)^2 = {}^2C_0 x^2 + {}^2C_1 x + {}^2C_2 x^0$$

the coefficients are 1, 2 and 1. If $n = 3$

$$(x+1)^3 = {}^3C_0 x^3 + {}^3C_1 x^2 + {}^3C_2 x + {}^3C_3 x^0$$

the coefficients are 1, 3, 3 and 1 and so on. It is concluded that the number of lines due to n equivalent spin 1/2 nuclei is n +1 and the intensity ratio is given by the coefficients of binomial expansion.

Coupling to nuclei I >1/2 leads to different relative intensities and multiplication. Let us consider the case AB, where A is a proton and B nucleus has spin I = 1. The number of orientations of spin B is 2I + 1 = 3 corresponding to m = 1, 0, -1. The spin a will experience different magnetic field resulting in three lines with line corresponding to m = 0 centred at δ_A. For the system AB_2 the possible configurations of B are given in Table 10.4. The intensity of the lines are in the ratio 1:2:3:2:1.

Table 10.4 Possible configurations and relative line intensities for coupling to two nuclei with I = 1.

Σm	Possible spin configurations (m_1 m_2,)	No. of spin combinations
2	(1,1)	1
1	(1,0), (0,1)	2
0	(1,-1), (0,0), (-1,1)	3
-1	(-1,0), (0,-1)	2
-2	(-1,-1)	1

Extending it further to AB_3, the possible values of m and spin configurations are given in Table 10.5. The seven lines are in the ratio 1:3:6:7:6:3:1. The results can be generalized and the number of lines are equal to 2nI + 1, where I is the spin, n is the number of nuclei. The intensity of lines are given by Pascal's triangle which is constructed by moving a box which can enclose 2I + 1 number of lines (Fig. 10.9). The box is moved along the line enclosing the first 1, then it is moved further so that now it encloses two numbers, then three numbers and so on. The numbers enclosed by the box are added at every step to give a number for the next line. For example, for I = 1 the box can enclose 2I + 1

Table 10.5 Possible configurations and relative line intensities for coupling to three nuclei with I = 1

Σm	Possible spin configurations (m_1, m_2, m_3)	No. of spin combinations
3	(1,1,1)	1
2	(1,1,0), (1,0,1), (0,1,1)	3
1	(1,0,0), (0,1,0), (0,0,1), (1,1,-1), (1,-1,1), (-1,1,1)	6
0	(0,0,0), (1,-1,0), (1,0,-1), (0,1,-1), (0,-1,1), (-1,1,0), (-1,0,1)	7
-1	(-1,0,0), (0,-1,0), (0,0,-1), (-1,-1,1), (-1,1,-1), (1,-1,-1)	6
-2	(-1,-1,0), (-1,0,-1), (0,-1,-1)	3
-3	(-1,-1,-1)	1

= 3 numbers. If we sweep the box through the line n = 1, it first encloses 1 thus giving 0 + 0 +1 = 1 for n = 2 line. In the next move the box has numbers 0, 1, 0 (that is 1 has moved in the middle partition) giving 0 + 1 + 0 = 1 for n = 2 line. In the last move in n = 1 line the box will have 1,0,0 giving 1 for n = 2 line. Thus AB$_2$ configuration has three lines with relative intensity 1:1:1. Now if the box is swept through n = 2 it first encloses 1, then 1 + 1 = 2, in the next move 1 + 1+ 1 = 3 and so on. Thus for AB$_3$ we have five lines with relative intensity ratio as 1:2:3:2:1.

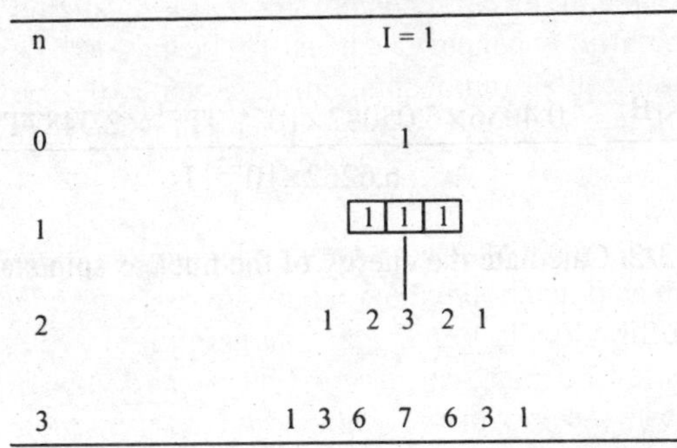

Fig. 10.9 Construction of Pascal's triangle for I = 1

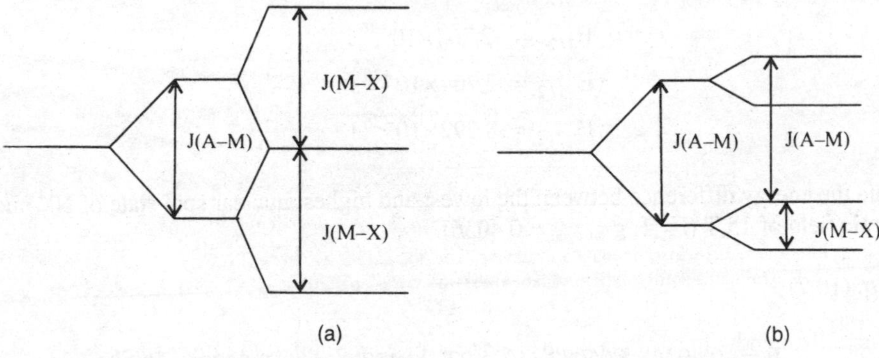

(a)   (b)

Fig. 10.10 Splitting of the H$_M$ proton due to coupling to H$_A$ and H$_X$ (a) J(A-M) = J(M-X) (b) J(A-M) ≠ J(M-X)

Consider a case when a nucleus is coupled simultaneously to chemically different groups of nuclei. Let us take an example where nucleus M is coupled to two chemically different spin 1/2 nuclei A and X. M is first split into a doublet by J (A-M). J (M-X) further splits this doublet line. If J (A-M) = J (M-X), a 1:2:1 triplet is observed. On the other hand, if J (A-M) $\neq$ J (M-X), then a doublet of doublet with all the lines having equal intensity arises as shown in Fig. 10.10.

Examples

1. Calculate the magnetic field needed to satisfy the resonance condition for unshielded proton in a 100   MHz radio frequency field. (g_{proton} = 5.585, β_N = 5.05082 x 10^{-27} JT^{-1}, h = 6.6262 x 10^{-34} Js).

    From Eq. (10.12)

$$B_0 = \frac{h\nu}{g_{proton}\,\beta_N} = \frac{6.6262\times10^{-34}\,Js\times100\times10^6\,Hz}{5.585\times5.05082\times10^{-27}\,JT^{-1}} \approx 2.3489T$$

2. Calculate the frequency needed to satisfy the resonance for N^{14} in a magnetic field of 2.3487 T ($g_{nitrogen}$ = 0.4036).

    From Eq. (10.12)

$$\nu = \frac{g_{nitrogen}\,\beta_N\,B_0}{h} = \frac{0.4036\times5.05082\times10^{-27}\,JT^{-1}\times2.3487T}{6.6262\times10^{-34}\,Js} \approx 7.22\;MHz$$

3. Cl^{35} has a nuclear spin of 3/2. Calculate the energy of the nuclear spin states in a magnetic field of 2T (g_{Cl} = 0.5472).

    From Eq. (10.7)

$$E_m = -g_{Cl}\beta_N B_0 m = -0.5472\times5.05082\times10^{-27}\,JT^{-1}\times2T\times m = -5.528\times10^{-26}\,J\times m$$

$$E_{3/2} = -8.292\times10^{-26}\,J$$
$$E_{1/2} = -2.764\times10^{-26}\,J$$
$$E_{-1/2} = 2.764\times10^{-26}\,J$$
$$E_{-3/2} = 8.292\times10^{-26}\,J$$

4. Calculate the energy difference between the lowest and highest nuclear spin state of N^{14} nucleus in a magnetic field of 15 T (I = 1, $g_{nitrogen}$ = 0.4036).

    From Eq. (10.7)

$$E = -0.4036\times5.05082\times10^{-27}\,JT^{-1}\times15T\times m = -3.05776\times10^{-26}\,J\times m$$

For m = 1

$$E_1 = -3.05776 \times 10^{-26} \, \text{J}$$

For m = -1

$$E_{-1} = 3.05776 \times 10^{-26} \, \text{J}$$

$$\Delta E = E_{-1} - E_1 = 6.1157 \times 10^{-26} \, \text{J}$$

5. Calculate the ratio of the number of protons in the upper to that in the lower spin state in a magnetic field of 2T at 10 mK.

From Eq. (10.15)

$$\frac{N_u}{N_l} = \exp\left(-\frac{g_{\text{proton}}\,\beta_N B_0}{k_B T}\right) = \exp\left(-\frac{5.585 \times 5.05082 \times 10^{-27}\,\text{JT}^{-1} \times 2\text{T}}{1.3807 \times 10^{-23}\,\text{JK}^{-1} \times 10 \times 10^{-3}\,\text{K}}\right) = \exp(-0.04086) \approx 0.9599$$

6. What is the value of γ for proton ($g_{\text{proton}} = 5.585$) ?

From Eq. (10.4)

$$\gamma_{\text{proton}} = \frac{2\pi\mu_N}{hI} = \frac{2.79255 \times 5.051082 \times 10^{-27}\,\text{JT}^{-1} \times 2 \times 3.14 \times 2}{6.6262 \times 10^{-34}\,\text{Js}} \approx 2.674 \times 10^8\,\text{s}^{-1}\text{T}^{-1}$$

7. What is the Larmor frequency for the proton at 10 T ($\gamma_{\text{proton}} = 2.674 \times 10^8\,\text{s}^{-1}\text{T}^{-1}$) ?

From Eq. (10.10)

$$\nu = \frac{\gamma_{\text{proton}} B_0}{2\pi} = \frac{2.674 \times 10^8\,\text{s}^{-1}\text{T}^{-1} \times 10\text{T}}{2 \times 3.14} = 0.4285 \times 10^9\,\text{Hz} = 428.5\,\text{MHz}$$

8. What is the difference in fractional population of C^{13} spins between the upper and lower states in a magnetic field of 2T at 298 K ? ($g_{\text{Carbon}} = 1.404$)

From Eq. (10.15)

$$\frac{N_u}{N_l} = \exp\left(-\frac{g_{\text{carbon}}\,\beta_N B_0}{k_B T}\right) = \exp\left(-\frac{1.404 \times 5.05082 \times 10^{-27}\,\text{JT}^{-1} \times 2\text{T}}{1.3807 \times 10^{-23}\,\text{JK}^{-1} \times 298\text{K}}\right)$$

$$\frac{N_u}{N_l} \approx 1 - 3.448 \times 10^{-6}$$

$$\frac{N_l - N_u}{N_l} \approx 3.448 \times 10^{-6}$$

9. A line in the H^1 NMR spectrum of a molecule had a width of 10 Hz. What is the effective relaxation time ?

According to Eq. (10.17)

$$T_2 = \frac{1}{\pi \Delta v_{1/2}} = \frac{1}{3.14 \times 10\,\text{Hz}} \approx 3.2\ \text{ms.}$$

10. The chemical shift for protons in CH_3F: 4.26 ppm, CH_3Cl: 3.05 ppm, CH_3Br: 2.68 ppm and CH_3I : 2.16 ppm. What might account for this trend ?

The halogens withdraw electron from the CH_3 group in the order $F > Cl > Br > I$. Thus the electron density and hence shielding at CH_3 group decreases in the same order. The difference in shielding between reference compound TMS increases thus increases the chemical shift.

11. The chemical shift of the CH_3 proton in acetaldehyde is 2.2 and that of CHO proton is 9.8. What is the difference in the local magnetic field between the two regions of molecule, when the external magnetic field is 7.0T.

From Eq. (10.22)

$$\delta = \frac{v_{\text{sample}} - v_{\text{ref}}}{v_{\text{ref}}}$$

$$2.2 = \frac{v_{CH_3} - v_{\text{ref}}}{v_{\text{ref}}}$$

$$9.8 = \frac{v_{CHO} - v_{\text{ref}}}{v_{\text{ref}}}$$

On subtracting

$$9.8 - 2.2 = 7.6 = \frac{v_{CHO} - v_{CH_3}}{v_{\text{ref.}}}$$

From Eq. (10.12)

$$v = \frac{g_H \beta_N B_0}{h} = \frac{5.585 \times 5.05082 \times 10^{-27}\ \text{JT}^{-1} \times 7\text{T}}{6.6262 \times 10^{-34}\ \text{Js}} = 298\ \text{MHz}$$

$$v_{CHO} - v_{CH_3} = 298 \times 7.6 = 2264.8\,\text{Hz}$$

From resonance condition (10.12)

$$h\nu_{CH_3} = g_H\beta_N B_{CH_3}$$

$$h\nu_{CHO} = g_H\beta_N B_{CHO}$$

and

$$h(\nu_{CHO} - \nu_{CH_3}) = g_H\beta_N(B_{CHO} - B_{CH_3}) = g_H\beta_N\left[(B_{loc})_{CH_3} - (B_{loc})_{CHO}\right]$$

$$(B_{loc})_{CH_3} - (B_{loc})_{CHO} = \frac{h(\nu_{CHO} - \nu_{CH_3})}{g_H\beta_N} = \frac{6.6262\times10^{-34}\,\text{Js}\times2264.8\text{Hz}}{5.585\times5.051\times10^{-27}\,\text{JT}^{-1}} \approx 53.19\,\mu\text{T}$$

12. The chemical shift of the CH_3 protons in acetaldehyde is 2.20 and that of CHO proton is 9.80. What is the splitting between the methyl and aldehyde proton resonance in a spectrometer operating at 350 MHz.

From Eq. (10.22)

$$\delta_{CHO} - \delta_{CH_3} = \frac{(\nu_{CHO} - \nu_{CH_3})\text{Hz}}{\nu(\text{MHz})}$$

$$\nu_{CHO} - \nu_{CH_3} = (9.8 - 2.2)\times350\,\text{Hz} = 2660\,\text{Hz}$$

13. At what frequency shift from TMS would a group of nuclei with $\delta = 2.3$ resonate in a spectrometer operating at 100 MHz?

From Eq. (10.22)

$$\delta = \frac{(\nu_i - \nu_r)\text{Hz}}{\nu(\text{MHz})}$$

$$\nu_i - \nu_r = 2.3\times100\,\text{Hz} = 230\,\text{Hz}$$

14. In a given solution the signal of benzene is 438 Hz higher in frequency than TMS. Calculate the chemical shift. The spectrometer is operating at 60 MHz.

From Eq. (10.22)

$$\delta = \frac{438}{60} = 7.3\,\text{ppm}$$

15. A proton spectrometer operating at 60 MHz was used to measure the frequency separation of the resonance of benzene and TMS, which was found to be 438 Hz, the benzene being to high frequency. What is the chemical shift of benzene on the δ scale? What would be the frequency separation and chemical shift be if the sample is measured with a spectrometer operating at 200 MHz?

From Eq. (10.22)

$$\delta = \frac{438\text{Hz}}{60\text{MHz}} = 7.3\text{ppm}$$

Since the chemical shift is independent of spectrometer frequency, therefore the chemical shift would also be 7.3 ppm at 200 MHz. The frequency separation between benzene signal and TMS signal would be from Eq. (10.22)

$$\nu_{\text{benzene}} - \nu_{\text{TMS}} = 7.3\text{ppm} \times 200\text{MHz} = 1460\text{Hz}$$

16. Which of the two chemically different types of protons in CH_3CHO resonate at higher frequency ?

There are two different kinds of protons in CH_3CHO, one is in CH_3 and other is in CHO. The proton of CHO is coordinated with more electronegative O. Therefore, proton of CHO is less screened than proton of CH_3. Hence the resonance of proton of CHO occurs at higher frequency.

17. Account for high resolution H^1 NMR spectrum of C_2H_5OH.

The low resolution spectrum of ethyl alcohol consists of three peaks corresponding to one OH proton at low end field, three methyl protons at the high field and two methylene protons in between these two fields. In high resolution spectrum the absorption peak of CH_3 protons will split into three peaks with relative intensity 1:2:1 (this is of AB_2 configuration with A methyl proton and B proton of CH_2). The absorption peak of methylene proton will primarily split into four peaks with intensities 1:3:3:1 due to interaction with the methyl protons (this is of AB_3 configuration with A protons of CH_2 and B protons of CH_3). Each of these four peaks will further split into two peaks of equal intensities due to coupling with the hydroxyl proton (of the form AB). Thus in a very high resolution spectrum, the methylene proton will exhibit an octet with relative intensities 1:1:3:3:3:3:1:1. Since proton of OH and CH_3 are far away from each other, therefore they do not affect each other. The high resolution spectrum of ethyl alcohol on a stick diagram is shown in Fig.10. . If the resolution is not very large, one often observe quintet with intensities 1:4:6:4:1 for CH_2 as if it has coupled with four protons (three from methyl and one from hydroxyl). The hydroxyl proton will split into three peaks of intensities 1:2:1 due to the coupling with methylene protons (AB_2 type).

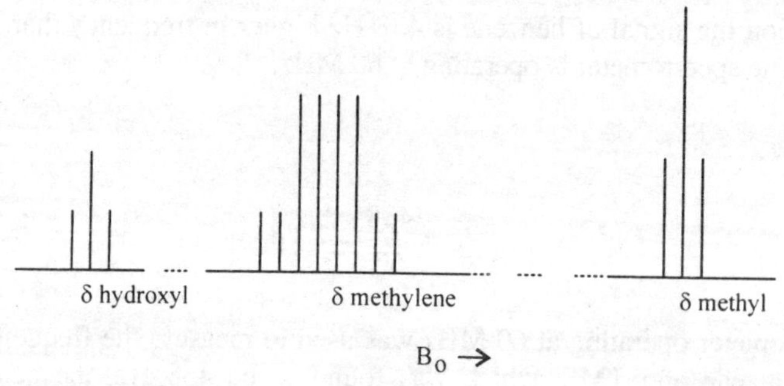

Fig. 10.11 High resolution H^1 NMR spectrum of ethyl alcohol

18. Discuss the effect of presence of trace of acid or alkali on the H^1 NMR spectrum of ethyl alcohol.

The hydroxyl proton of ethyl alcohol is not locked in any particular ethyl alcohol molecule. It is continuously exchanged from one to other molecule. The rate of this exchange is slow in pure ethyl alcohol. That is proton of H spent sufficient time on a molecule to interact with neighbouring CH_2 protons and thus causing splitting . However, on adding the trace of acid or alkali, the rate of exchange of proton of OH very much increased and it spends time on a molecule, which is not sufficient to allow its interaction with CH_2 group. As a result of this, splitting of OH and CH_2 due to each other does not take place and a single peak of OH is obtained. Similarly, the CH_2 proton show only splitting due to neighbouring CH_3 group and hence a quintet is replaced by the quartet.

19. What multiplicity and intensity distribution should be expected for H and D nuclei in the compound $CHDCl_2$.

The nuclear spin of H is 1/2 and that of D is 1. This is of AB type. Therefore the hydrogen peak would split into three components with intensity ratio 1:1:1 and D peak splits into two of equal intensity.

Problems

10.1   Calculate the magnetic field needed to satisfy the resonance for the Hg^{199} nucleus in a 17.83 MHz radio frequency field. ($g_{Hg} = 0.996$).

10.2   Calculate the frequency needed to satisfy the resonance for F^{19} in a magnetic field of 2.3487 T ($g_F = 5.255$).

10.3 S^{32} has a nuclear spin of 3/2. Calculate the energies of the nuclear spin states in a magnetic field of 7.50 T ($g_S= 0.4289$).

10.4 I^{127} has a nuclear spin of 5/2. Calculate the energies of the nuclear spin states in a magnetic field of 5 T ($g_I= 1.118$).

10.5   Calculate the energy difference between the lowest and highest nuclear spin state of B^{10} nucleus in a magnetic field of 10 T ($I = 3$, $g_B = 0.6002$).

10.6   Calculate the energy difference between the lowest and highest nuclear spin state of Cl^{35} nucleus in a magnetic field of 3 T ($I = 3/2$, $g_{Cl} = 0.4555$).

10.7   Calculate the ratio of the number of protons in the upper to that in the lower spin state in a magnetic field of 2T at 1 mK.

10.8   What is the value of γ for F^{19} ($I=1/2$, $g_F = 5.255$) ?

10.9   What is the value of γ for Na^{23} ($\mu_{Na} = 2.2171\ \beta_N$) ?

10.10   In a magnetic field of 2T what fraction of the protons have their spins line up with the field at 298 K.

10.11   A NMR spectrum gives a linewidth of 0.30 Hz. What is the effective transverse relaxation time?

10.12   The chemical shift of the CH_3 proton in acetaldehyde is 2.2 and that of CHO proton is 9.8. What is the difference in the local magnetic field between the two regions of molecule, when the external magnetic field is 1.5 T ?

10.13   The chemical shift of the CH_3 protons in acetaldehyde is 2.20 and that of CHO proton is 9.80. What is the splitting between the methyl and aldehyde proton resonance in a spectrometer operating at 60 MHz.

10.14   What is the shift of the resonance from TMS of a group of nuclei with $\delta = 9.8$ and operating frequency is 100 MHz.

10.15   Which of the two chemically different types of proton in CH_3COOH resonate at lower magnetic field or higher frequency ?

10.16   What high resolution spectrum can be expected for the protons in NH_4^+ ?

10.17   Explain the effect of D_2O on the NMR spectrum of ethyl alcohol.

10.18   Describe low and high resolution NMR spectrum of PF_3.

10.19   Sketch the low and high resolution H^1 NMR spectrum of $CHCl_2CH_2Cl$.

10.20   Sketch the low and high resolution H^1 NMR spectrum of ethyl group.

10.21   What multiplicity and intensity distribution should be expected for H^1 NMR spectrum of CH_3 and CH_2 proton in the compound $(CH_3)_3C-CH_2B$.

10.22   Sketch the low and high resolution H^1 NMR spectrum of acetaldehyde.

11

Mössbauer Spectroscopy

11.1 Isomer Nuclear Transitions

A nucleus has discrete energy levels. When a transition from upper level to lower energy level takes place, gamma rays are emitted. The time during which the nucleus remains in a state determine its mean lifetime for that state. Two nuclei with equal charge and mass number but in different excited states are called isomer nuclei. The lifetime of the excited state is ~ 10^{-6} to 10^{-8} s. The isomer transitions that are used in Mössbauer spectroscopy are shown in Fig. 11.1 for nuclei of iron and tin.

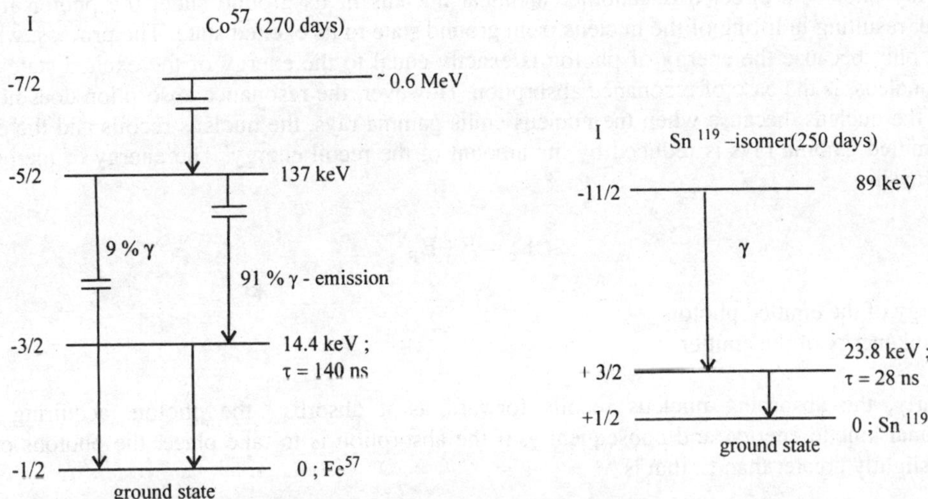

Fig 11.1 Isomer transitions of the Fe^{57} and Sn^{119} nuclei . Shown are transitions whose γ emission is used in the Mössbauer Spectroscopy : for Fe^{57} = 14.4 keV and for Sn^{119} = 23.8 keV; τ is transition time determining the natural line width ; I is nuclear spin in ground and excited states ; + and − sign indicate even and odd parity

The gamma emission of Fe^{57} nucleus with energy of 14.4 keV occurs as a result of isomer transition from the excited state to ground state of a Fe^{57} nucleus. Since in the case of a Fe^{57} an isomer with energy of 14.4 keV has a mean life time τ of only about 1.4 x10^{-7} s, in practice cobalt isotope Co^{57} with a half life of 270 days is taken as a source of gamma ray irradiation; then through electron capture transforms into an excited Fe^{57} isomer. In the case of tin, its isomer with a long lifetime (250 days) is used.

11.2 Resonance Fluorescence

(a) atomic resonance fluorescence

When light beam from sodium flame is focussed on a bulb containing vapours of sodium, a faint yellow glow is observed. The sodium atoms in the bulb are absorbing energy from the incident beam of yellow light (sodium D line) due to transition from lower $^2S_{1/2}$ energy level to upper $^2P_{1/2,\ 3/2}$ energy levels. Because of finite lifetime of upper levels, the absorbed energy is reemitted in all directions as a result of reverse transitions from $^2P_{1/2,3/2}$ to $^2S_{1/2}$. This lies at the base of resonance fluorescence (equal frequencies of primary and secondary emission). A comparison of the light which has passed through the bulb with that coming directly from the source shows that the result of passage through the sodium vapour is not simply weaken the sodium D lines, but to reduce the intensity of their wings. This is because of difference in temperature of the atoms in the flame and those in the bulb. Atoms in the flame, which are at higher temperature moves more rapidly, and therefore, the Doppler effect broadens the light, which they emit. The cooler atoms in the bulb absorb only the central portion of the broadened line.

(b) nuclear gamma resonance fluorescence

Consider a nucleus in its excited state whose energy is E. Nucleus emits gamma rays as a result of transition from excited state to the ground state. The energy of this gamma ray photon is $E_\gamma = h\nu$. If this gamma ray photon is directed on another identical nucleus in its ground state, the photon may be absorbed, resulting in lifting of the nucleus from ground state to its excited state. The process, which is possible only because the energy of photon is exactly equal to the energy of the excited state of the second nucleus, is the case of resonance absorption. However, the resonance absorption does not take place in the nucleus, because when the nucleus emits gamma rays, the nucleus recoils and the energy of the emitted gamma rays is reduced by the amount of the recoil energy. The energy of the emitted gamma rays is

$$E_e = E - E_R$$

E_e = energy of the emitted photon
E_R = recoil energy of the emitter

Similarly, the absorbing nucleus recoils forward as it absorbs the photon, acquiring some translational kinetic energy, and consequently, if the absorption is to take place, the photons energy must be slightly greater than E, that is

$$E_a = E + E_R$$

where E_a is the energy of the absorbing nucleus.

Fig.11.2 shows the position of E_e and E_a relative to the hypothetical recoil free situation. Since $E_e < E_a$, the emitted photon does not appear to have enough energy to excite the second nucleus. Therefore, resonance absorption is not usually observed in nuclear case.

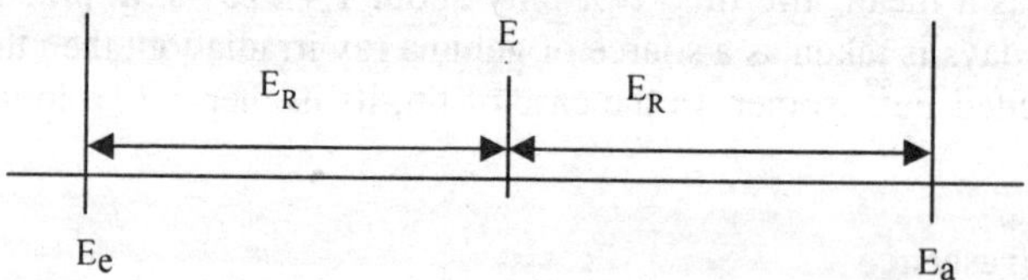

Fig. 11.2 Position of E_e and E_a relative to recoil free situation

Let us consider an isolated atom in the gas phase and the energy difference between the ground state E_g and excited state E_e is given by

$$E = E_e - E_g \tag{11.1}$$

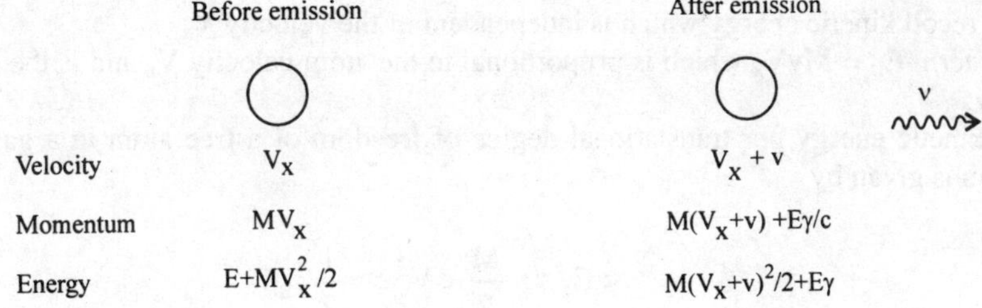

Fig. 11.3 The energy and momentum are conserved in the gamma emission process

Let the emitting atom of mass M is moving with a velocity V_x in the x-direction (Fig.11.3). The linear momentum of the atom before emission of gamma rays is MV_x. After emission of gamma ray, assumed in x-direction, the linear momentum of the system (gamma ray plus de-excited nucleus) must still equal to MV_x that is

$$MV_x = \frac{h\nu}{c} + M(V_x + v) \tag{11.2}$$

where $(V_x + v)$ is the velocity of the atom after emission of the gamma ray, v is vector and therefore, it can be negative. From Eq. (11.2)

$$v = -\frac{h\nu}{Mc} \tag{11.3}$$

recoil energy E_R is therefore by using Eq. (11.3) is

$$E_R = \frac{Mv^2}{2} = \frac{E^2}{2Mc^2} \tag{11.4}$$

before emission of gamma ray, the total energy above the ground state nucleus at rest is $(E + MV_x^2/2)$. After emitting the gamma ray of energy E_γ, the nucleus will have a new velocity $(V_x + v)$ due to recoil. The total energy of the system is $E_\gamma + M(V_x + v)^2/2$. From conservation of energy

$$E + \frac{1}{2}MV_x^2 = E_\gamma + \frac{1}{2}M(V_x + v)^2 \tag{11.5}$$

$$\delta E = E - E_\gamma = \frac{1}{2}Mv^2 + MvV_x \tag{11.6}$$

or

$$E = E_R + E_D \tag{11.7}$$

where δE is difference between energy of the nuclear transition E and energy of the emitted gamma ray photon E_γ . This difference depends
  (i)   on the recoil kinetic energy which is independent of the velocity V_x .
  (ii) on the term $E_D = MvV_x$ which is proportional to the atom velocity V_x and is the Doppler effect energy.

The mean kinetic energy per translational degree of freedom of a free atom in a gas with random thermal motion is given by

$$< E_k > = \frac{M}{2} < V_x^2 > \approx \frac{1}{2} k_B T \tag{11.8}$$

$<V_x^2>$ is the mean square velocity of the atom, k_B is the Boltzmann constant and T is the absolute temperature. From Eq. (11.8) we have

$$(< V_x^2 >)^{1/2} = \left(\frac{2 < E_k >}{M} \right)^{1/2} \tag{11.9}$$

using $E_D = MvV_x$ and Eq. (11.8) we have mean broadening

$$< E_D > = Mv(< V_x^2 >)^{1/2} = 2(E_R < E_k >)^{1/2} \tag{11.10}$$

The gamma ray distribution [Eq. (11.7)] is displaced by E_R and broadened by twice the geometric mean of the recoil energy and the average thermal energy [Eq. (11.10)]. The distribution is gaussian. Using Eq. (11.3) in Eq. (11.10) we have

$$< E_D > = E_\gamma \left(\frac{2 < E_k >}{Mc^2} \right)^{1/2} \tag{11.11}$$

for gamma rays of energy 10^4 eV, and a mass M = 100 amu, it is found that $E_R = 5.4 \times 10^{-4}$ eV and $<E_D> \sim 5 \times 10^{-3}$ eV at 300 K. Thus the resonance overlap for free atom resonance is small.

11.3 Mössbauer Effect

In 1958, Mössbauer used an Ir^{191} gamma ray source of 129 keV, for which the Doppler broadening at room temperature is about twice the value of E_R, therefore, the lines in Fig. 11.2 overlapped a little and resonance fluorescence could be observed. He expected that on cooling, the emitter and absorber, the absorption should decrease because of the decrease in Doppler width and thus in the overlap. However, he observed an increase in absorption. The absorption was as much as would be expected if there had been no recoil at all. The qualitative explanation of this fact is that at sufficiently low temperature an atom in a solid cannot recoil individually. The recoil momentum is absorbed by the crystal as a whole. The effective mass in Eq. (11.4) is therefore the mass of the crystal, which is so much larger than that of atom that the recoil energy is completely negligible. From Eq. (11.11), $<E_D>$ will also be negligible.

We know that the nucleus is not bound rigidly in crystal, but is free to vibrate. The recoil energy of a single nucleus can be taken up either by the whole crystal as discussed above or it can be transferred to the lattice by increasing the vibrational energy of the crystal. The vibrational energy levels of the crystal are quantized . Therefore, it can be excited only if the recoil energy corresponds closely with the allowed values, thus ensuring that the whole crystal recoils, leading to negligible recoil energy. Thus the necessary condition for the Mössbauer effect to occur is that the nucleus emitting the gamma ray photon should be in an atom, which has established vibrational integrity with the solid matrix. The vibrational energy of the lattice as a whole can change by discrete amounts 0, $\pm \hbar \omega$, $\pm 2 \hbar \omega$, ... If $E_R < \hbar \omega$, no transfer of energy will take place as either zero or $\hbar \omega$ units of vibrational energy but nothing intermediate can be transferred. When an average is taken over many emission processes, the energy transfer per event is exactly the free atom recoil energy. Let f be fraction of events which takes place without transferring energy to the lattice ($E_R < \hbar \omega$), then a fraction (1-f) will transfer one photon energy $\hbar\omega$, neglecting two, three etc. quantum transitions to a first approximation and therefore

$$E_R = (1-f)\frac{h}{2\pi}\omega \tag{11.12}$$

$$f = 1 - \frac{2\pi\, E_R}{h\omega} \tag{11.13}$$

Only those events give rise to the Mössbauer effect. f is often called Mössbauer coefficient. The probability that recoilless emission or absorption occurs is then given by general expression for the elastic scattering process

$$f = \exp[-4\pi^2 <s^2>/\lambda^2] \tag{11.14}$$

where λ is the wavelength of the gamma rays, $<s^2>$ is mean square thermal displacement of the appropriate atom in the direction of photon emission. Eq. (11.14) can be written as

$$f = \exp[-4\pi^2 E_\gamma^2 <s^2>/h^2c^2] \tag{11.15}$$

E_γ is gamma photon energy. $<s^2>$ is calculated on the basis of Einstein or Debye model. When $<s^2>$ is calculated on the basis of Debye model, f is called Debye Waller factor. The value of f becomes large for smaller values of E_γ and the smaller $<s^2>$. $<s^2>$ depends on the firmness of binding and on temperature. The reduction of force constant causes reduction of vibrational frequency of the lattice and hence increases in the probability of the recoil. The displacement of the nucleus must be smaller than wavelength. This is why Mössbauer effect is not observed in gases and non-viscous liquids. The smaller $<s^2>$ implies lower temperature and larger Debye or Einstein temperature. Because f decreases as E increases there is an upper limit to $E_\gamma \sim 150$ keV but in practice most Mössbauer nuclei have gamma ray energies < 100 keV. On the other hand gamma photon with $E_\gamma < 10$ keV are strongly absorbed by matter, so this energy forms an approximate lower limit.

The absorption process can be modulated by rigidly moving either the emitter or the absorber. If the emitter moves towards the absorber with a velocity v, the frequency ν of emitted gamma ray undergoes a Doppler shift.

$$v = v_0\left(1 + \frac{v}{c}\right) \tag{11.16}$$

where v_0 is the frequency of radiation from a stationary emitter. If the emitter and absorber are tuned in the beginning, the motion of the emitter (or absorber) causes detuning and reduces absorption. On the other hand, if the absorber and emitter are detuned at the beginning, the motion of emitter (or absorber) can be so adjusted as to bring in the desired tuning. If E_e and E_a are the energies of the emitter and absorber such that $E_e = hv_0$, $E_a = hv$, then using Eq.(11.16) we have

$$\frac{E_a - E_e}{E_a} = \frac{h(v - v_0)}{hv} = \frac{v}{c} \tag{11.17}$$

The above expression gives the velocity of emitter (or absorber) required to establish tuning.

Mössbauer spectra are measured by using spectrometer consisting of an emission source, a specimen (absorber), a detector for the resonant gamma rays transmitted through the absorber and a system capable of moving the emitter (or absorber). Emitter e.g. Fe^{57} is prepared by embedding the radioactive isotope, Co^{57}, in metallic iron, stainless steel and other host materials. A Mössbauer spectrum of a sample reveals resonance fluorescence of the same nucleus (the same isotope), which is incorporated in the emitter, and emits gamma rays. Therefore, by means of Fe^{57} emission it is possible to investigate only the iron making part of the absorber. Detection is done by gamma ray counter. The motion of the emitter relative to the absorber (required to equalize by means of Doppler effect), the energy of gamma ray quanta emitted by the emitter and absorbed by the specimen is measured. The number of counts in the detector is measured as a function of the velocity. The minimum of the transmission of gamma ray quanta corresponds to the maximum of the absorption by the absorber (specimen). This   corresponds to the Doppler velocity at which the resonance absorption occurs. The Mössbauer spectrum is therefore a velocity spectrum and represents the number of gamma rays quanta recorder by the counter at different Doppler velocities of emitter relative to the absorber.

11.4 Mössbauer Nuclei

For nuclei to be used for Mössbauer spectroscopy depend upon the following:
- (a)   The utilized radioactive isotope (parent nucleus) must have fairly long half-life.
- (b)   For a given element a source of gamma ray emission must exist - a parent nucleus whose decay gives rise to the appearance of isomer nuclei.
- (c)   The Mössbauer emission energy lies within the range of a few keV up to lower hundred of keV.
- (d)   The lifetime of the isomer excited level should be in the range $\sim 10^{-6} - 10^{-3}$ s. The lifetime determine the width Γ of lines and ratio $\Gamma/E \sim 10^{-10} - 10^{-14}$. Longer lifetime reduces Γ/E. Therefore, Doppler velocity becomes too low and measurable only with difficulty. Conversely, with rising Γ/E, the resonance selectivity diminishes.

Some of the properties of most important Fe^{57} nuclei of Mössbauer spectroscopy are given in Table 11.1

11.5 Isomer (Chemical) Shift

The Coulombic interaction between the nuclear charge and the electron charge alters the energy separation between the ground state and the excited state of the nucleus. Therefore, it causes a slight

Table 11.1 Properties of Fe^{57}

(a) Isotope abundance	:	2.17 %
(b) Energy of the first excited level (E)	:	14.4 keV
(c) Level lifetime(τ)	:	1.4×10^{-7} s
(d) Linewidth (Γ)	:	4.6×10^{-12} keV
(e) Ratio of linewidth to energy E	:	3×10^{-13}
(f) Ground state spin	:	1/2
(g) Excited state spin	:	3/2
(h) Ground state magnetic moment	:	0.095 nm
(i) Excited state magnetic moment	:	-0.15 nm
(j) Quadrupole moment of excited state	:	0.29 barn
(k) Recoil energy E_R	:	1.9×10^{-3} keV
(l) Typical Debye temperature	:	693K
(m) Doppler velocity due to natural linewidth	:	0.095 mm/s
(n) Recoilless fraction f (with Debye temperature equal to 693 K)	:	0.92 at 273 K, 0.79 at 573 K

shift in the position of the observed resonance line. This shift will be different in different chemical compounds (Fig.11.4). For this reason, it is generally known as the isomer shift or chemical shift or isomer chemical shift.

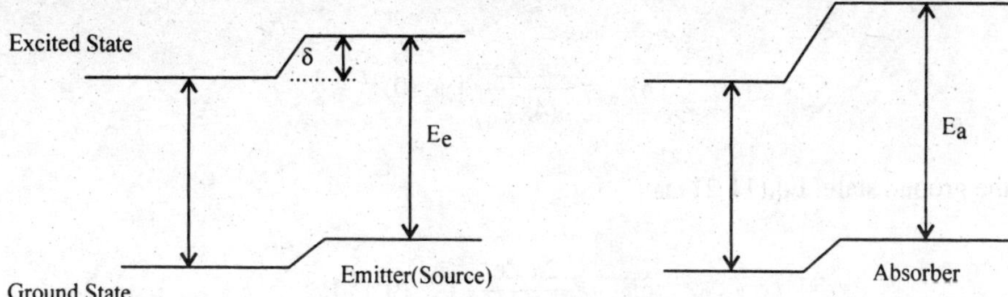

Fig. 11.4 Isomer Shift

The isomer shift is calculated by assuming the nucleus to be a uniformly charged sphere of radius R and the electronic charge density   is uniform over nuclear dimensions. At a distance r from the centre of the atomic nucleus the electrostatic potential is given by (for r > R)

$$V = \frac{Ze}{4\pi\,\varepsilon_0 r} \tag{11.18}$$

and if r < R

$$V' = \frac{Ze[3-(r^2/R^2)]}{(4\pi\,\varepsilon_0)2R} \tag{11.19}$$

Eq. (11.8) is for the hypothetical point nucleus. The s electrons ($l = 0$) also have a finite density within the atomic nucleus. The energy state of the nucleus is affected by the charge originating from the electrons in the sphere of radius R. The energy difference δE is given by

$$\delta E = \int_0^\infty \rho(V' - V)4\pi\, r^2 dr$$

$$\delta E = \int_0^R \frac{4\pi\, \rho\, Z\, e}{(4\pi\, \varepsilon_0)R}\left[\frac{3}{2} - \frac{r^2}{2R^2} - \frac{R}{r}\right] r^2 dr \tag{11.20}$$

$$\delta E = -\frac{2\pi\, Z\rho\, e\, R^2}{5(4\pi\, \varepsilon_0)} = \frac{2\pi\, Ze^2}{5(4\pi\varepsilon_0)}\,|\psi_s(0)|^2\, R^2 \tag{11.21}$$

where $\rho = -e\,|\psi_s(0)|^2$ is the charge density originating from the s electrons in the atomic nucleus.

The atomic nuclei in excited states usually have a different radius from those in the ground state; they may be either smaller or larger. The radius of excited state of I^{129} is greater than the radius of ground state of I^{129}. On the other hand the radius of the excited state of Fe^{57} is smaller than the radius of ground state of Fe^{57}. The energy difference δE will be different in excited and ground states. For excited state Eq. (11.21) is

$$\delta E_e = \frac{2\pi\, Ze^2}{5(4\pi\, \varepsilon_0)}\,|\psi_s(0)|^2\, R_e^2 \tag{11.22}$$

and for the ground state, Eq.(11.21) is

$$\delta E_g = \frac{2\pi\, Ze^2}{5(4\pi\, \varepsilon_0)}\,|\psi_s(0)|^2\, R_g^2 \tag{11.23}$$

On subtracting Eqs. (11.23) and (11.22) we obtain

$$\delta E_e - \delta E_g = \frac{2\pi\, Ze^2}{5(4\pi\, \varepsilon_0)}\,|\psi_s(0)|^2\, (R_e^2 - R_g^2) \tag{11.24}$$

by subtraction, the interaction energy resulting from the charges at a distance r > R from the centre of the   atomic nucleus is eliminated, as these interaction energies are approximately the same in the excited and ground states accordingly the integration in Eq. (11.20) is between the limit 0 and R. The magnitude of $|\psi_s(0)|^2$ depends somewhat on electron configuration, chemical environment and other factors varies from compound to compound.

An experiment usually determine the shift in gamma ray energies between absorber and the emitter and the shift δ is given by

$$\delta = E_{ab} - E_{em} = \frac{2\pi\, Ze^2}{5(4\pi\,\varepsilon_0)}\left[\,|\psi_s(0)|^2_{ab} - |\psi_s(0)|^2_{em}\,\right](R_e^2 - R_g^2) \tag{11.25}$$

$$R_e^2 - R_g^2 = (R_e - R_g)(R_e + R_g) \tag{11.26}$$

using $\Delta R = (R_e - R_g)$ and $R = (R_e + R_g)/2$ in Eq. (11.26)

$$R_e^2 - R_g^2 = \frac{2\Delta R R^2}{R} \tag{11.27}$$

From Eqs. (11.25) and (11.27) we have

$$\delta = E_{ab} - E_{em} = \frac{2\pi\, Ze^2}{5(4\pi\,\varepsilon_0)}\left[\,|\psi_s(0)|^2_{ab} - |\psi_s(0)|^2_{em}\,\right]2R^2\frac{\Delta R}{R}$$

$$\delta = k\frac{\Delta R}{R}\{\,|\psi_s(0)|^2_{ab} - |\psi_s(0)|^2_{em}\} \tag{11.28}$$

where $k = [4\pi\, R^2 Ze^2/(4\pi\,\varepsilon_0)5]$. The expression for the isomer shift has two factors (a) $\Delta R/R$ includes only nuclear parameters and is assumed to be constant for nuclei in different emitters and absorbers. (b) second factor $[\,|\psi_s(0)|^2_{ab} - |\psi_s(0)|^2_{em}\,]$ is an atomic parameter linked with the difference in the distribution of the s electron density in sample relative to the emitter. If $\Delta R/R$ is positive, and a positive isomer shift implies an increase in s electron density at the nucleus in going from emitter to absorber. If it is negative, the same (positive) shift signifies a decrease in s electron density. Electrons in 1s, 2s, 3s .... shell contribute to $|\psi_s(0)|^2$ but in decreasing amounts as the principal quantum number n increases. However, the inner shells are not markedly affected by chemical bonding. Therefore, the principal influence on the isomer shift will be by the outermost occupied s-orbitals. Shielding by other electrons effectively increases the s-radial function and decreases the s density at the nucleus.

The isomer shift is thus a measure of the s-electron density at the nucleus of absorber as compared with their density at the emitter nucleus. For the Fe^{57} nucleus, $\Delta R/R$ is negative and therefore a positive isomer shift corresponds to a decreased electron density on the nucleus of the absorber. The s-electron density, however, become subject to the effect produced by the 3d electrons of the iron. The effect manifests itself by shielding the s electrons on the nucleus. Therefore, an increase in the number of d electrons leads to a reduction of the s electron density at the nucleus, and consequently, to a greater isomer shift. For example, in comparison with metallic iron (configuration $3d^6 4s^2$ or $3d^7 4s^1$), the density of s electron diminishes in compounds of Fe^{2+} (electron configuration $3d^6$) or Fe^{3+} (electronic configuration $3d^5$) and a positive isomer shift is observed. Isomer shift for Fe^{2+} is 1.3 – 1.4 mm/s and for Fe^{3+} it is 0.4 – 0.5 mm/s. In Fe^{2+} compounds the number of 3d electrons is greater than in Fe^{3+} compounds, and therefore, the s electrons densities at the nucleus for Fe^{2+} is smaller and isomer shifts are generally larger.

11.6 Quadrupole Splitting

Any nucleus with a spin $I \geq 1$ has a non spherical charge distribution and hence a quadrupole moment. In Mössbauer spectra, the quadrupole splitting becomes manifest in the presence of two conditions

(a) When the Mössbauer nucleus displays the quadrupole moment in the ground or excited (isomer) states.

(b) The presence of electric field gradient (EFG) q at the site of nucleus position.

The asymmetric of the electric charge about the nucleus give rise to EFG. The EFG acting on the nucleus is given by

$$q = q_v (1-R) + q_L (1-\gamma_\infty) \qquad (11.29)$$

The aspherical distribution of lattice charges will contribute q_L, which is enhanced at the nucleus through polarisation of inner electron shell by a Sternheimer factor $(1-\gamma_\infty)$. The unfilled valence electron shells in transition elements produces a contribution q_v which is shielded from the nucleus by a factor $(1-R)$.

The interaction between the nuclear electric quadrupole moment Q and EFG is expressed by the Hamiltonian

$$H = Q \cdot \nabla E$$

and the energy is given by

$$E_Q = \frac{e^2 qQ[3m_I^2 - I(I+1)]}{4I(2I-1)} eV \qquad (11.30)$$

In the ground state, the Fe^{57} nucleus has a spin $I = 1/2$ and therefore have no quadrupole moment. In the excited state, Fe^{57} have a nuclear spin $I = 3/2$ and therefore have a quadrupole moment. The isomer level (excited level with $I = 3/2$) splits into two sublevels with $m_I = \pm 3/2, \pm 1/2$. Reversal of the sign of m_I will not change the nuclear charge distribution. Therefore EFG will not completely lift the four-fold degeneracy of the $I = 3/2$ state. The quartet will split into two doublets while $I = 1/2$ state will remain degenerate. For $I = 3/2$, the energy of the sublevels $m_I = \pm 3/2, \pm 1/2$ are

$$E_Q(\pm 3/2) = \frac{e^2 qQ}{4} eV$$

$$E_Q(\pm 1/2) = -\frac{e^2 qQ}{4} eV$$

and

$$\Delta = E_Q(\pm 3/2) - E_Q(\pm 1/2) = \frac{e^2 qQ}{2} eV \qquad (11.31)$$

The splitting of the level is shown in Fig. 11.5 . The transitions between these energy levels take place according to selection rule $\Delta m_I = 0, \pm 1$ and resulting in two lines. The separation between the transitions $m_I = \pm 1/2 \leftrightarrow \pm 1/2$ (σ transition) and $m_I = \pm 1/2 \leftrightarrow \pm 3/2$ (π transitions) is given by

Eq. (11.31). The relative probability of the transitions $m_I = \pm 1/2 \leftrightarrow \pm 1/2$ and $m_I = \pm 3/2 \leftrightarrow \pm 3/2$ are identical, but shows an angular dependence. In monocrystals, for transitions $m_I = \pm 3/2 \leftrightarrow \pm 1/2$ the

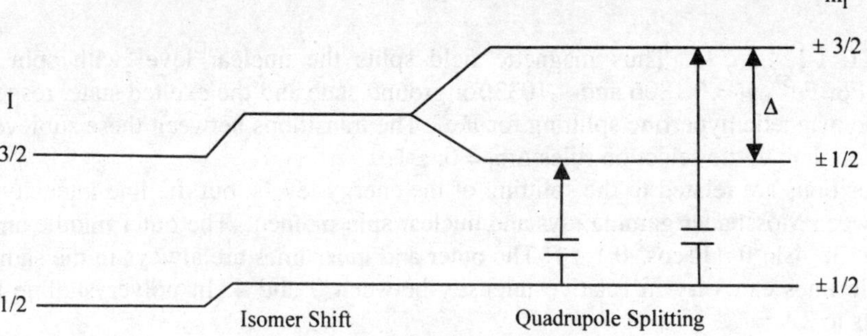

Fig. 11.5 Quadrupole splitting in Fe^{57}

angular dependence is of the form $3/2(1+\cos^2\theta)$ and for $m_I = 1/2 \leftrightarrow 1/2$ it is of the form $(1+3/2\sin^2\theta)$. In polycrystalline sample the intensities are equal.

11.7 Magnetic Hyperfine Structure

The energy levels of a nucleus, possessing a magnetic moment, are perturbed by the magnetic field. In the presence of magnetic field, the nuclear spin magnetic moment experiences a dipolar interaction with magnetic field i.e. Zeeman effect. The magnetic field at the nucleus is local one i.e. it does not extend over the whole of the crystal but arises due to interaction of the nucleus itself with its own electrons. The internal magnetic field at the nucleus is

$$B = B_{contact} + B_{orbital} + B_{dipole}$$

$B_{contact}$ arises due to interaction of the nucleus with s electrons. $B_{orbital}$ is due to orbital magnetic moment of the atom and $B_{dipolar}$ is dipolar field due to spin of these electrons. The field is almost completely linked with the interaction of the nucleus with s electrons, this being known as Fermi contact term. The s electrons are subject to the action of unpaired d electrons, which polarize the s electrons even of the occupied shells. The greatest number of unpaired d electrons e.g. five d electrons in Fe^{3+} causes a strong polarisation of the s shell and large magnetic field $\sim 40 - 60$ T. Decrease in the number of unpaired d electrons (e.g. four unpaired d electrons in Fe^{2+}) leads to decrease of the magnetic field to $\sim 20 - 40$T.

The nuclear magnetic moment $\boldsymbol{\mu_n}$ interact with the field and Hamiltonian is given by

$$H = -\mu_n \cdot B \tag{11.32}$$

and

$$\mu_n = \beta_N g_I I \tag{11.33}$$

where β_N is nuclear magneton.

From Eqs. (11.32) and Eq.(11.33) we have

$$H = -g_I \beta_N I \cdot B \tag{11.34}$$

and corresponding energy is

$$E = -g_I \beta_N m_I B \tag{11.35}$$

where m_I = I, I-1,........., -I. Thus magnetic field splits the nuclear level with spin I into 2I+1 components. For Fe^{57}, g_I = 0.1806 and -.1033 for ground state and the excited state, respectively. Fig. 11.6 show the magnetic hyperfine splitting for Fe^{3+}. The transitions between these sublevels gives rise to six lines according to the selection rules Δm_I = 0, ± 1.

The line positions are related to the splitting of the energy levels, but the line intensity is related to the angle between Mössbauer gamma rays and nuclear spin moment. The outer middle and inner lines are related by 3: $4\sin^2\theta$ /(1+$\cos^2\theta$) : 1. The outer and inner lines are always in the same proportion but the middle lines can vary in relative intensity between 0 and 4. In polycrystalline samples, the value averages to 2.

The splitting of the nuclear levels in the magnetic field is quite similar to the splitting of the levels in

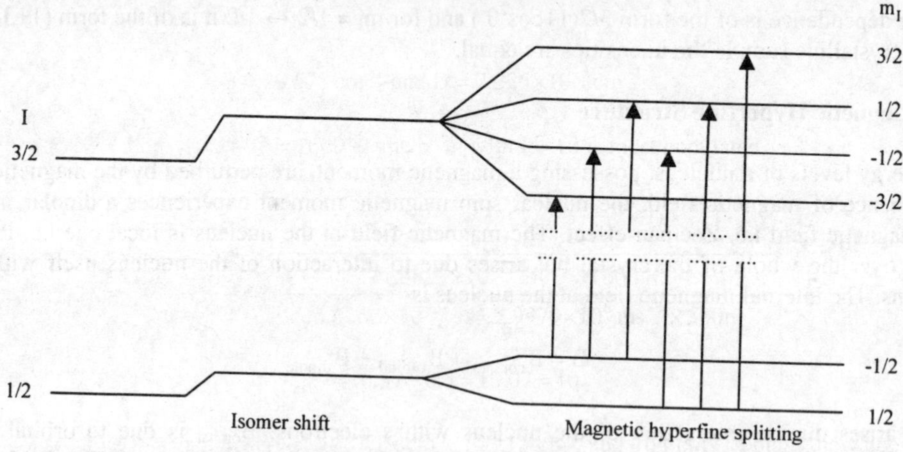

Fig. 11.6 Magnetic hyperfine splitting of the ground and first excited state of Fe^{57}

nuclear magnetic resonance (NMR) . However, there are differences between the NMR and Mössbauer effect as stated below

NMR	Mössbauer
1.The magnetic field in NMR is superimposed outer field generated by the magnet ~ 1 T.	1. Intensity of atomic magnetic field is ~ 20 – 60 T.
2.The ground state alone that comes under consideration	2. One considers both ground and excited states.
3.NMR results from transitions between neighbouring magnetic sublevels during absorption of a radio frequency quantum ~ MHz (10^{-7} –10^{-8} eV).	3. Gamma transitions from magnetic sublevels of the ground level to magnetic sublevels of the excited level occurs
4. Selection rules are Δm_I = ± 1.	4. Selection rules are Δm_I = 0, ±1

Examples

1. Calculate the recoil velocity of a free Mössbauer nucleus of mass 1.67×10^{-25} Kg when emitting a gamma ray of 0.1 nm wavelength. What is the Doppler shift of the gamma ray frequency to an outside observer?

From momentum conservation $Mv = \dfrac{h}{\lambda}$

$$v = \frac{h}{M\lambda} = \frac{6.6262 \times 10^{-34}\, js}{1.67 \times 10^{-25}\, Kg \times 0.1 \times 10^{-9}\, m} \approx 39.67 m/s$$

Doppler shift $\qquad\qquad \Delta v = \dfrac{v}{\lambda} = \dfrac{39.67 m/s}{0.1 \times 10^{-9}\, m} = 3.967 \times 10^{11}\, Hz$

2. The lifetime of Fe^{57} (excited) state is 1.5×10^{-7} s. The excited state is 14.4 keV above the ground state. Determine the linewidth Γ and ratio Γ /E.

We have $\quad \Gamma\tau \sim (h/2\pi)$

$$\Gamma \sim \frac{h}{2\pi\,\tau} = \frac{6.6262 \times 10^{-34}\, Js}{2 \times 3.14 \times 1.5 \times 10^{-7}\, s} = 0.703 \times 10^{-27}\, J = \frac{0.703 \times 10^{-27}}{1.6022 \times 10^{-19}}\, eV \approx 4.4 \times 10^{-9}\, eV$$

$$\frac{\Gamma}{E} = \frac{4.4 \times 10^{-9}\, eV}{14400\, eV} = 3.05 \times 10^{-3}$$

3. For a Fe^{57} nucleus with emission energy of E = 14400 eV, find the recoil energy.

We have

$$E_R = \frac{(h\,v)^2}{2Mc^2}$$

Considering the mass of Fe^{57} to be 57 amu and $h\,v = 14400$ eV we have

$$E_R = \frac{(14400 eV)^2}{2 \times 57 \times 931 \times 10^6\, eV} = 1.953 \times 10^{-5}\, eV$$

$\qquad\qquad$ (1 amu = 931×10^6 eV)

4. In a Mössbauer experiment gamma rays emitted in transitions from 0.129 MeV first excited state to the ground state of Ir^{191} (a) consider the recoil of the nucleus when it emits the gamma rays, and determine the downward shift in the energy of the gamma ray that results from the energy taken by the nuclear recoil (b) then compare this energy shift to the width of the first excited state of Ir^{191}, which has a measured lifetime of 1.4×10^{-10}s.

(a) Recoil energy is

$$E_R = \frac{(h\nu)^2}{2Mc^2}$$

Since the sum of gamma ray energy E and nuclear recoil energy must be equal to the energy available in gamma decay i.e. 0.129 MeV energy of the first excited state of the decaying nucleus. Therefore E is less than the energy of the first excited state by an amount E_R. This is the downward shift ΔE in the energy of the gamma ray due to nuclear recoil i.e.

$$\Delta E = -\text{ recoil energy } E_R = -\frac{E^2}{2Mc^2}$$

because M is so large, ΔE is very small compared to E, we may evaluate it approximately by setting E = 0.129 MeV. Using the relation 931 MeV = 1 amu, we have

$$\Delta E = -\frac{(1.29\times10^5 \text{eV})^2}{2\times191\times931\times10^6 \text{eV}} = -4.7\times10^{-2}\text{eV}$$

(b) If the lifetime τ of the first excited state of Ir^{191} is 1.4×10^{-10}s, the width Γ from $\Gamma\tau \sim \hbar$ is

$$\Gamma = h/2\pi\tau = \frac{6.62\times10^{-16}\text{eV}-\text{s}}{2\times3.14\times1.4\times10^{-10}\text{s}} = 7.5\times10^{-7}\text{eV}$$

clearly, the gamma rays emitted by the decay from the first excited state of Ir^{191} emitter nucleus cannot excite a Ir^{191} absorber nucleus from the ground state to its first excited state. The nuclear recoil shift of the gamma ray is larger by a factor of 10^4 than the width of the state it is supposed to excite. Therefore, the gamma ray is thrown completely out of resonance, and the resonant absorption is destroyed.

5.   If the relative velocity between source and absorber is 100 m/s for Fe^{57} having energy difference of ground and excited level 14400 eV, find the frequency shift.

We have $\nu = 14400 \times 2.418 \times 10^{14}$ Hz and

$$\Delta\nu = \frac{\nu\, v}{c} = \frac{14400\times2.418\times10^{14}\text{Hz}\times10^2\text{m/s}}{2.9979\times10^8\text{m/s}} \approx 1.1606\times10^{12}\text{Hz}$$

6.   A particular Mössbauer nucleus has spins 5/2 and 3/2 in its excited and ground states, respectively. Into how many lines will the gamma ray spectrum split if (a) the nucleus is under the influence of an internal electric field gradient, but no magnetic field is applied. (b) there is no electric field gradient at the nucleus but an external magnetic field is applied (c) both an internal electric field gradient and an external magnetic field are present.

(a) Under the action of internal electric field gradient, the excited state having I = 5/2 split into three levels with $m_I = \pm5/2, \pm3/2$ and $\pm1/2$. On the other hand ground state will split into two components with $m_I = \pm 3/2, \pm 1/2$. To obtain the number of allowed transitions the following procedure is adopted . The values of $\pm m_I$ for both initial and final states are written down in two rows with equal values of m_I directly below or   above each other. For the $I = 5/2 \leftrightarrow 3/2$ transitions they are as follow

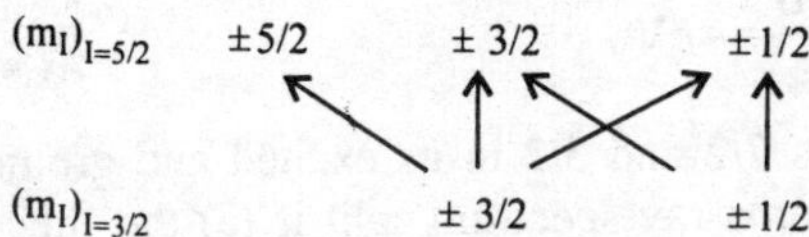

In this array, the vertical arrows indicate $\Delta m_I = 0$ transitions and the diagonal arrows indicates $\Delta m_I = \pm 1$ transitions. The total number of arrows indicates the number of allowed transitions. For $I = 5/2 \leftrightarrow 3/2$, the number of quadrupole transitions are five in numbers.

(b) In the presence of magnetic field, the degeneracy of the energy level of ground state and excited state is completely lifted. Excited state splits into five components with $m_I = 5/2, 3/2, 1/2, -1/2, -3/2, -5/2$ and ground state split into four components with $m_I = 3/2, 1/2, -1/2, -3/2$. To obtain the number of allowed transitions the following procedure is adopted . The values of m_I for both initial and final states, are written down in two rows with equal values of m_I directly below or above each other. For the $I = 5/2 \leftrightarrow 3/2$ transitions they are as follow

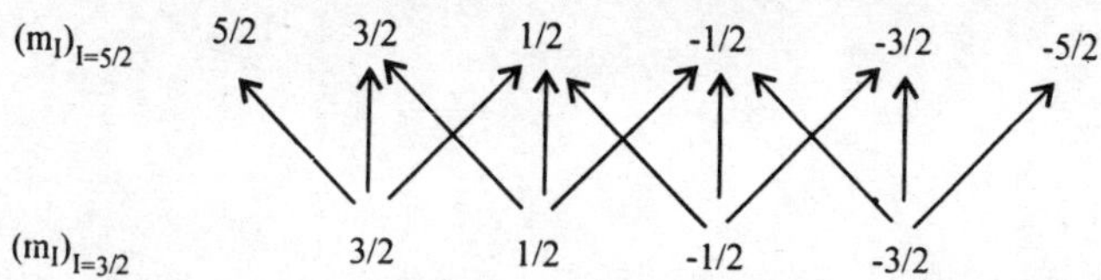

In this array, the vertical arrows indicate $\Delta m_I = 0$ transitions and the diagonal arrows indicates $\Delta m_I = \pm 1$ transitions. The total number of arrows indicates the number of allowed transitions. For $I = 5/2 \leftrightarrow 3/2$, the number of magnetic hyperfine transitions are 12 in numbers.

(c) Effect of electric field is only to change the spacing between the levels, thus again 12 lines will be observed.

Problems

11.1   A Zn^{67} nucleus has spins 5/2 and 1/2 in its excited and ground states, respectively. Into how many lines will the gamma ray spectrum split if (a) the nucleus is under the influence of an internal electric field gradient, but no magnetic field is applied. (b) there is no electric field gradient at the nucleus but an external magnetic field is applied (c) both an internal electric field gradient and an external magnetic field are present. For a Zn^{67} nucleus, the lifetime of the isomer state is 9400 ns. Determine the natural linewidth and recoil velocity.

11.2   Show that linewidth Γ_r is given by $\Gamma_r = \dfrac{273.8}{E_\gamma \tau_{1/2}}$ mm/s if E_γ is in keV and $\tau_{1/2}$ in ns. Γ_r is twice of Γ.

11.3 Show that recoil energy of the nucleus is $E_R = \dfrac{5.36942 \times E_\gamma^2}{M}$ eV If $E\gamma$ is in keV and M in amu.

11.4 Derive Eq. (11.19) .

11.5 Show that $\Gamma = \dfrac{4.562 \times 10^{-16}}{\tau_{1/2}s}$ eV .

11.6 A I^{127} nucleus has spins 7/2 and 5/2 in its excited and ground states, respectively. Into how many lines will the gamma ray spectrum split if (a) the nucleus is under the influence of an internal electric field gradient, but no magnetic field is applied. (b) there is no electric field gradient at the nucleus but an external magnetic field is applied (c) both an internal electric field gradient and an external magnetic field are present. For a I^{127} nucleus, the half lifetime of the isomer state is 1.86 ns. Determine the natural linewidth.

11.7 Sketch the Mössbauer energy levels for a nucleus which has the nuclear spin 5/2 and 7/2 in the ground and excited states. Into how many lines will the gamma ray spectrum splits if the nucleus is under the influence of magnetic field only.

12

Rotational Spectra of Diatomic Molecules

12.1 Molecular Spectra

The spectra of substances is in general assigned to one of the three categories :
   (i) The continuous spectra, which cannot be resolved into lines irrespective of the resolving power of the measuring instruments. These spectra are produced by the bodies heated to incandescent .
   (ii) The line spectra which are produced by atoms. The lines extend over the range of several hundred angstroms. These lines can be grouped into different series often overlapping. The separation of lines in each series is found to decrease as the wavelength decreases. These lines show further structure known as fine and hyperfine structure.
   (iii) Band spectra which are produced by the molecules and derive their names from the fact that in visible region under low spectroscopic resolution they appear as continuous band, bands of colour. The band consists of more or less broad wavelength regions and usually has at one end a sharp edge called a band head or band edge. At band head the intensity falls suddenly to zero, while it fades off more or less slowly towards the other end. According to the gradual fading of intensity take place toward high frequency or low frequency, the bands are said to be shaded or degraded to the violet or red. Bands are observed in some cases to have more than one head. Band spectra are observed both in emission and absorption. When band spectra are examined with spectroscopic instruments of high resolving power, they have been found to consist of a large number of discrete lines.

The spectra of a molecule can be divided into three spectral ranges from far infrared and microwave to extreme ultraviolet corresponding to different types of transitions between molecular quantum states. The molecules have spectrum in the far infrared region only when they possess a permanent electric dipole moment and this spectrum is caused by the transitions between the rotational states of the molecule. In the near infrared region the vibration rotation spectra is observed. This corresponds to radiation emitted in vibrational transition of molecules having electric dipole moment. The vibrational transitions are accompanied by change in rotational state. The bands in the visible and ultraviolet part of the spectrum are due to radiation emitted in transitions between electronic states. The electronic spectra have a fine structure determined by the rotational and vibrational state of the molecule during electronic transitions.

12.2 Classification of Molecules

The atoms in a molecule are generally arranged in three dimensions. It is convenient to revolve the rotation of three-dimensional body into rotational components about three mutually perpendicular directions through centre of gravity-the principal axis of rotation. The moment of inertia about three principal axes, called principal moment of inertia, is maximum or minimum. If the body has axes of symmetry, they coincide with the principal axes. In general three principal moments of inertia are different. Molecules may be classified into groups according to the relative values of their principal moment of inertia.

(i) linear molecules

In this type all the atoms of the molecules are arranged in a straight line such as CO, ICl or CO_2. Three directions of rotation may be taken as (a) about the bond axis, (b) end-over-end rotation in the plane of the paper (c) end-over-end rotation at right angles to the plane of the paper. The moment of inertia of (b) and (c) are same, that is, $I_b = I_c$, while that of (a), that is, I_a is very small. As an approximation we have $I_a = 0$, $I_b = I_c$.

(ii) symmetric tops

The molecule has two equal moment of inertia and third may be less or greater than the other two, that is, $I_b = I_c \neq I_a$. If $I_b = I_c > I_a$, the molecule is a prolate symmetric rotator , for example CH_3I (Fig.12.1). Since heavy iodine atom is on the a axis it makes no contribution to I_a , therefore $I_a < I_b = I_c$. If $I_b = I_c < I_a$ the molecule is referred to as oblate, for example, BCl_3 .

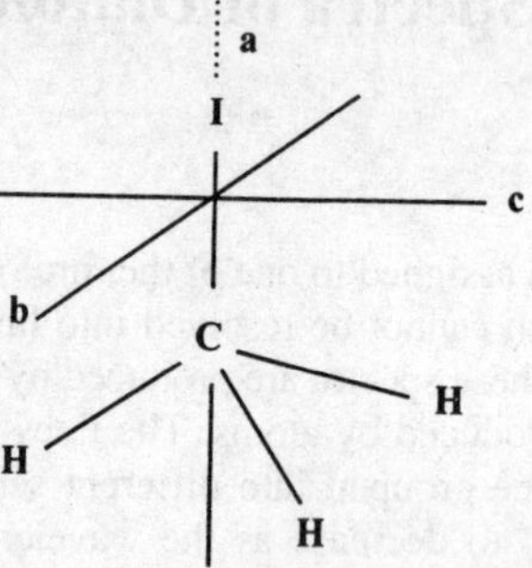

Fig. 12.1 Principal axes of inertia of methyl iodide

(iii) spherical tops

When a molecule has all three moment of inertia identical it is called a spherical top, that is, $I_a = I_b = I_c$. All regular tetrahedral molecules such as CH_4, SF_6, etc comes in this category. Fig.12.2 shows the principal axes of inertia of octahedral molecule SF_6 .

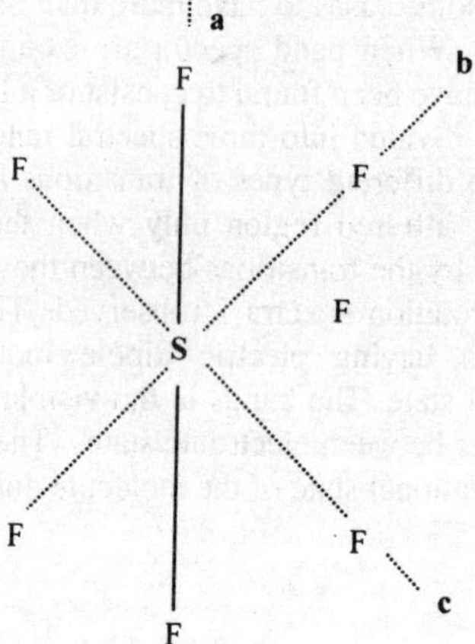

Fig. 12.2 The sulphur hexafluoride molecule

(iv) asymmetric tops

An asymmetric top molecule has no equal principal moment of inertia, that is $I_a \neq I_b \neq I_c$. Simple

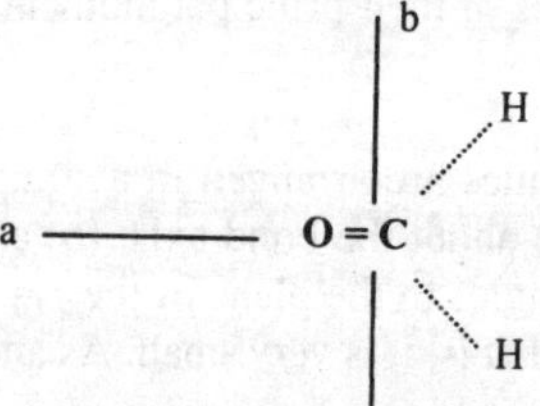

Fig. 12.3 Formaldehyde molecule

examples are H_2O, vinyl chloride, $CH_2=CHCl$, CHBrClF, H_2CO etc.. Fig.12.3 shows the formaldehyde molecule. The C=O bond of H_2CO is the principal a axis. For in planar molecule like formaldehyde, the moment of inertia about the axis perpendicular to the plane is equal to the sum of the two in - plane moments of inertia, that is, $I_c = I_a + I_b$.

12.3 The Rigid Rotator Model

In the rigid rotator model it is assumed that atoms of molecules are held rigidly at fixed distances from each other. Let m_1 and m_2 be the mass of the atoms separated by a distance r (Fig.12.4).

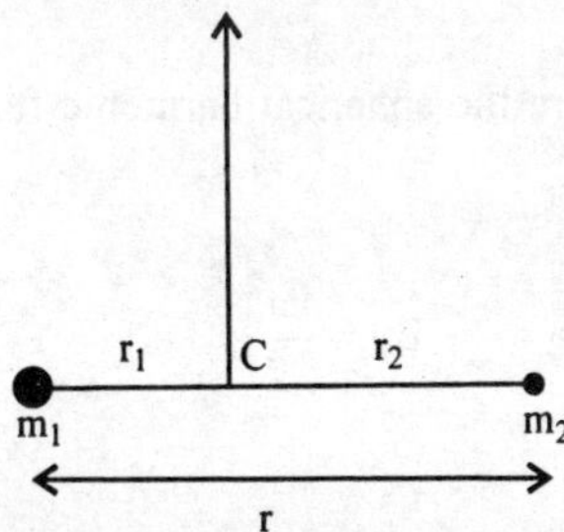

Fig. 12.4 Rotation of a diatomic molecule

The moment of inertia I about an axis passing through the centre of gravity C is

$$I = m_1 r_1^{\,2} + m_2 r_2^{\,2} \tag{12.1}$$

where r_1 and r_2 are the distances of atom of mass m_1 and m_2,respectively from the centre of gravity. The masses and distances of Fig.12.4 lead to the centre of gravity being located so that

$$m_1 r_1 = m_2 r_2 \tag{12.2}$$

and

$$r = r_1 + r_2 \tag{12.3}$$

From Eqs. (12.1) and (12.2) we have

$$r_1 = \frac{m_2 r}{m_1 + m_2} \tag{12.4}$$

$$r_2 = \frac{m_1 r}{m_1 + m_2} \tag{12.5}$$

Substituting the values of r_1 and r_2 from Eqs. (12.4) and (12.5), respectively into Eq.(12.1) gives

$$I = \frac{m_1^{\,2} m_2 r^2}{(m_1 + m_2)^2} + \frac{m_1 m_2^{\,2} r^2}{(m_1 + m_2)^2} = \frac{m_1 m_2 r^2}{m_1 + m_2} = \mu r^2 \tag{12.6}$$

where $\mu = \dfrac{m_1 m_2}{m_1 + m_2}$ is the reduced mass of the molecule. The moment of inertia is same as that of a mass μ

at a distance r from the axis passing through C.

The classical angular momentum $\mathbf{L}$ of the system is given by

$$L = I\omega_r \tag{12.7}$$

where ω_r is angular velocity of rotation and is related to the number of rotation per second, ν_{rot} (rotational frequency) by

$$\omega_r = 2\pi\nu_{rot} = 2\pi\nu_r \tag{12.8}$$

The eigenfunctions of the L^2 operator are the spherical harmonic functions $Y_{JM}(\theta, \varphi)$, which arise in the rotational kinetic energy E_r

$$E_r = \frac{I\omega_r^2}{2} \tag{12.9}$$

Using Eq. (12.7) in Eq. (12.9)

$$E_r = \frac{L^2}{2I} \tag{12.10}$$

The Schrödinger wave equation, Eq. (2.61), is

$$\nabla^2\Psi + \frac{8\pi^2\mu}{h^2}\left[E - V(r)\right]\Psi = 0$$

In spherical polar coordinates the above equation takes the form of Eq. (2.62)

$$\frac{1}{r^2}\frac{d}{dr}\left(r^2\frac{\partial\Psi}{\partial r}\right) + \frac{1}{r^2\sin\theta}\frac{d}{d\theta}\left(\sin\theta\frac{\partial\Psi}{\partial\theta}\right) + \frac{1}{r^2\sin^2\theta}\frac{\partial^2\Psi}{\partial\varphi^2} + \frac{8\pi^2\mu}{h^2}[E - V(r)]\Psi = 0$$

Using the form of L^2 operator from Eq. (2.115) the above equation can be written as

$$\frac{1}{r^2}\frac{d}{dr}\left(r^2\frac{\partial\Psi}{\partial r}\right) - \frac{4\pi^2 L^2\Psi}{h^2 r^2} + \frac{8\pi^2\mu}{h^2}[E_r - V(r)]\Psi = 0 \tag{12.11}$$

Since our rotator is rigid therefore, no change in potential energy is associated with rotation and thus the motion is that of molecule as a whole. The potential energy V(r) is constant which for convenience can be taken as zero. For rigid rotator, we take $r = r_e$, the equilibrium internuclear distance. Since r_e is constant, the derivative with respect to r can be dropped. The Eq. (12.11) reduces to

$$L^2\Psi = 2\mu r_e^2 E_r\Psi \tag{12.12}$$

The functions $Y_{JM}(\theta, \varphi)$ can be written[Eq. (2.82)]

$$Y_{JM}(\theta,\varphi) = [\frac{(2J+1)(J-M)!}{4\pi\,(J+M)!}]^{1/2}(-1)^M\,P_J^M(\cos\theta)\exp(iM\varphi) \qquad (12.13)$$

where quantum number J pertains to the magnitude of the angular momentum and M designates the z component of angular momentum. The forms of $Y_{JM}(\theta,\varphi)$ are obtained from Table 2.5 by replacing l by J and m by M.

The eigenvalues relations from Eqs. (2.116) and (2.117) are

$$L^2 Y_{JM}(\theta,\varphi) = (h/2\pi)^2\,J(J+1)\,Y_{JM}(\theta,\varphi) \qquad (12.14)$$

$$L_z Y_{JM}(\theta,\varphi) = M(h/2\pi)\,Y_{JM}(\theta,\varphi) \qquad (12.15)$$

where $J = 0,1,2,.......,\infty$ is called rotational quantum number and M = -J,-J+1, ...,J a total of 2J+1 values. Using Eq. (12.14) in Eq. (12.12) we have

$$(h/2\pi)^2\,J(J+1) = 2\mu\,r_e^2 E_r \qquad (12.16)$$

$$E_r = \frac{h^2 J(J+1)}{8\pi^2\,\mu\,r_e^2} = \frac{h^2 J(J+1)}{8\pi^2\,I} \qquad (12.17)$$

The rotational energy levels are given by Eq. (12.17) and degeneracy of each level is equal to 2J+1, the number of allowed values of M. This number is also called statistical weight. This degeneracy indicates that the energy does not depend on the orientation of the angular momentum. The degeneracy can be lifted in the presence of an electric or magnetic field. Eq. (12.17) can be obtained if L^2 of Eq. (12.15) is replaced by quantum mechanical value $(h/2\pi)^2\,J(J+1)$.

According to Eq. (12.17) we have a series of discrete energy levels whose energy increases with increasing J as shown in Fig.12.5. Since spectroscopic transitions are usually measured in wavenumber,

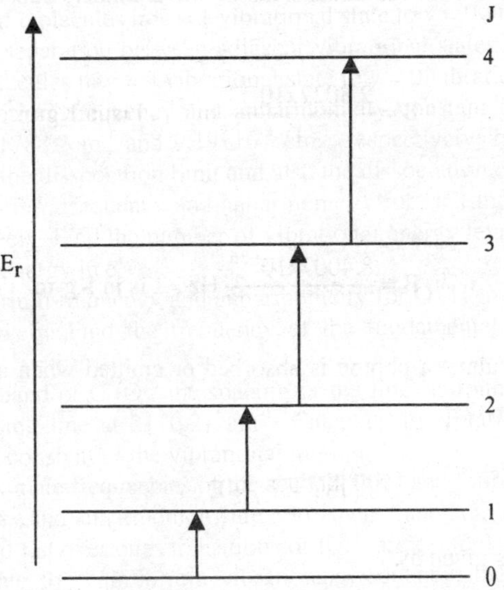

Fig. 12.5 Energy levels and allowed transitions for the rigid rotator

therefore Eq. (12.17) is written in terms of wavenumber and symbol $F(J) = \dfrac{E_r}{hc}$ is used for rotational term

$$F(J) = \frac{E_r}{hc} = \frac{hJ(J+1)}{8\pi^2 Ic} \tag{12.18}$$

In terms of frequency

$$F(J) = \frac{E_r}{h} = \frac{hJ(J+1)}{8\pi^2 I} \tag{12.19}$$

and

$$F(J) = BJ(J+1) \tag{12.20}$$

where

$$B = \frac{h}{8\pi^2 Ic} \text{ in wavenumbers} \tag{12.21}$$

$$B = \frac{h}{8\pi^2 I} \text{ in Hz} \tag{12.22}$$

where B is called rotational constant. Because of the nature of the experimental methods, microwave and millimeter wave spectroscopists use B with the dimensions of the frequency and infrared, visible and ultraviolet spectroscopists use B with the dimensions of wavenumber. Substituting the values of constants, Eq. (12.21) gives

$$B = \frac{2.8022 \times 10^{-44}}{I} \text{ m}^{-1} \text{ , I is in Kg-m}^2 \tag{12.23}$$

and Eq.(12.22) gives

$$B = \frac{8.4007 \times 10^{-36}}{I} \text{ Hz , I is in Kg-m}^2 \tag{12.24}$$

According to the Bohr postulate, a photon is absorbed or emitted when a transition take place between rotational state $E_r^{'}$ and $E_r^{''}$ such that

$$h\nu_r = E_r^{'} - E_r^{''} \tag{12.25}$$

The frequency of the photon is given by

$$\nu_r = \frac{E_r^{'} - E_r^{''}}{h}$$ (12.26)

where $E_r^{'}$ and $E_r^{''}$ are the energies of the upper and lower rotational states, respectively. Eq. (12.26) in wavenumbers is given by

$$\bar{\nu}_r = \frac{\nu_r}{c} = F(J') - F(J'')$$ (12.27)

where J' and J'' are the quantum numbers of upper and lower rotational states, respectively.

A molecule interacts appreciably with an electromagnetic field only if it has an electric or magnetic dipole moment. It is therefore required that molecule must possess a permanent electric dipole moment for the observation of rotational transitions. The rotational spectrum will be observed for hetronuclear molecules, which have electric dipole moment. This dipole moment changes during rotation and interacts with electromagnetic radiation. In the HBr molecule, for example, Br has an excess negative charge and H an excess positive charge, so that molecule is a small rotating dipole. Homonuclear molecules like H_2, N_2 and O_2 etc. do not show rotational spectrum while hetronuclear molecules like HCl, CO and DCl etc show rotational spectrum. Transitions between rotational states of molecules like H_2 can take place during collisions.

Pure rotational transitions may be observed in the microwave, millimetre wave or far infrared regions. They are electric dipole transitions and the allowed selection rules is

$$\Delta J = \pm 1$$ (12.28)

Using the selection rule in Eq. (12.27), the emitted or absorbed transitions of rigid rotator by Eq. (12.20), are

$$\bar{\nu}_r = F(J') - F(J'') = F(J+1) - F(J) = 2B(J+1)$$ (12.29)

where $J = 0,1,2,3, \ldots$ is used for J'' and B is in wavenumber .The allowed transitions are shown in Fig.12.5. If B is in unit of frequency, Eq. (12.29) reduced to

$$\nu_r = 2B(J+1)$$ (12.30)

The spectrum of the simple rigid rotator consists of a series of equally spaced spectral lines at 2B, 4B, 6B,..... The separation of successive line is constant and is equal to 2B. The lines extend to higher wavenumber as J increases. Whether the spectrum of a particular molecule lies in microwave, millimetre wave or far infrared region depends on the value of B and J.

In obtaining the rotational energy we have ignored the rotation about the symmetry axis, that is, bond axis. There are two reasons for this (a) the mass of an atom is located almost entirely in the nucleus whose radius is only $\sim 10^{-4}$ of the radius of atom itself, therefore the main contribution to moment of inertia of diatomic molecules about the symmetry comes from electrons. The mass of the electrons is only about 1/4000 of the total molecular mass and is concentrated in a region whose radius is about half the bond distance. The moment of inertia about the symmetry axis is thus very small and hence the value of B is large. According to Eq. (12.17) the energy levels are therefore widely spaced. To raise molecule from J = 0 state to J = 1 state requires large energy and such transition do not occur under normal spectroscopic conditions. (b) The molecule is in J = 0 state for rotation about symmetry axis, and they may be said to be not rotating. In such transitions, if occur, there is no dipole change and hence no rotational spectrum. In the rigid rotator model, rotator does not experiences any angular dependent forces (torques). In liquids, solids or even dense gases, where intermolecular interactions come into play, predictions of the theory will not apply.

12.4 Intensity Distribution in Rotational Spectra

Let m and n be the lower and upper rotational levels respectively. Under thermal equilibrium the population of upper and lower levels are N_n and N_m, respectively.

According to Boltzmann distribution law

$$\frac{N_n}{N_0} = \exp\left(-\frac{E_n}{k_BT}\right) \tag{12.31}$$

$$\frac{N_m}{N_0} = \exp\left(-\frac{E_m}{k_BT}\right) \tag{12.32}$$

where E_n and E_m are the energies of n and m rotational levels relative to $J = 0$ level. N_0 is the population of J = 0 level. From Eqs. (12.31) and (12.32)

$$\frac{N_n}{N_m} = \exp\left(-\frac{E_n - E_m}{k_BT}\right) = \exp\left(-\frac{\Delta E}{k_BT}\right) \tag{12.33}$$

N, the total number of molecules per unit volume, will be

$$N = N_0 + N_1 + N_2 + \ldots\ldots + N_i = N_0 \sum_i \exp\left(-\frac{E_i}{k_BT}\right) \tag{12.34}$$

where the summation is taken over all the levels. If the ith level has degeneracy g_i, the total population will be

$$N = N_0 \sum_i g_i \exp\left(-\frac{E_i}{k_BT}\right) \tag{12.35}$$

Number of molecules in the level J with degeneracy $g_J = 2J + 1$ and energy E_J is

$$N_J = g_J N_0 \exp\left(-\frac{E_J}{k_BT}\right) \tag{12.36}$$

$$N_J = (2J+1)N_0 \exp\left(-\frac{F(J)hc}{k_BT}\right) = (2J+1)N_0 \exp\left(-\frac{BJ(J+1)hc}{k_BT}\right) \tag{12.37}$$

N_J increases linearly with J because of the factor 2J+1 but decreases due to the exponential factor. For small values of J, the increase in N_J due to 2J+1, degeneracy factor, dominates the decrease due to exponential factor, but at high J the opposite is the case. J_{max}, the J value of the energy level with the highest population is obtained when

$$\frac{dN_J}{dJ} = 0$$

Putting the value of N_J in the above and differentiating

$$\left[\frac{2k_BT-(2J+1)^2 Bhc}{k_BT}\right]\exp\left(-\frac{BhcJ(J+1)}{k_BT}\right)=0$$

$$(2J+1)^2=\frac{2k_BT}{hc}$$

$$J_{max}\approx\sqrt{\frac{k_BT}{2hcB}}-\frac{1}{2} \qquad (12.38)$$

The approximation is because J_{max} must of course be an integer. If B has dimension of frequency, Eq. (12.38) is

$$J_{max}\approx\sqrt{\frac{k_BT}{2hB}}-\frac{1}{2} \qquad (12.39)$$

J_{max}, increases with increasing T and decreasing B. From Eqs. (12.35) and (12.37)

$$\frac{N_J}{N}=\frac{(2J+1)\exp\left(-\dfrac{hcBJ(J+1)}{k_BT}\right)}{\sum\limits_i g_i\exp\left(-\dfrac{E_i}{k_BT}\right)} \qquad (12.40)$$

where

$$\sum\limits_i g_i\exp\left(-\frac{E_i}{k_BT}\right)=1+3\exp\left(-\frac{2hcB}{k_BT}\right)+5\exp\left(-\frac{6hcB}{k_BT}\right)+...... \qquad (12.41)$$

For large T or small B, the summation in Eq. (12.41) can be replaced by integral

$$\sum\limits_i g_i\exp\left(-\frac{E_i}{k_BT}\right)=Q_r=\int\limits_0^\infty (2J+1)\exp\left(-\frac{hcBJ(J+1)}{k_BT}\right)dJ \qquad (12.42)$$

Putting $J(J+1)=x$, $(2J+1)\,dJ=dx$ and $A=(hcB/k_BT)$ in Eq. (12.42)

$$Q_r=\int\limits_0^\infty \exp(-Ax)dx=\frac{1}{A}=\frac{k_BT}{hcB}$$

$$(12.43)$$

The Eq. (12.40) takes the form

$$N_J=\frac{NhcB(2J+1)}{k_BT}\exp\left(-\frac{hcBJ(J+1)}{k_BT}\right)$$

$$(12.44)$$

The intensity of absorption I_{abs} for a transition from lower state m to an upper state n is the net result of stimulated absorption and stimulated emission. For rotational transitions the energy separation is very small and therefore spontaneous emission can be ignored. Further, population of state n is appreciable as compared to state m. The rate R_{mn} of stimulated absorption of energy is given by

$$dN_m/dt = (N_m/N)B_{nm}\,\rho(\nu_{nm})h\,\nu_{nm} \tag{12.45}$$

$$dN_n/dt = (N_n/N)B_{nm}\,\rho(\nu_{nm})h\,\nu_{nm} \tag{12.46}$$

Net absorption is the difference of the Eqs.(12.45) and (12.46)

$$I_{abs} = \frac{dN_m}{dt} - \frac{dN_n}{dt} = \frac{(N_m - N_n)}{N} B_{nm}\,\rho(\nu_{nm})\,h\nu_{nm} \tag{12.47}$$

From Eq. (12.33)

$$\frac{N_m - N_n}{N_m} = 1 - \exp\!\left(-\frac{h\,\nu_{nm}}{k_B T}\right) \tag{12.48}$$

Substituting Eq. (12.48) in Eq. (12.47)

$$I_{abs} = \frac{N_m}{N} B_{nm}\,\rho(\nu_{nm})\,h\nu_{nm}\left[1 - \exp\!\left(-\frac{h\nu_{nm}}{k_B T}\right)\right] \tag{12.49}$$

For small energy difference between rotational levels or for large T

$$\exp\!\left(-\frac{h\,\nu_{nm}}{k_B T}\right) \cong 1 - \frac{h\,\nu_{nm}}{k_B T}$$

and Eq.(12.49) becomes

$$I_{abs} \cong \frac{N_m}{Nk_B T} B_{nm}\,\rho(\nu_{nm})\,h^2\nu_{nm}^2 \tag{12.50}$$

It can be shown that

$$B_{nm} = \frac{8\pi^3\mu^2(J+1)}{(4\pi\varepsilon_0)3h^2(2J+1)} \tag{12.51}$$

where μ is electric dipole moment. Substituting Eq. (12.51) in Eq. (12.50)

$$I_{abs} \cong \frac{8\pi^3\mu^2\nu_{nm}^2\,\rho(\nu_{nm})N_m(J+1)}{3(4\pi\varepsilon_0)Nk_B T(2J+1)} \tag{12.52}$$

Substituting the value of Nm/N from Eq. (12.44) and putting $J = J''$ for m and $J = J'$ for n in Eq.(12.52)

$$I_{abs} \cong \frac{8\pi^3\, \mu^2 v_{J'J}^2\, \rho(v_{J'J})hcB(J+1)}{3(4\pi\varepsilon_0)(k_B T)^2}\, \exp\left[-\frac{hcBJ(J+1)}{k_B T}\right] \tag{12.53}$$

From Eq. (12.53) it is seen that the absorption strength is proportional to the frequency of transition and square of electric dipole moment, both of which vary with J. Further, intensity of absorption is proportional to J +1 instead of 2J + 1.

12.5 Symmetry Properties of the Rotational Levels

The wavefunction Ψ for a state of molecule is

$$\Psi = \psi_e \psi_n \tag{12.54}$$

where ψ_e is the wavefunction for the electronic state and wavefunction ψ_n describes the stationary state corresponding to the motion of nuclei. If nuclear motion is analysed into rotational and vibrational parts, ψ_n can be written as

$$\psi_n = \psi_v \psi_r \tag{12.55}$$

where ψ_v and ψ_r are respectively the wavefunctions for vibrational and rotational states. From Eqs. (12.54) and (12.55)

$$\Psi = \psi_e \psi_v \psi_r \tag{12.56}$$

Consider the inversion of all the coordinates involved in ψ_e , i.e., x $\rightarrow$ -x, y $\rightarrow$ -y and z $\rightarrow$ -z. Inversion of the electronic part of the wavefunction can be done by rotating the entire molecule by 180^0 about an axis perpendicular to the internuclear axis and then performing a reflection through a plane perpendicular to the rotation axis and passing through internuclear axis. On rotating the molecule by 180^0, the coordinates of the electrons remains unchanged relative to nuclei and therefore electronic wavefunction remain unaltered. The effect of reflection operation depends on the wavefunction describing the state. For most of the diatomic molecules, the ground state wavefunction is positive both above and below any plane passing through the bond axis. The electronic wavefunctions, that are symmetric relative to such planes, are designated by a plus sign and most of the ground state electronic wavefunction are of this type. The ψ_v depends on the magnitude of internuclear distance and hence remains unaffected by inversion operation. The wavefunction ψ_r depends on θ and φ and is represented by $Y_{JM}(\theta, \varphi)$. The polar coordinates equivalent to the inversion implied by x $\rightarrow$ -x, y $\rightarrow$ -y and z $\rightarrow$ -z is obtained by replacing r $\rightarrow$ r, $\theta \rightarrow \pi - \theta$ and $\varphi \rightarrow \pi + \varphi$. On applying this to Y_{11} and Y_{21}

$$Y_{11} = \sqrt{\frac{3}{8\pi}}\sin\theta\,\exp(i\varphi) \xrightarrow[\varphi\rightarrow\pi+\varphi]{\theta\rightarrow\pi-\theta} -\sqrt{\frac{3}{8\pi}}\sin\theta\,\exp(i\varphi) = -Y_{11} \tag{12.57}$$

$$Y_{21} = \sqrt{\frac{15}{8\pi}}\sin\theta\cos\theta\,\exp(i\varphi) \xrightarrow[\varphi\rightarrow\pi+\varphi]{\theta\rightarrow\pi-\theta} \sqrt{\frac{15}{8\pi}}\sin\theta\,\cos\theta\,\exp(i\varphi) = Y_{21} \tag{12.58}$$

In general, for even values of J, the function Y_{JM} remains unchanged by inversion while for odd values of J, Y_{JM} changes sign. According to Eq. (12.56) the sign of function Ψ depends on the sign of ψ_r as ψ_v and ψ_e

remains invariant. Thus a rotational level is called positive or negative depending on whether the Ψ remains unchanged or changes sign for such operator. The rotational levels are positive or negative depending on whether J is even or odd as shown in the Fig.12.6. A positive level combines only with negative and vice versa. Transitions between two positive levels or between two negative levels are forbidden.

12.6 Isotope Effect

The internuclear distance and force constant in diatomic molecules are determined by the electronic structure. For rigid rotator where no vibration occurs, the interatomic distances and force constant are equal in isotopic molecule. Since the masses of isotopes are different , therefore , the corresponding

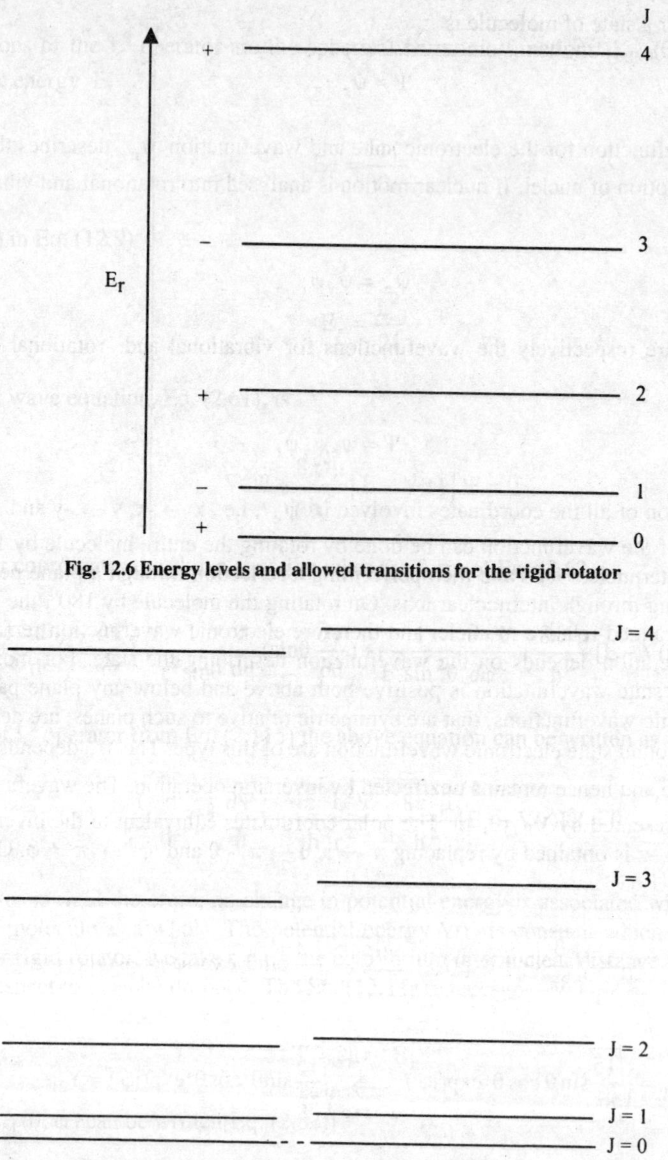

Fig. 12.6 Energy levels and allowed transitions for the rigid rotator

Fig. 12.7 Rotational levels of two isotopic molecules. To the right for lighter isotope

rotational constants are also different. Eq. (12.20) can be written for isotopic molecules as

$$F_{isot} = B_{isot} J(J+1) \tag{12.59}$$

where F_{isot} is rotational term for isotope and B_{isot} is rotational constant of the isotopic molecule. Fig.12.7 show the rotational levels of two isotopic molecules. The ratio of B and B_{isot} from Eqs. (12.21) or (12.22) is

$$\frac{B}{B_{isot}} = \frac{I_{isot}}{I} = \frac{\mu_{isot} r_e^2}{\mu\, r_e^2} = \frac{\mu_{isot}}{\mu} \tag{12.60}$$

where μ_{isot} and I_{isot} are the reduced mass and moment of inertia, respectively of isotopic molecule. For lighter isotope, the rotational constant is large in comparison with the heavier isotope. Therefore, the rotational lines of lighter isotopes are more separated than the heavier ones. By measuring the values of rotational constants, mass of the unknown isotope can be calculated.

12.7 The Nonrigid Rotator

The general features of pure rotational spectra of most molecules can be explained fairly well on the basis of rigid rotator model. However, a quantitative fit of the spectrum requires the refinement of this model. According to the rigid rotator model of the diatomic molecules, the interval between adjacent rotational lines is 2B. However, it is observed that the separation between adjacent lines decreases with increase in J.

In the rigid rotator model it is assumed that the atoms of the molecules are always at a fixed distance between each other. However, the bond between atoms can stretch and contract in a vibrational motion; it cannot be entirely rigid. As a result of flexible bond between the atoms, the rotational motion tends to throw the atoms outward from the centre of mass due to centrifugal forces. Classically, this effect increases with the speed of rotation and quantum mechanically with the value of J. The centrifugal distortion stretches the bond and hence tends to increase the moment of inertia I of the molecule. As rotational constant B is inversely proportional to I, the interval between successive energy levels is thus reduced from the rigid rotator value as the value of J increases.

In the nonrigid rotator model of the molecule it is assumed that the two nuclei of diatomic molecules are connected by a massless spring. It is assumed that motion of two masses can be described by a single particle of reduced mass μ as was done in the rigid rotator model. Consider the particle of mass μ rotating about a fixed point with an angular velocity ω_r and in the absence of rotation, the particle is at a distance r_e from the fixed point. As the particle rotate the length increases to r_c. The centrifugal force acting on the body is $\mu r_c \omega_r^2$. Assuming the vibration of molecule as that of simple harmonic oscillation, this centrifugal force is balanced by the restoring force $k\,(r_c - r_e)$ so that molecule remains bound. The centrifugal force is given by

$$F_c = \mu\, \omega_r^2 r_c = \frac{(I\omega_r)^2}{\mu\, r_c^3} = \frac{L^2}{\mu\, r_c^3} \tag{12.61}$$

where $L = I\omega_r$ is the angular momentum.

Equating the centrifugal and restoring force

$$k(r_c - r_e) = \frac{L^2}{\mu\, r_c^3}$$

$$r_c - r_e = \frac{L^2}{\mu\, k\, r_c^3} \approx \frac{L^2}{\mu\, k\, r_e^3} \tag{12.62}$$

and

$$r_c = r_e\left(1 + \frac{L^2}{\mu\, k\, r_e^4}\right)$$ (12.63)

total energy of the rigid rotator is

$$E_{nr} = \frac{I\omega_r^2}{2} + \frac{k(r_c - r_e)^2}{2} = \frac{L^2}{2\,\mu\, r_c^2} + \frac{k(r_c - r_e)^2}{2}$$ (12.64)

Substituting the value of $r_c - r_e$ and r_c from Eqs. (12.62) and (12.63), respectively in Eq.(12.64) we get

$$E_{nr} = \frac{L^2}{2\mu\, r_e^2} - \frac{L^4}{2\mu^2\, k\, r_e^6} + \ldots\ldots\ldots$$ (12.65)

Replacing L^2 by its quantum mechanical value $(h/2\pi)^2 J(J+1)$ in Eq.(12.65)

$$E_{nr} = \frac{h^2 J(J+1)}{8\pi^2 \mu\, r_e^2} - \frac{h^4 J^2 (J+1)^2}{32\pi^4 \mu^2 k\, r_e^6}$$ (12.66)

Using $\overline{\omega} = \dfrac{1}{2\pi c}\sqrt{\dfrac{k}{\mu}}$ we have $k = 4\pi^2 c^2 \mu\, \overline{\omega}^2$. Substituting this value of k in Eq. (12.66) and dividing E_{nr} by hc we have

$$F(J) = \frac{E_{nr}}{hc} = \frac{hJ(J+1)}{8\pi^2 Ic} - \left(\frac{h}{8\pi^2 Ic}\right)^3 \frac{4J^2(J+1)^2}{\overline{\omega}^2}$$

$$F(J) = BJ(J+1) - \frac{4B^3}{\overline{\omega}^2} J^2(J+1)^2 = BJ(J+1) - DJ^2(J+1)^2$$ (12.67)

where

$$D = \frac{4B^3}{\overline{\omega}^2}$$ (12.68)

D is called centrifugal constant and $\overline{\omega}$ is in wavenumber. D is always positive for diatomic molecules and depends on the vibrational frequency $\overline{\omega}$. The transition wavenumber or frequency is now

$$\overline{v}_r\ (\text{or}\ v_r) = F(J') - F(J'')$$ (12.69)

The selection rule is $\Delta J = \pm 1$. Substituting the value of F (J) from Eq. (12.67) in Eq. (12.69) with J' as J+1 and J" as J

$$\overline{v}_r(v_r) = F(J+1) - F(J) = 2B(J+1) - 4D(J+1)^3$$ (12.70)

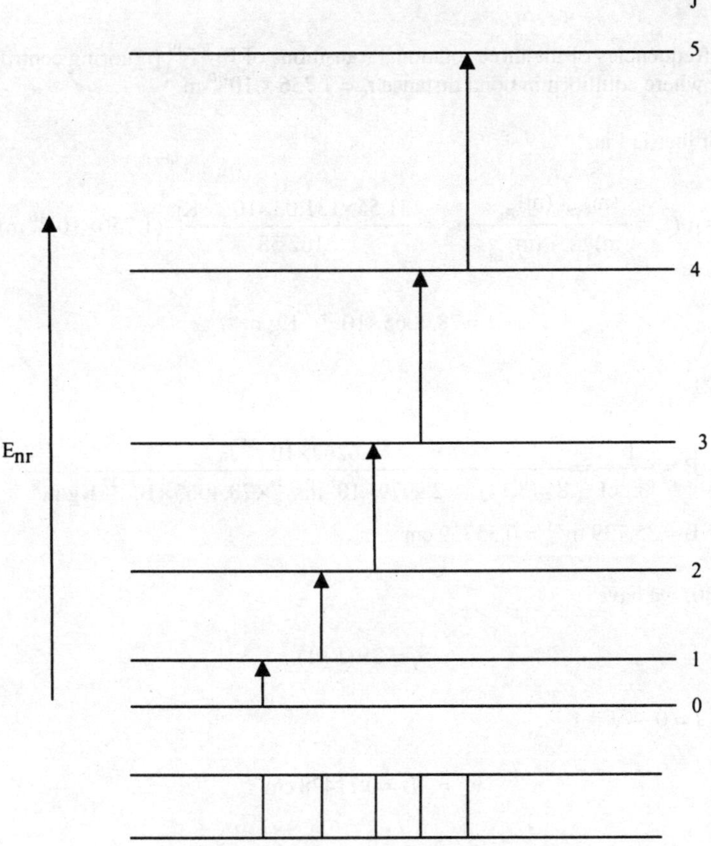

Fig. 12.8 Energy levels and allowed rotational transitions of non-rigid rotator

Fig.12.8 shows the energy levels of a nonrigid rotator along with the allowed transitions. The observed lines are no longer equidistant as for rigid rotator. The separation between rotational lines decreases slightly with increasing J. The effect is very small as B >> D. When very accurate experimental data are used even Eq. (12.67) is not satisfactory. This is due to the fact that we have considered vibrations as that of simple harmonic oscillator and ignored the higher order terms in J (J+1). If we consider vibrations as that of an anharmonic oscillator, the Eq. (12.67) takes the form

$$\bar{v}_r(v_r) = BJ(J+1) - DJ^2(J+1)^2 + HJ^3(J+1)^3 + KJ^4(J+1)^4$$

H, K etc. are small constants and are negligible in comparison with D. From the value of centrifugal constant the value of J of a line can be determined in an observed spectrum. Further, approximate value of vibrational frequency of a diatomic molecule can be determined by Eq. (12.68). The value of $\bar{\omega}$ would be approximate because the calculations assume simple harmonic motion, which is only an approximation to the truth. Further D is small and is not very precisely determined. Dividing large B^3 by small D leads only to approximate value of $\bar{\omega}$.

Examples

1. Calculate the frequencies of the three rotational transitions of $Br^{79}F^{19}$ (ignoring centrifugal distortion) in MHz and cm^{-1} where equilibrium bond distance $r_e = 1.756 \times 10^{-10}$ m.

The moment of inertia I is

$$I = \mu\, r^2 = \frac{(m)_{F^{19}}\, (m)_{Br^{79}}}{(m)_{F^{19}} + (m)_{Br^{79}}}\, r^2 = \frac{31.55 \times 131.03 \times 10^{-27}\ Kg}{162.58} (1.756 \times 10^{-10}\, m)^2$$

$$I = 78.4065 \times 10^{-47}\ Kg\ m^2$$

From Eq. (12.21)

$$B = \frac{h}{8\pi^2 cI} = \frac{6.6262 \times 10^{-34}\ Js}{8 \times (3.14)^2 \times 2.9979 \times 10^8\ ms^{-1} \times 78.4065 \times 10^{-47}\ Kg\ m^2}$$

$$B = 35.739\ m^{-1} = 0.35739\ cm^{-1}$$

From Eq. (12.30) we have

$$\bar{v}_r = 2B(J+1)$$

For transition $J = 0 \rightarrow J = 1$

$$\bar{v}_r = 2B = 0.71478\ cm^{-1}$$

For transition $J = 1 \rightarrow J = 2$

$$\bar{v}_r = 4B = 1.42956\ cm^{-1}$$

For transition $J = 2 \rightarrow J = 3$

$$\bar{v}_r = 6B = 2.14434\ cm^{-1}$$

In terms of MHz

$$v_r = \bar{v}_r c$$

For transition $J = 0 \rightarrow J = 1$,

$$v_r = 0.71478\ cm^{-1} \times 2.9979 \times 10^{10}\ cm/s \approx 21428\ MHz$$

Similarly, $v_r = 42857$ MHz for transition $J = 1 \rightarrow J = 2$ and $v_r = 64285$ MHz for transition $J = 2 \rightarrow J = 3$.

2. Find the value of rotational constant for the molecule $Br^{79}F^{19}$ if the most intense spectral line at 300 K is for the transition $J = 17 \rightarrow J = 18$.

From Eq. (12.38)

$$J_{max} = \sqrt{\frac{k_B T}{2hcB}} - \frac{1}{2}$$

$$17 = \sqrt{\frac{1.3807 \times 10^{-16}\, \text{erg K}^{-1} \times 300\, \text{K}}{2 \times 6.6262 \times 10^{-27}\, \text{erg s} \times 2.9979 \times 10^{10}\, \text{cm s}^{-1}\, B}} - \frac{1}{2}$$

$$B = \frac{3 \times 1.3807 \times 10^{-14}\, \text{erg}}{2 \times 6.6262 \times 10^{-27}\, \text{erg s} \times 2.9979 \times 10^{10}\, \text{cm s}^{-1} \times (17.5)^2} \approx 0.3404\, \text{cm}^{-1}$$

3. The moment of inertia of HBr^{79} is $3.30 \times 10^{-47}\, \text{kg m}^2$. Calculate the energies of first three rotational levels of the molecules in eV and in units of cm^{-1}. Find internuclear distances in atomic units and in angstroms.

The rotational constant B is

$$B = \frac{h}{8\pi^2 cI} = \frac{6.6262 \times 10^{-34}\, \text{Js}}{8 \times (3.14)^2 \times 2.9979 \times 10^8\, \text{ms}^{-1} \times 3.30 \times 10^{-49}\, \text{Kg m}^2} = 849.14 \text{m}^{-1} = 8.4914 \text{cm}^{-1}$$

The energy of the Jth rotational level is given by Eq. (12.17)

$$E_J = \frac{h^2 J(J+1)}{8\pi^2 I} = \frac{(6.6262 \times 10^{-34}\, \text{Js})^2}{8 \times (3.14)^2 \times 3.30 \times 10^{-47}\, \text{Kg m}^2}\, J(J+1) = (1.68677 \times 10^{-22}\, \text{J}) J(J+1)$$

$$E_J = \frac{1.68677 \times 10^{-22}}{1.60219 \times 10^{-19}}\, J(J+1)\, \text{eV} = 1.0527 \times 10^{-3}\, \text{eV}$$

For J = 1, $E_1 = 2 \times 1.0527 \times 10^{-3}\, \text{eV} \approx 2.10 \times 10^{-3}\, \text{eV}$,
For J = 2, $E_2 = 2 \times 3 \times 1.0527 \times 10^{-3}\, \text{eV} \approx 6.316 \times 10^{-3}\, \text{eV}$
For J = 3, $E_3 = 3 \times 4 \times 1.0527 \times 10^{-3}\, \text{eV} \approx 12.63 \times 10^{-3}\, \text{eV}$

$$\bar{v}_J = \frac{E_J}{hc} = \frac{1.686707\, \text{J}}{6.6262 \times 10^{-34}\, \text{Js} \times 2.9979 \times 10^8\, \text{ms}^{-1}}\, J(J+1)$$

$$\bar{v}_J = 8.4909\, J(J+1)\, \text{cm}^{-1}$$

$$\bar{v}_1 = 16.98\, \text{cm}^{-1},\ \bar{v}_2 = 50.94\, \text{cm}^{-1},\ \bar{v}_3 = 101.89\, \text{cm}^{-1}$$

4. The three alternative lines in the measured rotational spectrum of CO are at $7.68991907\, \text{cm}^{-1}$, $15.378662\, \text{cm}^{-1}$ and $23.065043\, \text{cm}^{-1}$. Assign the lines to their appropriate J" to J' transition. Determine the rotational constant.

The separation between alternate lines is 4B. From the average of the distances we obtain B equal to 1.92189 cm^{-1}. From Eq. (12. 30)

$$\bar{\nu}_r = 2B\,(J+1)$$

$$\bar{\nu} = 7.68991907 \text{ cm}^{-1} = 2\times1.92189(J+1)\text{ cm}^{-1}$$

$$J = 1$$

Thus the rotational line corresponding to 7.68991907 cm^{-1} is due to transition $J = 1 \rightarrow J = 2$. Since the lines are alternate therefore, the rotational line at 15.378662 cm^{-1} is due to transition $J = 3 \rightarrow J = 4$ and rotational line at 23.065043 xcm^{-1} is due to transition $J = 5 \rightarrow J = 6$.

5. The three alternative lines in the measured rotational spectrum of a molecule are at 3.845033.9 cm^{-1}, 11.5345096 cm^{-1} and 19.222223 cm^{-1}. Find the value of J, which has maximum population at 300K.

   Since the lines are alternate therefore the spacing between the two lines is 4B. From the difference of these lines we obtain an average value of B equal to 1.92215 cm^{-1}. Substituting this value of B in Eq. (12. 30)

$$J_{max} = \sqrt{\frac{k_B T}{2hcB}} - \frac{1}{2} = \sqrt{\frac{1\cdot3807\times10^{-16}\,\text{erg K}^{-1}\times300\text{K}}{2\times6\cdot6262\times10^{-27}\,\text{erg s}\times2\cdot9979\times10^{10}\,\text{cm s}^{-1}}} - \frac{1}{2}$$

$$J_{max} \approx 6.864 = 7$$

6. The $J = 0 \rightarrow J = 1$ rotational absorption line occurs at 1.153×10^{11} Hz in $C^{12}O^{16}$ and at 1.102×10^{11} Hz in $C^{x}O^{16}$. Find the mass number of unknown carbon isotope.

   We have

$$\bar{\nu}_r = 2B = \frac{2h}{8\pi^2 Ic} = \frac{h}{4\pi^2\mu r^2 c}$$

For $C^{12}O^{16}$

$$(\bar{\nu}_r)_c = \nu_1 = 1\cdot153\times10^{11}\,\text{Hz} = \frac{h}{4\pi^2\mu_1 r^2 c}$$

For $C^{x}O^{16}$

$$(\bar{\nu}_r)_c = \nu_2 = 1\cdot102\times10^{11}\,\text{Hz} = \frac{h}{4\pi^2\mu_2 r^2 c}$$

where μ_1 and μ_2 are the reduced masses of $C^{12}O^{16}$ and $C^{x}O^{16}$ molecules, respectively.

$$\mu_1 = \frac{12\times15.994191}{12+15.994191}\text{ a.u.} \quad \text{and} \quad \mu_2 = \frac{x\times15.994191}{x+15.994191}\text{ a.u.}$$

$$\frac{\nu_1}{\nu_2} = \frac{1\cdot153}{1\cdot102} = 1.0462795 = \frac{\mu_2}{\mu_1} = \frac{x\times(12+15.99491)}{12\times(x+15.99491)}$$

$$x = 13$$

The isotope of carbon is C^{13}.

7. A $Hg^{200}Cl^{35}$ molecule emits a 4.4 cm photon when it undergoes a rotational transition from $J = 1 \rightarrow J = 0$. Find the interatomic distance in this molecule.

Reduced mass of the molecule is

$$\mu = \frac{199.968316\times34.968853}{199.968316+34.968853}\ a.u. = \frac{6992.6626}{234.937169}\ a.u. = 29.763969\ a.u.$$

The wavelength of the transition $\lambda = 4.4$ cm. Thus $2B = \dfrac{1}{4.4}$ cm^{-1} and from Eq. (12.21)

$$B = \frac{1}{8.8}\ cm^{-1} = 11.3635\ m^{-1} = \frac{h}{8\pi^2\mu r^2 c}$$

$$r^2 = \frac{6.6262\times10^{-34}\ J\ s}{8\times(3.14)^2\times11.3635\ m^{-1}\times29.763969\times1.66053\times10^{-27}\ Kg\times2.9979\times10^8\ ms^{-1}}$$

$$r^2 = 5.0193\times10^{-20}\ m^2$$

$$r \approx 0.2240\ nm$$

8. For HCl molecule compute the ratio of the number in the ground state to number in the excited state at 290 K. The rotational constant is 10.44 cm^{-1}.

From Eq. (12.31)

$$\frac{N_{J=1}}{N_{J=0}} = 3\exp\left(-\frac{2\times10.44\times100 m^{-1}\times6.6262\times10^{-34}\ Js\times2.9979\times10^8\ ms^{-1}}{1.3807\times10^{-23}\ JK^{-1}\times290K}\right)$$

$$\frac{N_{J=1}}{N_{J=0}} = 2.7065$$

$$\frac{N_{J=0}}{N_{J=1}} \approx 0.37$$

9.   The rotational constant of HCl is 10.593 cm^{-1} and centrifugal distortion constant is 5.3 x 10^{-4} cm^{-1}. Calculate the vibrational frequency.

From Eq. (12.68)

$$\overline{\omega}^2 = \frac{4B^3}{D} = \frac{4\times(10.593\text{ cm}^{-1})^3}{5.3\times10^{-4}\text{ cm}^{-1}}$$

$$\overline{\omega} = 2995\text{ cm}^{-1}$$

10. For HF molecule the rotational transition corresponding to $J = 4 \rightarrow J = 5$ occur at 204.62 cm^{-1} and $J = 5 \rightarrow J = 6$ occur at 244.93 cm^{-1}. From this calculate the value of rotational and centrifugal constants.

From Eq. (12.70)

$$\overline{\nu}_r(J = 4 \rightarrow J = 5) = 204.62\text{ cm}^{-1} = 10\,B - 500\,D$$

$$\overline{\nu}_r(J = 5 \rightarrow J = 6) = 244.93\text{ cm}^{-1} = 12\,B - 864\,D$$

On solving these two equations

$$B = 20.578\text{ cm}^{-1} \text{ and } D = 2.325\times10^{-3}\text{cm}^{-1}$$

11. Show that for $B = 5$ cm^{-1} and at $T = 1600$ K, the level with $J = 10$ is most populated.

From Eq. (12.38)

$$J_{max} = \sqrt{\frac{k_BT}{2hcB}} - \frac{1}{2} = \sqrt{\frac{1.3807\times10^{-23}\text{ JK}^{-1}\times1600\text{ K}}{2\times6.6262\times10^{-34}\text{ Js}\times2.9979\times10^{8}\text{ ms}^{-1}\times500\text{m}^{-1}}} - \frac{1}{2}$$

$$J_{max} \approx 10.57 - 0.5 \approx 10.07 = 10$$

12. Three consecutive lines are observed at 82.19 cm^{-1}, 123.15 cm^{-1} and 164 cm^{-1}. Obtain the value of the rotational constant and assigned the J values between which transition occur for the rotational line at 164 cm^{-1}.

From Eq. (12.29)

$$82.19\text{ cm}^{-1} = 2B\,(J+1)$$
$$123.15\text{ cm}^{-1} = 2B(J+2)$$
$$164.00\text{ cm}^{-1} = 2B(J+3)$$

From first two equations

$$\frac{82.19}{123.15} = \frac{J+1}{J+2}$$

$$82.19(J+2) = 123.15(J+1)$$

$$J = 1$$

Substituting this value of J in first of the equation we obtain $B = 20.55$ cm^{-1} . The transition corresponding to rotational line at 164 cm^{-1} corresponds to $J = 3 \rightarrow J = 4$.

13. The moment of inertia of HCl molecule is 2.71×10^{-47} Kg-m^2 and rotational constant is 1034 m^{-1}. Calculate the frequency and period of rotation in $J = 1$ and $J = 3$ state.

From Eqs. (12.7) and (12.14)

$$L = I\omega_r = (h/2\pi) \sqrt{J(J+1)}$$

$$\omega_r = 2\pi \, \nu_{rot} = \frac{h}{2\pi \, I} \sqrt{J(J+1)}$$

$$\nu_{rot} = \frac{h}{4\pi^2 I} \sqrt{J(J+1)} = 2Bc\sqrt{J(J+1)}$$

$$\nu_{rot} = 2 \times 1034 \text{ m}^{-1} \times 2.9979 \times 10^8 \text{ ms}^{-1}\sqrt{J(J+1)} \approx 6.1998 \times 10^{11} \text{s}^{-1} \sqrt{J(J+1)}$$

For $J = 1$

$$\nu_{rot} = 6.1998 \times 10^{11} \text{s}^{-1} \sqrt{2} \approx 8.77 \times 10^{11} \text{Hz}$$

$$T_{rot} = \frac{1}{\nu_{rot}} = \frac{1}{8.77 \times 10^{11} \text{Hz}} \approx 1.14 \times 10^{-12} \text{s}.$$

For $J = 3$

$$\nu_{rot} = 6.1998 \times 10^{11} \text{s}^{-1} \sqrt{12} \approx 21.5 \times 10^{11} \text{Hz}$$

$$T_{rot} = \frac{1}{\nu_{rot}} = \frac{1}{21.5 \times 10^{11} \text{Hz}} \approx 4.65 \times 10^{-13} \text{s}.$$

Problems

12.1   Calculate the value of rotational constant for $P^{31}O^{16}$ molecule when $r_e = 0.1402$ nm.

12.2   The rotational constant for $I^{127}Cl^{35}$ is 0.090 cm^{-1}. Calculate the value of r_e.

12.3   The three alternative lines in the measured rotational spectrum of HF are at 82.19 cm^{-1}, 164.0 cm^{-1} and 244.93 cm^{-1}. Assign the lines to their appropriate J" to J' transition. Determine the rotational constant.

12.4   For HCl molecules compute the ratio of number in the $J = 12$ state to the number in the ground state at 300 K. The rotational constant is 10.44 cm^{-1}.

12.5   The rotational constant of HF is 20.561 cm^{-1} and centrifugal distortion constant is 2.13×10^{-3}cm^{-1}. Calculate the vibrational frequency.

12.6   Three consecutive lines in the rotational spectrum of H^1Br^{79} are observed at 84.544 cm^{-1}, 101.355 cm^{-1} and 118.112 cm^{-1}. Assign the lines to their appropriate $J" \rightarrow J'$ transitions, then deduce values for rotational and centrifugal constants hence evaluate bond length and approximate vibrational frequency of the molecule.

$$\boxed{13}$$

Vibrational Spectra of Diatomic Molecules

13.1 Simple Harmonic Oscillator Approximation

In this model it is assumed that atoms of diatomic molecule are vibrating along the direction of the bond. As a result of this there is a periodic lengthening and shortening of the bond length. The two atoms of the molecule with masses m_1 and m_2 can be considered to be joined by a spring of the spring constant k as shown in Fig.12.4. In the absence of external force, the linear momentum of the system is conserved. The oscillations of the atoms therefore cannot affect the motion of the centre of mass. The two atoms vibrate back and forth relative to their centre of mass in opposite direction. According to Hook's law force exerted by the two atoms of a molecule on each other when they are displaced from their equilibrium position, is proportional to the change in the inter-nuclear distance, that is, $k\,(r-r_e)$. For the atom of mass m_1, from Newton's second law we have

$$m_1 \frac{d^2 r_1}{dt^2} = -k(r - r_e) \tag{13.1}$$

For the atom of mass m_2

$$m_2 \frac{d^2 r_2}{dt^2} = -k(r - r_e) \tag{13.2}$$

where r is the distance between the two atoms. Substituting the values of r_1 and r_2 from Eqs. (12.4) and (12.5), respectively into Eqs. (13.1) and (13.2)

$$\frac{m_1 m_2}{m_1 + m_2} \frac{d^2 r}{dt^2} = \mu \frac{d^2 r}{dt^2} = -k(r - r_e) \tag{13.3}$$

where μ is reduced mass . The Eq. (13.3) is identical with the equation of simple harmonic oscillator, that is,

$$m \frac{d^2 x}{dt^2} = -kx$$

except that mass m of the particle is replaced by the reduced mass μ and displacement x by $r-r_e$. The vibrational frequency of the molecule is

$$\nu_{osc} = \frac{1}{2\pi} \sqrt{\frac{k}{\mu}} \tag{13.4}$$

Here only one frequency, whose magnitude depends on the masses of atoms of molecules and force constant, is possible. Since the restoring force is $-k\,(r-r_e)$ therefore the potential energy is given by

$$V = \frac{1}{2}k(r - r_e)^2 = \frac{1}{2}kx^2 \qquad (13.5)$$

where $x = r - r_e$. A plot of V versus $(r- r_e)^2$, which is a parabola, is shown in Fig.13.1

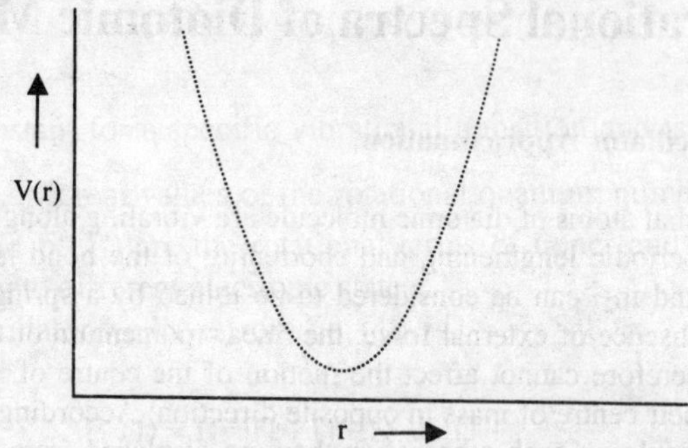

Fig. 13.1 Harmonic potential curve

The Schrödinger wave equation in one dimension for a particle of mass μ is

$$\frac{d^2\psi}{dx^2} + \frac{8\pi^2\mu}{h^2}[E - V]\psi = 0 \qquad (13.6)$$

where E is the energy and V is given by Eq.(13.5). Substituting the value of V in Eq. (13.6)

$$\frac{d^2\psi}{dx^2} + \frac{8\pi^2\mu}{h^2}[E - \frac{1}{2}kx^2]\psi = 0 \qquad (13.7)$$

Let

$$\lambda = \frac{8\pi^2\mu}{h^2}E \qquad (13.8)$$

and

$$\alpha^2 = \frac{4\pi^2\mu k}{h^2} \qquad (13.9)$$

Eq. (13.7) is then written as

$$\frac{d^2\psi}{dx^2} + (\lambda^2 - \alpha^2 x^2)\psi = 0 \qquad (13.10)$$

with the boundary condition $\psi \to 0$ as $|x| \to \infty$. Let us suppose αx to be very large in particular $\alpha x \gg 1$ and $\alpha x \gg \lambda$. Eq. (13.10) becomes

$$\frac{d^2\psi}{dx^2} - \alpha^2 x^2 \psi = 0 \tag{13.11}$$

The equation is satisfied asymptotically by the exponential function

$$\psi = \exp(\pm\alpha x^2/2) \tag{13.12}$$

Of the two asymptotic solutions exp $(-\alpha x^2/2)$ and exp $(\alpha x^2/2)$, the latter is unsatisfactory as a wavefunction, since it tends rapidly to infinity with increasing value of x. Let

$$\psi = \exp(-\alpha x^2/2)f(x) \tag{13.13}$$

The Eq. (13.10) becomes

$$f'' - 2\alpha f' + (\lambda - \alpha)f = 0 \tag{13.14}$$

Let

$$\xi = \sqrt{\alpha}\, x \tag{13.15}$$

and

$$f(x) = H(\xi) \tag{13.16}$$

Using Eqs. (13.15) and (13.16) in Eq.(13.14)

$$H''(\xi) - 2\xi H'(\xi) + \left(\frac{\lambda}{\alpha} - 1\right)H(\xi) = 0 \tag{13.17}$$

The Eq. (13.17) can be solved by assuming a power series of the form

$$H(\xi) = \sum_{s=0}^{\infty} a_s \xi^s \tag{13.18}$$

Substituting Eq. (13.18) in Eq. (13.17)

$$\sum_{s=0}^{\infty} s(s-1)a_s \xi^{s-2} + \sum_{s=0}^{\infty}\left(\frac{\lambda}{\alpha} - 1 - 2s\right)a_s \xi^s = 0 \tag{13.19}$$

In order for Eq. (13.19) to vanish for all values of ξ, that is, for H (ξ) to be a solution of Eq. (13.17) the coefficients of individual power of ξ must vanish separately, that is,

$$1.2a_2 + \left(\frac{\lambda}{\alpha} - 1\right)a_0 = 0$$

$$2.3a_3 + \left(\frac{\lambda}{\alpha} - 1 - 2\right)a_1 = 0 \qquad \text{etc.}$$

In general

$$(s+1)(s+2)a_{s+2} + \left(\frac{\lambda}{\alpha} - 1 - 2s\right)a_s = 0$$

or

$$\frac{a_{s+2}}{a_s} = \frac{\left(2s + 1 - \dfrac{\lambda}{\alpha}\right)}{(s+1)(s+2)} \tag{13.20}$$

This expression is called recursion formula. Using this, a_2, a_3, ... can be calculated in terms of a_0 and a_1, which are arbitrary. If a_0 is set equal to zero only odd powers appear; with $a_1=0$, the series contains only even powers. For large values of s in Eq. (13.20)

$$\frac{a_{s+2}}{a_s} \underset{s\to\infty}{=} \frac{\left(2s + 1 - \dfrac{\lambda}{\alpha}\right)}{(s+1)(s+2)} = \frac{2s}{s^2} = \frac{2}{s} \tag{13.21}$$

It is seen that this behaviour is the same as that of the series for $\exp(\xi^2)$

$$\exp(\xi^2) = 1 + \xi^2 + \frac{\xi^4}{2!} + \cdots\cdots + \frac{\xi^s}{(s/2)!} + \frac{\xi^{s+2}}{[(s/2)+1]!} + \cdots\cdots$$

The ratio of the coefficients of ξ^{s+2} and ξ^s is 2/s for large values of s. The H (ξ) behaves like $\exp(\xi^2)$. The function

$$\psi = \exp(-\alpha x^2/2)f(x) = \exp(-\xi^2/2)H(\xi)$$

behaves like an $\exp(\xi^2/2)$, which increases with increasing value of ξ^2 or increasing value of x , thus making it unacceptable as a wavefunction. Therefore, we choose the values of the energy parameter, which will cause the series for H (ξ) to break off after a finite number of terms. This yields a satisfactory wavefunction, because the negative exponential factor $\exp(-\xi^2/2)$ will cause the function to approach zero for large value of ξ. The value of λ which causes the series to break off after the vth term as is seen from the Eq. (13.21) is

$$\lambda = (2v + 1)\alpha \tag{13.22}$$

$$\frac{8\pi^2 \mu E}{h^2} = (2v+1)\frac{2\pi \mu \omega_{osc}}{h}$$

$$E = \left(v + \frac{1}{2}\right)(h/2\pi)\,\omega_{osc} \tag{13.23}$$

$$E = (v + \frac{1}{2}) h \nu_{osc} \tag{13.24}$$

where $\omega_{osc} = 2\pi\nu_{osc}$. The possible energy levels of the molecule are restricted to the infinite discrete set of values given by the Eq. (13.23) with v = 0, 1, 2,........ .
Thus the wavefunction (eigenfunction) is

$$\psi_v(x) = N_v \exp(-\xi^2/2) H_v(\xi) \tag{13.25}$$

where $H_v(\xi)$ is Hermite polynomial of vth degree in ξ and N_v is a constant which is adjusted so that ψ_v is normalized, so that ψ_v satisfies the equation

$$\int_{-\infty}^{+\infty} \psi_v^* \psi_v \, dx = 1 \tag{13.26}$$

The value of N_v which makes the above condition true, is

$$N_v = \left[\left(\frac{\alpha}{\pi} \right)^{\frac{1}{2}} \frac{1}{2^v \, v!} \right]^{\frac{1}{2}} \tag{13.27}$$

The first few Hermite polynomials are given in Table 13.1

Table 13.1 Hermite polynomials

$H_0(\xi)$	=	1
$H_1(\xi)$	=	2ξ
$H_2(\xi)$	=	$4\xi^2 - 2$
$H_3(\xi)$	=	$8\xi^3 - 12\xi$
$H_4(\xi)$	=	$16\xi^4 - 48\xi^2 + 12$

From Eqs. (13.25), (13.27) and Table 13.1, ψ_v is obtained for various values of v as given in Table 13.2.

Table 13.2 Wavefunctions of simple harmonic oscillator

ψ_0	=	$(\alpha/\pi)^{1/4} \exp(-\xi^2/2)$
ψ_1	=	$(\alpha/4\pi)^{1/4} 2\xi \exp(-\xi^2/2)$
ψ_2	=	$(\alpha/64\pi)^{1/4} (4\xi^2 - 2) \exp(-\xi^2/2)$
ψ_3	=	$(\alpha/2304\pi)^{1/4} (8\xi^3 - 12\xi) \exp(-\xi^2/2)$

According to Eq. (13.24) the energy level diagram consists of a series of equidistant levels. Fig.13.2 shows the simple harmonic oscillator potential energy curve with the energy levels indicated by the horizontal lines for each value of v. The points where this line intersects the curve represent E = V and hence, kinetic energy is zero. These points are called turning points of vibration. At the mid-point of vibrational energy level all the energy of the nuclei is kinetic. For v = 0 level, which has an energy E =

$(1/2)h\nu_{osc}$, is called zero-point level and $(1/2)h\nu_{osc}$ the zero point energy. This is the minimum vibrational energy, that the molecule may have at 0 K, and is the consequence of uncertainty principle.

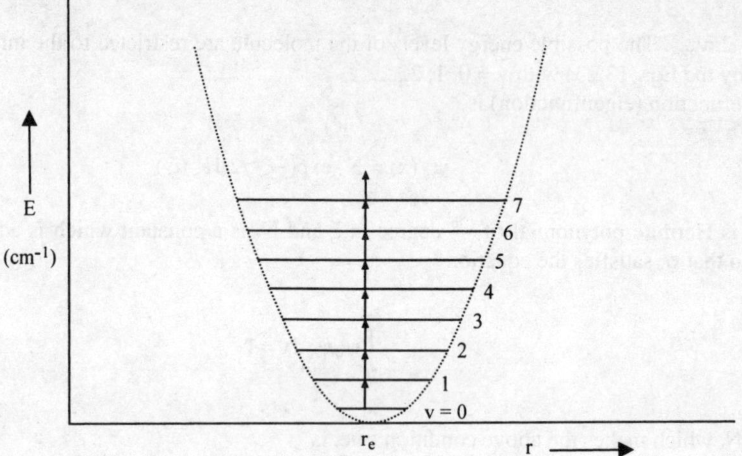

Fig. 13.2 The vibrational energy levels and allowed transitions between them for a diatomic molecule

The vibrational term G (v) from Eq. (13.24) is

$$G(v) = \frac{E(v)}{hc} = \frac{\nu_{osc}}{c}\left(v + \frac{1}{2}\right) = \overline{\omega}\left(v + \frac{1}{2}\right) \qquad (13.28)$$

where ν_{osc}/c is designated by $\overline{\omega}$. $\overline{\omega}$ is the vibrational frequency measured in wavenumbers.

Quantum mechanically, the emission of radiation takes place as a result of a transition of the oscillator from a higher to a lower state and absorption take place by the reverse process. The wave number of emitted or absorbed light is given by

$$\overline{\omega} = G(v') - G(v'') \qquad (13.29)$$

where v' and v" are the vibrational quantum numbers of the upper and lower vibrational states, respectively.

The transition moment for a transition between lower and upper vibrational states with wavefunctions $\psi_{v''}$ and $\psi_{v'}$ of harmonic oscillator, respectively is given by

$$\mathbf{R}_{v'v''} = \int \psi_{v'}^{*}\,\boldsymbol{\mu}\,\psi_{v''}\,dx \qquad (13.30)$$

where $x = r - r_e$, $\boldsymbol{\mu}$ is electric dipole moment. For homonuclear molecules $\boldsymbol{\mu}$ is zero. On the other hand $\boldsymbol{\mu}$ is non-zero for hetronuclear diatomic molecules and varies with $r - r_e$. The variation in $\boldsymbol{\mu}$ can be expressed as

$$\boldsymbol{\mu} = \boldsymbol{\mu}_e + \left(\frac{d\boldsymbol{\mu}}{dx}\right)_e x + \frac{1}{2!}\left(\frac{d^2\boldsymbol{\mu}}{dx^2}\right)_e x^2 + \cdots\cdots \qquad (13.31)$$

Substituting Eq. (13.31) in Eq. (13.30)

$$\mathbf{R}_{v'v''} = \boldsymbol{\mu}_e \int \psi_{v'}^{*}\psi_{v''}\,dx + \left(\frac{d\boldsymbol{\mu}}{dx}\right)_e \int \psi_{v'}^{*}\,x\,\psi_{v''}\,dx + \cdots\cdots \qquad (13.32)$$

The first term in Eq. (13.32) is zero as μ_e is constant and simple harmonic oscillator wavefunctions are orthogonal. The second term is non zero if v' and v" differ by one. This constitutes the vibrational selection rule

$$\Delta v = \pm 1 \qquad (13.33)$$

Eq. (13.29) with the selection rule $\Delta v = \pm 1$ gives

$$G(v+1) - G(v) = \overline{\omega}\,(v+\frac{3}{2}) - \overline{\omega}\,(v+\frac{1}{2}) = \overline{\omega} \qquad (13.34)$$

The frequency of the radiated light is equal to the frequency $\overline{\omega}$ of the oscillator. Since the spacing between the energy levels is equal, therefore in harmonic oscillator approximation all the transitions obeying the selection rule are coincident at a wavenumber $\overline{\omega}$. Each vibrational transition gives rise to a band in the spectrum. The word 'line' is reserved for describing a transition between rotational levels. The intensities of the absorption spectrum decrease rapidly with increasing v". The number of molecules N_v in the v vibrational level relative to v = 0 vibrational level is given by

$$\frac{N_v}{N_0} = \exp\left(-\frac{G_0(v)hc}{k_B T}\right) \qquad (13.35)$$

where N_0 is the number of molecules in v = 0 vibrational state and G_0 (v) is measured with respect to v = 0 level. N_v decreases with increase in v value and hence the intensity also decreases. All bands with v" $\neq$ 0 are called hot bands , because the populations of the lower levels of such transitions and therefore intensities of the transitions are increased at higher temperatures.

13.2 The Anharmonic Oscillator

The vibrational absorption spectrum of a diatomic molecule is dominated by $\Delta v = \pm 1$ transitions called fundamental transitions. But sometimes $\Delta v = \pm 2, \pm 3, \ldots\ldots$ transitions are also observed. These are called overtones and usually the first overtone, $\Delta v = \pm 2$ is stronger than the second overtone, $\Delta v = \pm 3$ which in turn is more intense than third overtone and so on.

In a diatomic molecule, the atoms oscillate about their equilibrium position, at which the internuclear distance is r_e. The potential energy V(r) near equilibrium position can be expressed in Taylor series, with r-r_e = x as

$$V(x) = V(0) + \frac{dV}{dx}\bigg|_0 x + \frac{1}{2}\frac{d^2V}{dx^2}\bigg|_0 x^2 + \frac{1}{3!}\frac{d^3V}{dx^3}\bigg|_0 x^3 + \cdots\cdots \qquad (13.36)$$

The energy at x = 0, that is, V (0) can be taken as zero. The first derivative term is also absent because the slope is zero at the minimum in V (x). The second derivative term $d^2V/dx^2\big|_0 = k$ and

$$\frac{1}{2}\frac{d^2V}{dx^2}\bigg|_0 x^2 = \frac{1}{2}kx^2 \qquad (13.37)$$

where $\dfrac{d^2V}{dx^2}\bigg|_0 = k$ is force constant.

In Sec. 13.1, the vibrational energy levels and eigenfunctions were obtained assuming that potential energy is of the form given by the third term of Eq. (13.36) and ignoring all higher terms. The potential energy curve corresponding to this term is shown in the Fig.13.2. This potential is a good approximation to the true potential for small displacement x and becomes poorer as the potential energy increases. One big failure of the simple harmonic oscillator approximation is that it does not allow dissociation of molecule, because the potential increases to infinity as x tends to infinity.

For large value of r, the diatomic molecule dissociates and neutral atoms are formed. The atoms at large distances do not influence each other ; force constant is zero and correspondingly the potential energy remains constant even if r is increased to infinity. At small values of r, the positive charge on the nuclei causes mutual repulsion which increasingly oppose their approaching each other. This causes potential energy curve to be steeper than for a harmonic oscillator as shown in Fig.13.3. The main features of the potential energy curve is therefore (a) as $r \rightarrow \infty$, V(r) tends asymptotically to a constant value so that $V(\infty) - V(r_e) = D_e$, the dissociation energy, (b) as $r \rightarrow 0$, V(r) goes to infinity, or to a large positive value.

The presence of overtones can be explained on the basis of anharmonicity, that is, deviation from harmonic behaviour. There are two ways by which overtones can appear and they may be called mechanical and electrical anharmonicity. Mechanical anharmonicity is usually referred to mean that V(r) is not a quadratic function but rather a more realistic function discussed above. This anharmonicity leads to decrease in spacing between adjacent vibrational levels as quantum number v increases. In the Taylor series expansion of electric moment [Eq. (13.31)], the second derivative of the dipole moment permits transitions $\Delta v = \pm 2$ and higher order derivatives allow $\Delta v = \pm 3, \pm 4,..$ overtones. This effect is called electric anharmonicity.

A simple modification of the harmonic oscillator potential function of Eq. (13.5) to take account of anharmonicity is to include a term $-g(r - r_e)^3$ to give

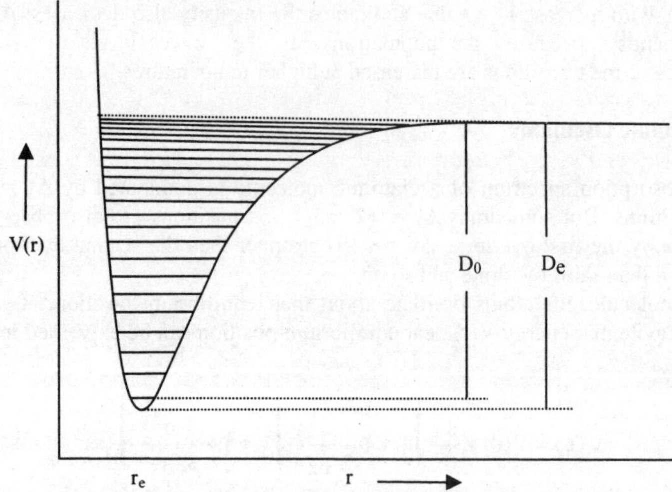

Fig. 13.3 Potential energy function and vibrational energy levels for a diatomic molecule

$$V = f(r - r_e)^2 - g(r - r_e)^3 \qquad (13.38)$$

where f = k /2 and g << f. The effect of the additional term is to steepen the harmonic oscillator curve for r < r_e and make it more shallow for r > r_e. Thus use of this function is limited only to region where r – r_e is small. For a better approximation quartic and higher terms in r – r_e have to be added in Eq. (13.38). Eq. (13.38) for the potential can be used in the Schrödinger Eq. (13.5) to deduce the energy levels of the anharmonic oscillator. The solution is obtained by the perturbation method and leads to energy level expression that can be written as

$$E_v = hc\overline{\omega}_e \left(v+\frac{1}{2}\right) - hc\overline{\omega}_e x_e \left(v+\frac{1}{2}\right)^2 + hc\overline{\omega}_e y_e \left(v+\frac{1}{2}\right)^3 \tag{13.39}$$

where $\overline{\omega}_e$ is the oscillating frequency . The anharmonic correction terms $\overline{\omega}_e x_e$ and $\overline{\omega}_e y_e$ are related to the coefficients g and h. Further, $\overline{\omega}_e \gg \overline{\omega}_e x_e \gg \overline{\omega}_e y_e$. $\overline{\omega}_e y_e$ is generally negligible in comparison with $\overline{\omega}_e x_e$. The vibrational term G (v) is obtained by dividing Eq. (13.39) by hc

$$G(v) = \frac{E_v}{hc} = \overline{\omega}_e \left(v+\frac{1}{2}\right) - \overline{\omega}_e x_e \left(v+\frac{1}{2}\right)^2 + \overline{\omega}_e y_e \left(v+\frac{1}{2}\right)^3 \tag{13.40}$$

The sign of x_e is usually positive, so the result of the anharmonic correction is to bring the vibrational energy levels closer and closer together as v increases. The last term of Eq. (13.40) is small and can be ignored. On comparing Eqs. (13.28) and (13.40), ignoring term in y_e,

$$\overline{\omega} = \overline{\omega}_e \left[1 - x_e \left(v+\frac{1}{2}\right)\right]$$

Thus anharmonic oscillator behaves like the harmonic oscillator but with oscillation frequency that decreases with increasing v. $\overline{\omega} = \overline{\omega}_e$ provided v = -1/2 (a hypothetical case) that is, the molecule would be at the equilibrium point with zero vibrational energy. Thus $\overline{\omega}_e$ is defined as the hypothetical equilibrium oscillation frequency of the anharmonic oscillator.

The zero point energy is obtained by putting v = 0 in Eq. (13.40)

$$G(0) = \frac{\overline{\omega}_e}{2} - \frac{\overline{\omega}_e x_e}{4} + \frac{\overline{\omega}_e y_e}{8} \tag{13.41}$$

which differs slightly from the one obtained for the harmonic oscillator.

The term values relative to G (0), from Eqs. (13.40) and (13.41) are

$$G_0(v) = G(v) - G(0) = \left(\overline{\omega}_e - \overline{\omega}_e x_e + \frac{3}{4}\overline{\omega}_e y_e\right)v - \left(\overline{\omega}_e x_e - \frac{3}{2}\overline{\omega}_e y_e\right)v^2 + \overline{\omega}_e y_e v^3 \tag{13.42}$$

$$G_0(v) = \overline{\omega}_0 v - \overline{\omega}_0 x_0 v^2 + \overline{\omega}_0 y_0 v^3 \tag{13.43}$$

where

$$\overline{\omega}_0 = \overline{\omega}_e - \overline{\omega}_e x_e + \frac{3}{4}\overline{\omega}_e y_e \tag{13.44}$$

$$\overline{\omega}_0 x_0 = \overline{\omega}_e x_e - \frac{3}{2}\overline{\omega}_e y_e \tag{13.45}$$

$$\overline{\omega}_0 y_0 = \overline{\omega}_e y_e \tag{13.46}$$

The separation between successive terms, or first difference is (neglecting cubic term)

$$\Delta G(v+\tfrac{1}{2}) = G_0(v+1) - G_0(v) = \overline{\omega}_0 - \overline{\omega}_0 x_0 - 2\overline{\omega}_0 x_0 v$$

$$= \overline{\omega}_e - 2\overline{\omega}_e x_e - 2\overline{\omega}_e x_e v \tag{13.47}$$

The second difference (neglected cubic term) is

$$\Delta^2 G_{v+1} = \Delta G(v+\tfrac{3}{2}) - \Delta G(v+\tfrac{1}{2})$$

$$= [G_0(v+2) - G_0(v+1)] - [G_0(v+1) - G_0(v)]$$

$$= -2\overline{\omega}_0 x_0 = -2\overline{\omega}_e x_e \tag{13.48}$$

The wavenumber for the transition between vibrational levels (neglecting y_e term) using Eq. (13.43) is

$$\overline{v} = G_0(v') - G_0(v'') = \overline{\omega}_0(v'-v'') - \overline{\omega}_0 x_0(v'^2 - v''^2) \tag{13.49}$$

At normal temperature, before absorption of energy , majority of the molecules are in $v'' = 0$ state. The selection rules for anharmonic oscillator are $\Delta v = \pm 1, \pm 2, \pm 3, \pm 4, \ldots\ldots$. For $v'' = 0$ and $v' = 1, 2, 3, \ldots$

$$\overline{v}_{0\rightarrow 1} = \overline{\omega}_0 - \overline{\omega}_0 x_0 = \overline{\omega}_e - 2\overline{\omega}_e x_e = \overline{\omega}_e(1-2x_e) \tag{13.50}$$

$$\overline{v}_{0\rightarrow 2} = 2\overline{\omega}_0 - 4\overline{\omega}_0 x_0 = 2\overline{\omega}_e - 6\overline{\omega}_e x_e = 2\overline{\omega}_e(1-3x_e) \tag{13.51}$$

$$\overline{v}_{0\rightarrow 3} = 3\overline{\omega}_0 - 9\overline{\omega}_0 x_0 = 3\overline{\omega}_e - 12\overline{\omega}_e x_e = 3\overline{\omega}_e(1-4x_e) \tag{13.52}$$

If the wavenumbers for two transitions are known then x_e, $\overline{\omega}_e$ and $\overline{\omega}_e x_e$ can be determined. At about absolute zero degree, molecules have no rotational energy but are merely vibrating with zero point energy, that is, molecules are in $v = 0$ level. Now the energy required to separate the stable molecule AB, initially in the $v = 0$ level , into unexcited atoms A and B is known as dissociation energy D_0 . It is given by

$$D_0 = [G_0(1) - G_0(0)] + [G_0(2) - G_0(1)] + \ldots\ldots = \sum_v \Delta G_{v+1/2} \tag{13.53}$$

From Fig.13.3

$$D_e = D_0 + G(0) = D_0 + \frac{\overline{\omega}_e}{2} - \frac{\overline{\omega}_e x_e}{4} + \frac{\overline{\omega}_e y_e}{8} \tag{13.54}$$

When anharmonic oscillator receives more energy than D_e , the molecule dissociates and above D_e there is no discrete vibrational level . Thus D_e must be the maximum value of G (v), that is,

$$D_e = G_{max}(v) \tag{13.55}$$

when the molecule dissociates. Using Eq. (13.40) and ignoring small $w_e y_e$ terms we have

$$\frac{d}{dv}[G(v)]_{v=v_c} = \overline{\omega}_e - 2\overline{\omega}_e x_e (v_c + \frac{1}{2}) = 0$$

$$v_c = \frac{1}{2x_e} - \frac{1}{2} \tag{13.56}$$

Substituting the value of v_c from Eq. (13.56) in Eq. (13.40) and ignoring y_e terms,

$$G_{max}(v) = \frac{\overline{\omega}_e}{4x_e} \tag{13.57}$$

From Eq. (13.55)

$$D_e = G_{max}(v) = \frac{\overline{\omega}_e}{4x_e} \tag{13.58}$$

The value of D_e can be determined from the values of $\overline{\omega}_e$ and x_e.

13.3 Potential Energy Functions

A potential energy curve is a graphical representation of the potential energy V(r) versus bond length or the change in potential energy of the molecule as a function of the distortion of the bond of the molecule from its equilibrium distance. V(r) is commonly represented by some empirical function whose validity can be tested by comparing the calculated energy levels with the experimental values. One of the most popular approximations for V(r) is due to P.M.Morse. The Morse function is an empirical potential that has the general appearance of the anharmonic potential for a real molecule. It is given by

$$V(r) = D_e[1 - \exp(-a(r - r_e))]^2 \tag{13.59}$$

where D_e is the dissociation energy and a is a constant for a given molecular electronic state. The function has the following features (i) V(r) goes to zero at r_e (ii) V(r) approaches D_e for large r and (iii) V(r) is very large as r tends to zero. Differentiating Eq. (13.59) with respect to r and putting $r = r_e$,

$$\frac{dV}{dr}\Big|_{r=r_e} = 0, \frac{d^2V}{dr^2}\Big|_{r=r_e} = 2a^2 D_e \tag{13.60}$$

From Eqs.(13.37) and (13.60)

$$\frac{d^2V}{dr^2}\Big|_{r=r_e} = k = 2a^2 D_e \tag{13.61}$$

But $k = \mu(\omega_{osc})^2 = \mu (2\pi v_{osc})^2$, and Eq.(13.61) becomes

$$4\pi^2 v_{osc}^2 \mu = 2a^2 D_e$$

or

$$\nu_{osc} = \frac{a}{2\pi}\sqrt{\frac{2D_e}{\mu}}$$
(13.62)

Using the Morse function the term values are found to be

$$G(v) = a\sqrt{\frac{D_e h}{2\pi^2 \mu c}}(v+\frac{1}{2}) - \frac{ha^2}{8\pi^2 \mu c}(v+\frac{1}{2})^2$$
(13.63)

with

$$a = \sqrt{\frac{2\pi^2 \mu c}{D_e h}}\,\overline{\omega}_e$$
(13.64)

Many other potential functions have been proposed in the literature, the important among them are (i) the Rydberg function (ii) the Lippincolt function (iii) Lippincolt and Schroeder function (iv) Rosen and Morse function (v) Hulburt and Hirshfelder function etc.

13.4 The Vibrating Rotator

Let us consider the case in which diatomic molecule is simultaneously executing both vibrational and rotational motion. When we neglect the interaction between vibration and rotation, the total term value T is given by the sum of rotational term values F (J) [Eq.(12.67)] and vibrational term values G(v) [Eq.(13.40)], ignoring terms containing $\omega_e y_e$,

$$T = G(v) + F(J) = \overline{\omega}_e(v+\frac{1}{2}) - \overline{\omega}_e x_e(v+\frac{1}{2})^2 + BJ(J+1) - DJ^2(J+1)^2$$
(13.65)

The selection rules for v and J are given by Eqs. (13.33) and (12.28), respectively. The correction term because of anharmonic oscillator (that is, $\overline{\omega}_e x_e$) and non-rigid rotator (that is, D) would be very small and ignoring them , the term value is

$$T = \overline{\omega}_e(v+\frac{1}{2}) + BJ(J+1)$$
(13.66)

Using selection rules, we have for $\Delta J = 1$, that is J→J+1 (R branch)

$$\overline{\nu} = \overline{\nu}_0 + 2B(J+1) = R(J)$$
(13.67)

where J = 0, 1, 2, ... and $G(v') - G(v'') = \overline{\nu}_0$
For $\Delta J = -1$, that is, J→J-1 (P branch)

$$\overline{\nu} = \overline{\nu}_0 - 2BJ = P(J)$$
(13.68)

where J = 1, 2, ...

In these expressions the value of J to be inserted is the rotational quantum number of the lower vibrational state. These transitions and the resulting absorption lines are shown in Fig.13.4. There should be a set of absorption lines spaced by a constant amount 2B on the high frequency side of the band centre, which occurs at, $\bar{v}_0$ and a corresponding set on the low frequency side with a similar constant spacing. Furthermore , there should be a gap in the band centre corresponding to the absent

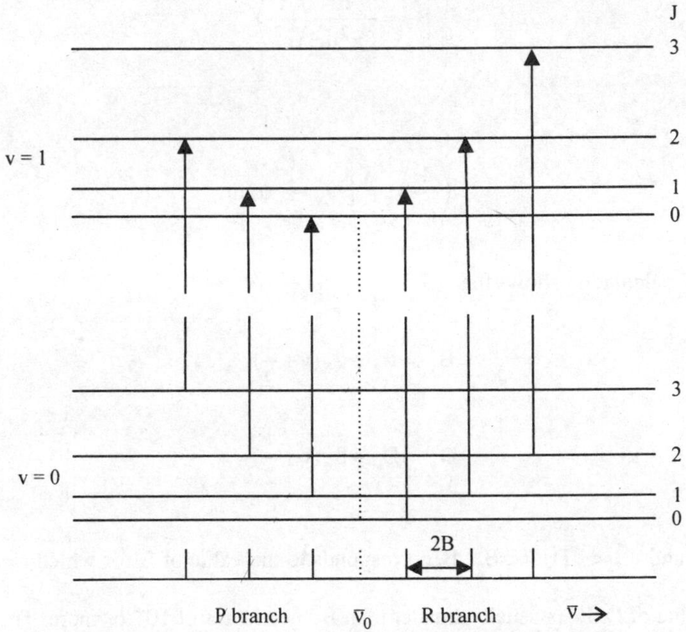

**Fig. 13.4 Rotational - vibrational energy levels using Eq.(13.66) and allowed transitions
between them.The rotational constant is same in both the vibrational states**

$\Delta J = 0$ transitions all of which would have a frequency $\bar{v}_0$. Low frequency set of lines is labelled as P branch of rotational vibrational band and high frequency set of lines as R branch. However, experimentally it is found that spacing between components of P and R branch are not equal.

The energy level expression given by Eq. (13.66) is based on the assumption that the energy of the molecule can be treated in terms of separate contribution from the rotation of the molecule and the vibration of the molecule. The fact that the spectral transitions derived on this basis are not in complete accord with the observed bands, that is, spacing of the components is not constant, can be attributed to coupling between the rotation and vibration of the molecule. During vibration, the internuclear distance r is no longer constant, but changes with time. Period of rotation $\sim 10^{-12}$ sec. And period of vibration is $\sim 10^{-14}$ sec. Hence a molecule can vibrate some hundred times during a rotation. Thus

$$B = \frac{h}{8\pi^2 \mu c r^2}$$

is no longer valid and may be replaced by

$$B = \frac{h}{8\pi^2 \mu c} \left\langle \frac{1}{r^2} \right\rangle \tag{13.69}$$

where $\left\langle \dfrac{1}{r^2} \right\rangle$ is a time average over one vibrational period. The transition to quantum mechanical treatment can in this case be affected by replacing the time average by quantum mechanical average over the wavefunction

$$B_v = \frac{h}{8\pi^2 \mu c} \left\langle \frac{1}{r^2} \right\rangle_v \tag{13.70}$$

where

$$\left\langle \frac{1}{r^2} \right\rangle_v = \int_0^\infty \psi_v^* \frac{1}{r^2} \psi_v \, dr \tag{13.71}$$

Quantum mechanical calculation shows that

$$B_v = B_e - \alpha_e (v + \frac{1}{2}) \tag{13.72}$$

$$D_v = D_e - \beta_e (v + \frac{1}{2}) \tag{13.73}$$

in general $\alpha_e \ll B_e$, and $\beta_e \ll D_e \ll B_e$. B_e corresponds to the value of B for which $r = r_e$ and $D_e = \dfrac{4B_e^3}{\overline{\omega}_e^2}$ is the equilibrium value of D. D_v is usually smaller than B_v by a factor of 10^{-4} or more. The centrifugal force, which is proportional to the square of the angular momentum stretches the bond against the restoring force and the work consumed in this process, must be subtracted from energy calculated in the rigid rotator approximation. The effect upon rotational terms given by Eq. (12.67) may be represented by

$$F_v(J) = B_v J(J+1) - D_v J^2 (J+1)^2 \tag{13.74}$$

B_v implies that rotational term is a function of vibrational quantum number. Since the higher energy vibrational state has a greater vibrational amplitude, it will have a large effective value of r, hence $I_1 > I_0$ and therefore $B_1 < B_0$. The term values of a vibrating rotator in the presence of interaction between rotational and vibrational motion of the molecule can be written as

$$T = G(v) + F_v(J) = \overline{\omega}_e (v + \frac{1}{2}) - \overline{\omega}_e x_e (v + \frac{1}{2})^2 + B_v J(J+1) - D_v J^2 (J+1)^2 \tag{13.75}$$

The energy level diagram, which results from the above equation, is shown in Fig. 13.5. For the lowest vibrational state (v = 0) the rotational constant B_0 is to be used. B_0 is somewhat smaller than B_e which corresponds to the unrealizable completely vibrationless state. This is the first a series of vibrational levels ; the distance between these levels decreases slightly. For each vibrational level v there is a series of rotational levels. The distance between rotational levels increases with quantum number J.

The infrared spectrum consists of a series of bands due to transitions in which vibrational quantum number v changes. Under high resolution, each of these bands is found to be composed of a large number of closely spaced lines, forming the rotational fine structure, which are due to the change in the rotational

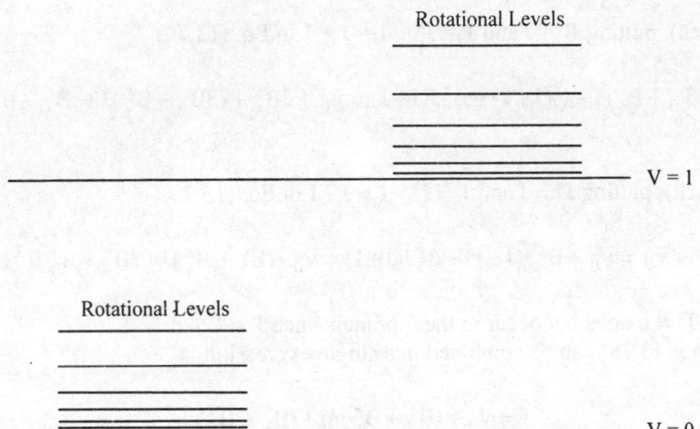

Fig. 13.5 Rotational energy levels for the first two vibrational states of a diatomic molecule

quantum number J. Fig. 13.6 shows the rotational energy levels associated with vibrational level v' (upper) and v" (lower) between which a vibrational transition allowed by selection rule $\Delta v = \pm 1$ .The rotational selection rule governs the transitions between the two sets of levels is $\Delta J = \pm 1$.

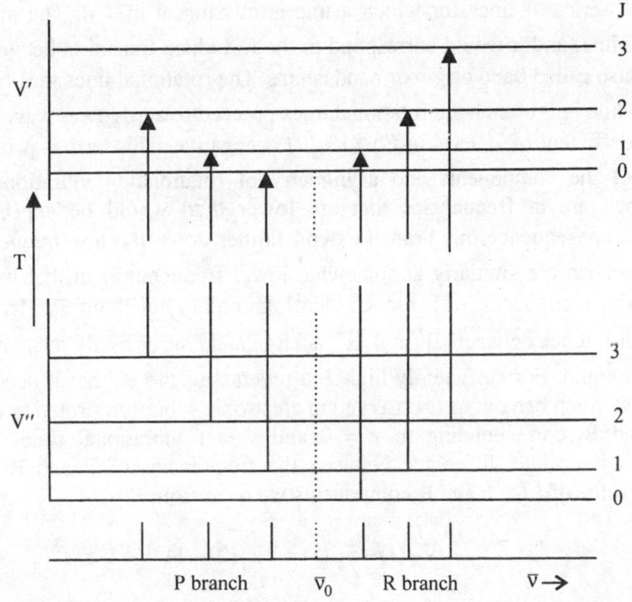

Fig. 13.6 Some transitions between the rotational-vibrational energy levels of a diatomic molecule.The rotational constant is smaller in the upper vibrational state

Consider a transition from v' to v" (neglecting centrifugal distortion, that is, D_v)

$$\overline{v} = T(v') - T(v'') = G(v') - G(v'') + F_v(J') - F_v(J'') = \overline{v}_0 + B'_v J'(J'+1) - B''_v J''(J''+1) \tag{13.76}$$

where $G(v') - G(v'') = \overline{v}_0$ is the frequency of pure vibrational transition without taking account of rotation $(J' = J'' = 0)$

For $\Delta J = 1$ (R branch), putting $J'' = J$ and $J' = J'' + 1 = J + 1$ in Eq. (13.76)

$$\bar{v}_R = \bar{v}_0 + B'_v(J+1)(J+2) - B''_v J(J+1) = \bar{v}_0 + 2B'_v + (3B'_v - B''_v)J + (B'_v - B''_v)J^2 \qquad (13.77)$$

where $J = 0, 1, 2, 3, \ldots\ldots$
For $\Delta J = -1$ (P branch), putting $J'' = J$ and $J' = J'' - 1 = J - 1$ in Eq. (13.76)

$$\bar{v}_P = \bar{v}_0 + B'_v J(J-1) - B''_v J(J+1) = \bar{v}_0 - (B'_v + B''_v)J + (B'_v - B''_v)J^2 \qquad (13.78)$$

with $J = 1, 2, 3,\ldots\ldots$. $J'' = 0$ does not occur in the P branch since $J' = J'' - 1 = -1$.
The Eqs. (13.77) and (13.78) can be combined in a single expression

$$\bar{v} = \bar{v}_0 + (B'_v + B''_v)m + (B'_v - B''_v)m^2 \qquad (13.79)$$

where

$$m = 1, 2, 3, \ldots\ldots \text{ for R branch (that is, } m = J + 1)$$
$$m = -1, -2, -3 ,\ldots\ldots\text{for P branch (that is, } m = -J)$$

Thus we have a single series of lines for which a line is missing at $m = 0$. The missing line at $\bar{v} = \bar{v}_0$ is called zero line (null line) and it would correspond to the forbidden transition between rotational levels $J' = 0$ and $J'' = 0$. $\bar{v}_0$ is also called band origin or band centre. The rotational lines in R branch extend towards higher wavenumbers while in P branch the rotational lines proceed toward lower wavenumbers.

For $B'_v < B''_v$, the coefficient of J^2 term in Eq. (13.79) is negative. This term is primarily responsible for the unequal spacing of the components and asymmetry of rotational – vibrational band .The high J component of P branch are at frequencies that are lower than would be expected on the basis of $B'_v = B''_v = B$ and as a consequence this branch extend farther down the low frequency side. The high J components of the R branch are similarly at somewhat lower frequency than that would be expected for $B'_v = B''_v = B$. The high J components of R branch therefore tend to bunch up. For large Δv (2, 3, 4, .., that is 2-0, 3-0,4-0,..) the difference between B'_v and B''_v is large and consequently the convergence of the lines in the structure is more rapid. For sufficiently high J, a reversal of the R branch occurs. The band is then said to have a band head. Such behaviour is observed in electronic-vibration - rotation band.

The value of B_0 and B_1 corresponding to $v = 0$ and $v = 1$ vibrational states, respectively can be determined by taking appropriate difference between the frequencies of P and R branch components. Starting with the same value of J for P and R components we have from Eqs. (13.77) and (13.78)

$$\bar{v}_R(J) - \bar{v}_P(J) = 2B_1(2J+1) \qquad (13.80)$$

A plot of $\bar{v}_R(J) - \bar{v}_P(J)$ versus $(J+1/2)$ gives the value of $4B_1$. An R component starting from the J level and a P component starting from J+2 level end on the same J level in the upper state, thus from Eqs. (13.77) and (13.78)

$$\bar{v}_R(J) - \bar{v}_P(J+2) = 2B_0(2J+3) \qquad (13.81)$$

A plot of $\bar{v}_R(J) - \bar{v}_P(J+2)$ versus $(J+3/2)$ gives the value of $4B_0$. The Eqs. (13.80) and (13.81) are called combination relations.

Examples

1. In the near infrared spectrum HCl has a single intense band at 2885.9 cm^{-1}. If this represents a vibration spectrum, obtain the vibrational frequency.

    We have

    $$\nu_{osc} = \overline{\omega}c = 2885.9 \text{ cm}^{-1} \times 2.9979 \times 10^{10} \text{ cm s}^{-1} \approx 8.65 \times 10^{13} \text{ Hz}$$

2. The vibrational frequency of H^1Cl^{35} is 8.97 x 10^{13} Hz. Find the force constant for the H-Cl bond.

    The reduced mass of H^1Cl^{35} is

    $$\mu = \frac{m_H m_{cl}}{m_H + m_{cl}} = \frac{1.673 \times 58.06 \times 10^{-27} \text{ Kg}}{1.673 + 58.06} = 1.6261 \times 10^{-27} \text{ Kg}$$

    From Eq. (13.4)

    $$\nu_{osc} = 8.97 \times 10^{13} \text{ Hz} = \frac{1}{2\pi}\sqrt{\frac{k}{\mu}} = \frac{1}{2 \times 3.14}\sqrt{\frac{k}{1.6261 \times 10^{-27} \text{ Kg}}}$$

    $$k = (2 \times 3.14 \times 8.97 \times 10^{13} \text{ Hz})^2 \times 1.6261 \times 10^{-27} \text{ Kg} = 516 \text{ N / m}$$

3. Calculate the fundamental frequency of DCl if the fundamental vibrational frequency of HCl is 2990 cm^{-1} assuming same force constant.

    From Eq. (13.4)

    $$\overline{\omega} = \frac{\nu_{osc}}{c} = \frac{1}{2\pi c}\sqrt{\frac{k}{\mu}}$$

    $$\frac{\overline{\omega}_{DCl}}{\overline{\omega}_{HCl}} = \frac{\overline{\omega}_{DCl}}{2990 \text{ cm}^{-1}} = \sqrt{\frac{\mu_{HCl}}{\mu_{DCl}}} = \sqrt{\frac{(m_D + m_{Cl}) \times m_H}{(m_H + m_{Cl}) \times m_D}} = \sqrt{\frac{(2+35) \times 1}{(1+35) \times 2}} = \sqrt{\frac{37}{72}} = \sqrt{0 \cdot 5139}$$

    $$\frac{\overline{\omega}_{DCl}}{2990 \text{ cm}^{-1}} = 0 \cdot 71687$$

    $$\overline{\omega}_{DCl} = 0.71687 \times 2990 \text{ cm}^{-1} \approx 2143.4 \text{ cm}^{-1}.$$

4. The fundamental vibrational band of Cl_2 molecule is at 2940.8 cm^{-1}. Each atom of the molecule has an atomic weight 35. Determine the corresponding fundamental vibration band of Cl_2 molecule in which one atom has atomic weight 35 and the other 37. What is separation of spectral lines?

    From Eq. (13.4)

$$\overline{\omega}_{osc} = \frac{\nu_{osc}}{c} = \frac{1}{2\pi c}\sqrt{\frac{k}{\mu}}$$

$$\overline{\omega}_{Cl^{35}-Cl^{35}} = \frac{1}{2\pi c}\sqrt{\frac{k}{\mu_{Cl^{35}-Cl^{35}}}} = 2940.8 \text{ cm}^{-1}$$

$$\overline{\omega}_{Cl^{35}-Cl^{37}} = \frac{1}{2\pi c}\sqrt{\frac{k}{\mu_{Cl^{35}-Cl^{37}}}}$$

$$\frac{2940.8 \text{ cm}^{-1}}{\overline{\omega}_{Cl^{35}-Cl^{37}}} = \sqrt{\frac{\mu_{Cl^{35}-Cl^{37}}}{\mu_{Cl^{35}-Cl^{35}}}} = \sqrt{\frac{70\times37}{35\times72}} \approx \sqrt{1.02778} = 1.01379$$

$$\overline{\omega}_{Cl^{35}-Cl^{37}} = \frac{2940.8 \text{ cm}^{-1}}{1.01379} = 2900.8 \text{ cm}^{-1}$$

The separation of the spectral lines is 2940.8 cm^{-1} - 2900.8 cm^{-1} = 40 cm^{-1}.

5. Calculate the ratio of the number of molecules in $v = 1$ vibrational state to $v = 0$ vibrational state at 298K if the spacing between levels is 2×10^{-13} erg/mole.

From Eq. (13.35)

$$\frac{N_1}{N_0} = \exp\left(-\frac{2\times10^{-13}\text{ erg mol}^{-1}}{1.38\times10^{-16}\text{ erg deg}^{-1}\text{ mol}^{-1}\times298 \text{ K}}\right) = \exp(-4.86334) = 7.7\times10^{-3} \approx 0.008$$

Thus less than 1% of the molecules are in the $v = 1$ state.

6. For HCl molecule, the separation between adjacent vibrational states is 2885.9 cm^{-1}. Calculate the ratio of the number of molecules in $v = 1$ vibrational state to $v = 0$ vibrational state at 1000 K.

From Eq. (13.35)

$$\frac{N_{v=1}}{N_{v=0}} = \exp\left(-\frac{G_0(1)hc}{k_BT}\right) = \exp\left(-\frac{2885.9 \text{ cm}^{-1}\times6.6262\times10^{-27}\text{ erg}-\text{s}\times2.9979\times10^{10}\text{ cm s}^{-1}}{1.3807\times10^{-16}\text{ erg deg}^{-1}\text{ mol}^{-1}\times1000 \text{ K}}\right)$$

$$\frac{N_{v=1}}{N_{v=0}} = \exp(-13.847) = 9.72\times10^{-7}$$

7. Find the ratio of HCl molecules in the first excited rotational state to those in the first excited vibrational state at 1000 K. The rotational constant is 1.32×10^{-3} eV and separation between adjacent vibrational levels is 2990 cm^{-1}.

From Eqs. (12.37) and (13.35)

$$\frac{N_{J=1}}{N_{v=1}} = 3\frac{\exp(-E_{J=1}/k_BT)}{\exp(-E_{v=1}/k_BT)} = 3\exp[(E_{v=1}-E_{J=1})/k_BT]$$

$$2990 \text{ cm}^{-1} = (2990/8065) \text{ eV} = 0.370738 \text{ eV}$$

$$\frac{N_{J=1}}{N_{v=1}} = 3\exp\left[\frac{(0\cdot370738-2\times1.32\times10^{-3})\times1.6\times10^{-19}}{1.3807\times10^{-23}\text{ Jmol}^{-1}\text{ deg}^{-1}\times1000\text{K}}\right] = 3\exp\left(\frac{0\cdot368098}{0.086294}\right)$$

$$\frac{N_{J=1}}{N_{v=1}} = 3\exp(4.265642) = 213.6 \approx 214$$

8. The equilibrium vibrational frequency and anharmonicity constant for HI molecule are 2309.5 cm^{-1} and 0.0172 cm^{-1}, respectively. Calculate the fundamental and first overtone transition.

The fundamental transition from Eq. (13.50)

$$\overline{v}_{0\rightarrow1} = \overline{\omega}_e(1-2x_e) = 2309.5\,(1-2\times0.0172)\text{ cm}^{-1} = 2230.06 \text{ cm}^{-1}$$

the first overtone transition is

$$\overline{v}_{0\rightarrow2} = \overline{\omega}_e(1-3x_e) = 2309.5\,(1-3\times0.0172)\text{ cm}^{-1} = 4380.66 \text{ cm}^{-1}$$

9. The fundamental and fist overtones transitions of HBr are centred at 2559.08 cm^{-1} and 5027.54 cm^{-1}, respectively. Evaluate the equilibrium vibrational frequency, the anharmonicity constant, zero point energy and the force constant of the molecule.

From Eqs. (13.50) and (13.51)

$$\frac{\overline{\omega}_e(1-2x_e)}{2\overline{\omega}_e(1-3x_e)} = \frac{2559\cdot08 \text{ cm}^{-1}}{5027\cdot54 \text{ cm}^{-1}}$$

$$5027\cdot54(1-2x_e) = 2559\cdot08\times2(1-3x_e)$$

$$x_e = 0\cdot0171$$

From Eq. (13.50)

$$2559.08 \text{ cm}^{-1} = \overline{\omega}_e(1-2\times0.0171)\text{ cm}^{-1} = 0\cdot9658\,\overline{\omega}_e$$

$$\overline{\omega}_e = 2649.7 \text{ cm}^{-1}$$

Zero point energy from Eq. (13.41), ignoring $\overline{\omega}_e y_e$ term

$$G(0) = \frac{\overline{\omega}_e}{2} - \frac{\overline{\omega}_e x_e}{4} = \frac{2649\cdot7 \text{ cm}^{-1}}{2} - \frac{2649\cdot7\times0\cdot0171 \text{ cm}^{-1}}{4} = 1313.52 \text{ cm}^{-1}$$

Mass of $Br^{79} = 131.03 \times 10^{-27}$ Kg and mass of $H^1 = 1.673 \times 10^{-27}$ Kg. The reduced mass of H^1Br^{79} is

$$\mu = \frac{1 \cdot 673 \times 131 \cdot 03 \times 10^{-27}\,\text{Kg}}{(1 \cdot 673 + 131 \cdot 03)} \approx 1.6519 \times 10^{-27}\,\text{Kg}$$

Force constant from Eq. (13.4)

$$k = (2\pi c \overline{\omega}_e)^2 \mu = (2 \times 3 \cdot 14 \times 2 \cdot 998 \times 10^8\,\text{ms}^{-1} \times 1313 \cdot 52 \times 100\,\text{m}^{-1})^2 \times 1 \cdot 6519 \times 10^{-27}\,\text{Kg}$$

$$k = 411 \cdot 1\,\text{N/m}$$

10. The value of equilibrium frequency and anharmonicity for $O^{16}H^1$ molecule in the ground electronic state are 3735.21 cm^{-1} and 82.81 cm^{-1}, respectively. Find the number of vibrational energy levels below the dissociation limit and the dissociation energy in eV.

From Eq. (13.56)

$$v_{\max} = \frac{1}{2x_e} - \frac{1}{2} = \frac{\overline{\omega}_e}{2\overline{\omega}_e x_e} - \frac{1}{2} = \frac{3735 \cdot 21}{2 \times 82 \cdot 81} - \frac{1}{2} = 22 \cdot 05 = 22$$

From Eq. (13.58)

$$D_e = \frac{\overline{\omega}_e}{4x_e} = \frac{\overline{\omega}_e^2}{4\overline{\omega}_e x_e} = \frac{(3735 \cdot 21)^2}{4 \times 82 \cdot \cdot 81}\,\text{cm}^{-1} = 42119 \cdot 9\,\text{cm}^{-1} = \frac{42119 \cdot 9}{8065}\,\text{eV} = 5 \cdot 22\,\text{eV}$$

$$D_0 = \overline{\omega}_e v - \overline{\omega}_e x_e v(v+1) = 22 \times 3735 \cdot 21\,\text{cm}^{-1} - 82 \cdot 81 \times 22 \times 23\,\text{cm}^{-1}$$

$$D_0 = 82174 \cdot 62\,\text{cm}^{-1} - 41901 \cdot 86\,\text{cm}^{-1} = 40272 \cdot 76\,\text{cm}^{-1} = \frac{40272 \cdot 76}{8065}\,\text{eV} = 4.99\,\text{eV}$$

11. The value of equilibrium frequency and anharmonicity constant for C_2 molecule in the ground electronic state are 1641.4 cm^{-1} and 7.11x10^{-3} cm^{-1}, respectively. Find the number of vibrational energy levels below the dissociation limit and the dissociation energy in kJ.

From Eq. (13.56)

$$v_{\max} = \frac{1}{2x_e} - \frac{1}{2} = \frac{10^3}{2 \times 7 \cdot 11} - \frac{1}{2} = 69 \cdot 82 = 69$$

From Eq. (13.58)

$$D_e = \frac{\overline{\omega}_e}{4x_e} = \frac{1641 \cdot 4}{4 \times 7 \cdot 11 \times 10^{-3}}\,\text{cm}^{-1} = 57706 \cdot 57\,\text{cm}^{-1}$$

$$D_e = 57706 \cdot 57 \times 11.958\,\text{kJmol}^{-1}$$

$$D_0 = 56888 \cdot 8 \text{ cm}^{-1} = 56888 \cdot 8 \times 11 \cdot 958 \text{ Jmol}^{-1} = 680.28 \text{ kJmol}^{-1}$$

12. Spectrum of HCl shows a very intense absorption at 2886 cm^{-1}, a weaker one at 5668 cm^{-1} and a very weak at 8347 cm^{-1}. Calculate the equilibrium frequency of the molecule, anharmonicity constant, force constant, D_e and D_o.

From Eqs. (13.50) and (13.51)

$$\frac{1-2x_e}{2(1-3x_e)} = \frac{2886}{5668}$$

$$5668(1-2x_e) = 2886 \times 2(1-3x_e)$$

$$x_e = 0 \cdot 01739$$

From Eq. (13.50)

$$2886 \text{ cm}^{-1} = \overline{\omega}_e (1 - 2 \times 0.01739)$$

$$\overline{\omega}_e = 2990.05 \text{ cm}^{-1} \approx 2990 \text{ cm}^{-1}$$

From Eq. (13.4)

$$k = (2\pi c\overline{\omega}_e)^2 \mu = (2 \times 3.14 \times 2.998 \times 10^8 \text{ ms}^{-1} \times 2990 \times 100 \text{m}^{-1})^2 \mu$$

$$k = 316 \cdot 902 \times 10^{27} \text{ s}^{-2} \times \mu$$

substituting the value of reduced mass from example 1 in to the above equation

$$k = 316 \cdot 902 \times 10^{27} \text{ s}^{-2} \times 1 \cdot 626 \times 10^{-27} \text{ Kg} = 515.28 \text{ Nm}^{-1}$$

From Eq. (13.58)

$$D_e = \frac{\overline{\omega}_e}{4x_e} = \frac{2990 \text{ cm}^{-1}}{4 \times 0 \cdot 0174} \approx 42960 \text{ cm}^{-1}$$

From Eq. (13.54) ignoring term in $\overline{\omega}_e y_e$

$$D_0 = D_e - \frac{\overline{\omega}_e}{2} + \frac{\overline{\omega}_e x_e}{4}$$

$$D_0 = 42960 \text{ cm}^{-1} - \frac{2990}{2} \text{ cm}^{-1} + \frac{2990 \times 0 \cdot 01739}{4} \text{ cm}^{-1} \approx 41478 \text{ cm}^{-1}$$

13. In the fundamental band of HCl, R (0) = 2906.25 cm^{-1}, P (1) = 2860.09 cm^{-1}. Find the missing line of the band centre and the rotational constant of the molecule.

From Eqs. (13.67) and (13.68)

$$\bar{v}_0 = \frac{R(0)+P(1)}{2} = \frac{2906\cdot25+2860\cdot09}{2} \, cm^{-1} = 2883\cdot17 \, cm^{-1}$$

and

$$B = \frac{R(0)-P(1)}{4} = \frac{2906\cdot25-2860\cdot09}{4} \, cm^{-1} = 11.54 \, cm^{-1}$$

14. In the fundamental band of HCl, the spacing of the lines is found to be constant with a value of 21.2 cm^{-1}. The band is centred on a missing line at 2990 cm^{-1}. Calculate the bond length in nm and the energy of the second excited state of the vibrational motion of the molecule in eV.

The spacing between the lines is 2B. Thus, 2B = 21.2 cm^{-1} or B = 10.6 cm^{-1}. The reduced mass of HCl from example 1 is = 1.626x10^{-27}Kg. From Eq. (12.21)

$$r = \sqrt{\frac{h}{8\pi^2 cB\mu}} = \sqrt{\frac{6\cdot6262\times10^{-34} \, Js}{8(3.14)^2 \times2.9979\times10^8 \, ms^{-1} \times(10.6\times100 \, m^{-1})\times1.626\times10^{-27} \, Kg}}$$

$$r = 0\cdot1274 \, nm$$

From Eq. (13.28)

$$E_v \, (cm^{-1}) = (v+\frac{1}{2})\overline{\omega}_e$$

$$E_{v=2} = (2+\frac{1}{2})\overline{\omega}_e = \frac{5}{2}\overline{\omega}_e = 2\cdot5\times2990 \, cm^{-1} = 7475 \, cm^{-1}$$

$$E_{v=2} = \frac{7475}{8065} \, eV = 0.9268 \, eV$$

15. Assume the following data for H^{1}Cl^{35}. Reduced mass = 1.62626 x 10^{-27} Kg, bond length 0.1275 nm, fundamental frequency = 2991.3 cm^{-1}. Calculate the wavenumber of lines R(2) and P(2).

The moment of inertia of HCl is

$$I = \mu\gamma_e^2 = 2\cdot6437\times10^{-47} \, Kg \, m^2$$

The rotational constant

$$B = \frac{h}{8\pi^2 I c} = \frac{6.6262 \times 10^{-34}\,\text{Js}}{8 \times (3.14)^2 \times 2.6437 \times 10^{-47}\,\text{Kg m}^2 \times 2.998 \times 10^8\,\text{ms}^{-1}} = 10.59\,\text{cm}^{-1}$$

From Eqs.(13.77) and (13.78)

$$R(2) = \overline{\omega}_e + 6B = (2991.3 + 6 \times 10.59)\,\text{cm}^{-1} = 3054.84\,\text{cm}^{-1}$$

$$P(2) = \overline{\omega}_e - 4B = (2991.3 - 4 \times 10.59)\,\text{cm}^{-1} = 2948.9\,\text{cm}^{-1}$$

16. The rotational constant for the $v = 0$ state of the molecule is $10\,\text{cm}^{-1}$ and for $v = 1$ state is $9.5\,\text{cm}^{-1}$. Estimate the rotational constant in the state $v = 2$.

From Eq. (13.72)

$$B_{v=1} = 9.5\,\text{cm}^{-1} = B_e - \alpha_e\left(1 + \frac{1}{2}\right) = B_e - 1.5\alpha_e$$

$$B_{v=0} = 10\,\text{cm}^{-1} = B_e - \alpha_e\left(0 + \frac{1}{2}\right) = B_e - 0.5\alpha_e$$

From these two equations, we obtain $B_e = 10.25\,\text{cm}^{-1}$ and $\alpha_e = 0.5\,\text{cm}^{-1}$. From Eq. (13.72)

$$B_{v=2} = B_e - \alpha_e\left(v + \frac{1}{2}\right) = 10.25\,\text{cm}^{-1} - 0.5\left(2 + \frac{1}{2}\right) = (10.25 - 0.5 \times 2.5)\,\text{cm}^{-1} = 9.0\,\text{cm}^{-1}$$

Problems

13.1   For HCl molecule, the separation between adjacent vibrational states is $2885.9\,\text{cm}^{-1}$. Calculate the ratio of the number of molecules in $v = 1$ vibrational state to $v = 0$ vibrational state at 300 K.

13.2   For I_2 molecule, the separation between adjacent vibrational states is $200\,\text{cm}^{-1}$. Calculate the ratio of the number of molecules in $v = 3$ vibrational state to $v = 0$ vibrational state at 300 K.

13.3   The value of equilibrium frequency and anharmonicity constant for C_2 molecule in an excited electronic state are $1788.2\,\text{cm}^{-1}$ and $9.19 \times 10^{-3}\,\text{cm}^{-1}$, respectively. Find the number of vibrational energy levels below the dissociation limit and also the dissociation energy in kJ.

13.4   The value of equilibrium frequency and anharmonicity for HCl molecule are a $2989.74\,\text{cm}^{-1}$ and $52.05\,\text{cm}^{-1}$, respectively. Find the number of vibrational energy levels below the dissociation limit and the dissociation energy in eV.

13.5   The value of equilibrium frequency and anharmonicity for $O^{16}H^1$ molecule are a $3735.21\,\text{cm}^{-1}$ and $82.81\,\text{cm}^{-1}$, respectively. Find the frequency of the fundamental, first and second overtone of $O^{16}H^1$.

13.6   In the fundamental band of $C^{12}O^{16}$ the spacing of the lines is found to be $3.86\,\text{cm}^{-1}$. The band is centred on the missing line at $2170.21\,\text{cm}^{-1}$. Calculate the rotational constant, the internuclear separation and force constant of the vibrational motion.

13.7   Calculate the approximate frequencies of the second, third and fourth harmonics of NaF in Hz. The equilibrium frequency and anharmonicity are $536.10\,\text{cm}^{-1}$ and $3.83\,\text{cm}^{-1}$, respectively.

13.8   The fundamental and fist overtones transitions of ICl^{35} are centred at $381.28\,\text{cm}^{-1}$ and $759.64\,\text{cm}^{-1}$, respectively. Evaluate the equilibrium vibrational frequency, the anharmonicity constant, zero point energy and the force constant of the molecule.

13.9   The vibrational frequency of H^1Cl^{35} is 2990.6 cm^{-1}. Without calculating the bond force constant, estimate the frequency of D^2Cl^{35}.

13.10   Calculate the value of force constant of the bond in a CO molecule if the spacing between the vibrational levels are 8.45 x 10^{-2} eV.

13.11   The wavenumber of the fundamental vibrational motion of the molecule HBr is 2650 cm^{-1}. calculate the energy of the lowest and first excited states in eV. Also calculate the force constant in S.I. units.

13.12   The molecule of HCl show a strong absorption line of wavelength 3.4662 x 10^{-4} cm. assuming that this is due to vibrational motion, calculate the force constant of the HCl bond.

13.13   The equilibrium vibration frequency of I_2 is 215 cm^{-1} and anharmonicity constant is 0.003. At 300 K calculate the intensity ratio of first overtone and fundamental transition.

13.14   How does the increase in the vibrational energy of a molecule affect its moment of inertia ?

14

Electronic Spectra of Diatomic Molecules

14.1 The Born-Oppenheimer Approximation

A molecule is formed by binding two or more atoms in such a way that the total energy is lower than the sum of the energy of the constituents. The electrons of the inner shell of atoms forming molecules remain localized about each nucleus. On the other hand valence electrons are distributed throughout the molecule and charge distribution of these electrons provide the binding force. The bonds are normally of ionic or covalent nature. The force experienced by the electrons and nuclei is of comparable intensity. Since the nuclei are about two thousand times heavier than the electrons, therefore the motion of the nuclei is much slower than that of electrons. Hence, it can be assumed that nuclei occupy nearly fixed position in the atom.

The Hamiltonian for the molecule can be written as

$$H = T_N + T_e + V_{ee} + V_{eN} + V_{NN} \tag{14.1}$$

where $\ \ T_N\ \ $ = kinetic energy operator of all the nuclei,

$T_e\ \ $ = kinetic energy operator of all the electrons

V_{ee} = potential energy operator for coulombic interaction between electron-electron

V_{eN} = potential energy operator for coulombic attraction of all electrons and all nuclei

V_{NN} = coulombic repulsion between nucleus nucleus

If ψ is an eigenfunction of H then

$$H\Psi(r,R) = E\Psi(r,R) \tag{14.2}$$

The coordinates of the electrons and nuclei cannot be separated out because of term in Coulombic interaction between electrons and nuclei. Born and Oppenheimer, however, were able to show that an approximate solution of Eq. (14.1) can be obtained by first solving Eq. (14.1) for the electrons alone with nuclear positions fixed. The operator T_N is then neglected assuming nuclear mass is infinite, there by making potential V_{NN} constant. The Hamiltonian is then

$$H_E = T_e + V_{ee} + V_{eN}$$

and Schrödinger wave equation is

$$H_E\psi_E(r,R) = E_E(R)\psi_E(r,R) \tag{14.3}$$

where r is electronic coordinate and R is internuclear distance. V_{eN} depends on the position of the nuclei. From Eq. (14.2) $E_e(R)$ can be obtained for different values of internuclear distances. The potential energy can be obtained by adding V_{NN} to $E_e(R)$, that is,

$$V(R) = E_e(R) + V_{NN} = E_e(R) + \frac{Z_a Z_b e^2}{4\pi\varepsilon_0 R} \qquad (14.4)$$

where Z_a and Z_b are the atomic numbers of two nuclei a and b , respectively. The value E_E decreases as the nuclei approaches each other because attraction between electrons and nuclei increases and this is opposed by the repulsion between the two nuclei. If a band is formed there will be a minimum distance at which these two balance and this is called equilibrium bond length. A further decrease in the separation of nuclei causes energy to rise sharply. The electronic state is then a bound state. The curve representing the variation of $E_E + V_{NN}$ is usually referred to as potential curve. The potential energy curve for each electronic state of the molecule has a different shape and in the case of stable states, the minima of potential curve falls at different equilibrium internuclear distances. If the potential curve has no minimum, the electronic state is unstable. In obtaining potential energy curve, we have ignored the nuclear kinetic energy term of Eq.(14.1). Therefore, V(R) does not depend on the mass of the nuclei. If V(R) is found for a given molecule, it applies to all isotopic variation for example V(R) is same for H_2, HD and D_2 and hence also have same bond length within Born - Oppenheimer approximation.

The variation of E_E as a function of R simply provides a part of the potential field in which nuclei move. V_{NN} provides the other part of this field. For the nuclear motion, the Hamiltonian is

$$H_N = T_N + V_{NN} + E_E \qquad (14.5)$$

and Schrödinger wave equation is

$$H_N \psi_N(R) = (T_N + V_{NN} + E_E)\psi_N = E_N \psi_N(R) \qquad (14.6)$$

where ψ_N is a function of R only and represents the stationary state of the nuclei. The total wavefunction is

$$\Psi = \psi_E(r, R)\psi_N(R) \qquad (14.7)$$

The total energy E is given by

$$E = E_E + E_N \qquad (14.8)$$

The wavefunction ψ_N, in a first approximation, can be expressed as

$$\psi_N = \psi_v \psi_r \qquad (14.9)$$

where ψ_v is the vibrational eigenfunction of a linear oscillator. Thus to a first approximation

$$\Psi = \psi_e \psi_v \psi_r \qquad (14.10)$$

The total energy E of the molecule apart from the translational energy may be regarded as the sum of rotational energy E_r, vibrational energy E_v and the electronic energy E_e, that is,

$$E = E_e + E_v + E_r \qquad (14.11)$$

where E_e is defined as the energy of stable electronic state corresponding to the minimum of V. The minimum of lowest electronic state is chosen as zero point of the energy scale. This choice is different from that of atoms. The total energy in terms of wavenumbers is

$$T = \frac{E}{hc} = \frac{E_e}{hc} + \frac{E_v}{hc} + \frac{E_r}{hc} = T_e + G(v) + F(J) \tag{14.12}$$

The electrons move faster than nuclei; thus the nuclei feel the potential energy of the averaged electronic distribution. This forms the basis of Born-Oppenheimer approximation. However, this approximation can breakdown if electronic energy spacings are not large compared to vibrational spacings. This breakdown results in a situation where (i) time scales of nuclear and electronic motions are not separable and (ii) it is not possible to separate nuclear and electronic energies.

14.2 Classification of Electronic States

In diatomic molecules, the electrons move in strong electrostatic field of nuclei which is cylindrically symmetric about the bond axis. The orbital angular momentum **L** precesses very rapidly about the direction of electrostatic field and only axial component of L is a constant of motion. The axial component is characterized by the quantum number M_L where M_L = L, L-1, ..., -L. Since internuclear field is of electrical nature, therefore energy is not changed by the exchange of $M_L \leftrightarrow -M_L$. The absolute value of M_L is designated by the symbol Λ (Fig.14.1) where

$$\Lambda = |M_L| = 0,1,2,3, ...,L \tag{14.13}$$

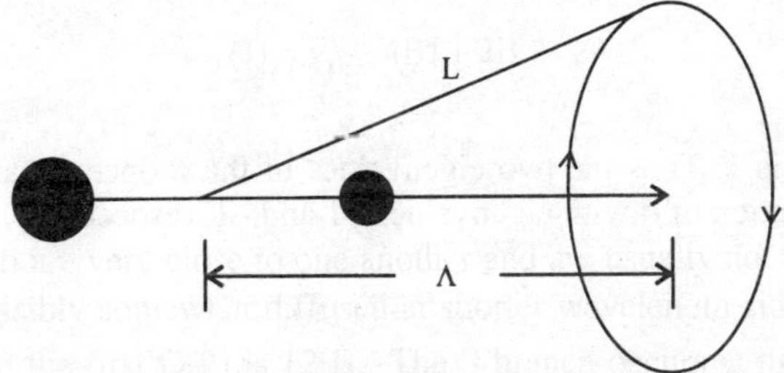

Fig. 14.1 The vector diagram showing the coupling of L about the electric field along the internuclear axis producing the axial component Λ

The symbols for the states are written as follows:

Λ	0,	1,	2,	3,	4,	....
Symbol of state	Σ,	Π,	Δ,	Φ,	Γ,	....

All the states are doubly degenerate except Σ state because of $M_L \leftrightarrow - M_L$ symmetry. The individual electronic spins vectorially add up to produce integral or half integral spin quantum number depending on whether there is even or odd number of electrons. If S is total spin, then multiplicity is 2S+1. The multiplicity of the term is designated by a superscript. The term symbol for a state is written as $^{2S+1}\Lambda$. For $\Lambda > 0$, the orbital motion of the electrons produces a magnetic field along the bond axis and **S** precesses about the magnetic field direction (Fig.14.2). The quantized components M_S of **S**, about the magnetic field direction, have value $\hbar M_S$. The quantum number M_S is designated by the symbol Σ

which can take values S, S – 1, S – 2, ..., – S. Thus, Σ can take 2S+1 values. For the electronic state Σ (Λ=0), there is no resultant magnetic field and therefore, M_S is not defined. These states have only one component, whatever be the multiplicity. Spin-orbit coupling can lift the degeneracy of $^{2S+1}\Lambda$ state. The symbol Ω is used to designate the quantum number of z component of total (spin plus orbit) angular momentum in diatomic molecules if coupling between **L** and **S** is weak. The quantum number Ω is written as a subscript to the term symbol. The revised term symbol is $^{2S+1}\Lambda_\Omega$ with Ω = Λ+S, Λ+S – 1 ,........, Λ-S. For example, a triplet Δ state is split by spin orbit coupling into three doubly degenerate levels.

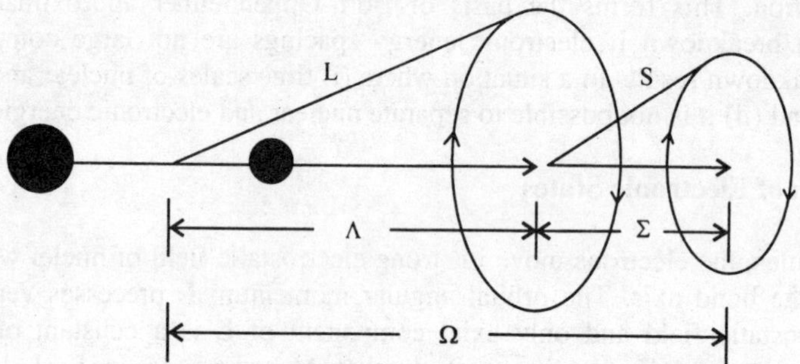

Fig. 14.2 Vector addition of $\Lambda + \Sigma$ producing the total electronic angular momentum Ω

In classification of molecular electronic state, symmetry properties of electronic functions must be considered besides quantum numbers Σ, Λ and Ω. The reflection operator σ acing twice in succession on electronic wavefunction must give the original wavefunction, that is,

$$\sigma^2 \psi_e = (+1)\psi_e \qquad (14.14)$$

The eigenvalue of σ^2 is 1. Thus the two eigenvalues of the σ operator are +1 and –1. If ψ_e^+ and ψ_e^- represent the eigenfunction of σ with eigenvalues +1 and –1, respectively, then

$$\sigma\psi_e^+ = (+1)\psi_e^+ \qquad (14.15)$$

$$\sigma\psi_e^- = (-1)\psi_e^- \qquad (14.16)$$

The function ψ^+ remains unchanged, but ψ^- changes sign under reflection operation. The electronic states which are doubly degenerate ($\Lambda > 0$) may be distinguished as + or -, for example, Δ^+, Δ^- etc. The electronic state Σ is not degenerate but can still be classified as Σ^+ or Σ^- . Homonuclear diatomic molecules also have inversion symmetry through the midpoint of the bond. With the midpoint taken as origin of the cartesian coordinate system and under inversion operation (x_i, y_i, z_i) $\rightarrow$ ($-x_i$, $-y_i$, $-z_i$) of electronic function twice in succession gives the same electronic function then

$$i^2 \psi_e = (+1)\psi_e \qquad (14.17)$$

The eigenvalues of operator **i** are therefore +1 and – 1. The wavefunction either remains unchanged or changed sign under inversion. The wavefunction is called even or g (gerade) if it remains unchanged upon inversion. On the other hand if wavefunction changes its sign upon inversion it is called odd or u (ungerade). The symbols g and u are written as subscripts to the term value, for example, Σ_g, Σ_u , etc.

In considering diatomic molecules we are concerned with cases in which S, Σ, Λ and Ω adequately represent the quantized electronic angular momentum. In an isolated molecule, however, the total angular momentum is due to vectorial coupling of nuclear angular momentum **I**, electronic spin angular momentum **S** and electronic orbital angular momentum **L**. The four important cases for coupling that occur are described as Hund's cases (a) through (d). The most important are (a) and (b)

Hund's case (a):   In this case (Fig.14.3) the interaction between nuclear rotation and electronic angular momentum is assumed to be weak. The spin orbit coupling is also weak. **L** and **S** precess

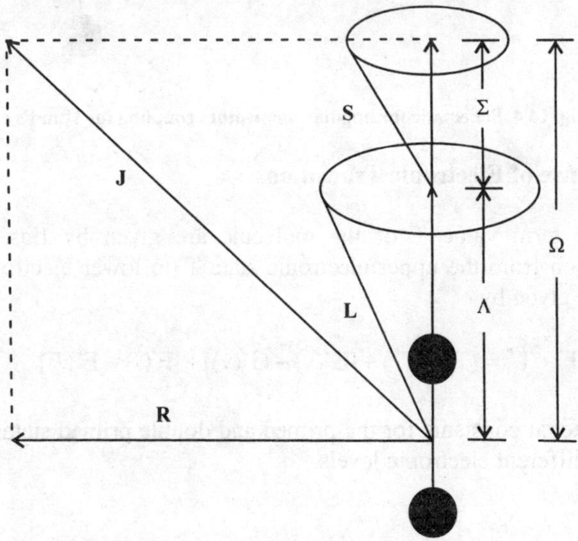

Fig. 14.3 Schematic for angular momentum coupling in Hund's case a

independently about the bond axis in diatomic molecules. Along the bond direction the projections of **L** and **S**, that is, Λ and Σ, respectively add together to give Ω. In this case

$$\mathbf{J} = \mathbf{R} + \Omega \qquad (14.18)$$

with a quantum number $J = R + \Omega$

Hund's case (b):   In this case (Fig.14.4) it is assumed that coupling of S and Λ is weak. **S** couples to the nuclear rotational axis. Here the magnetic field is weak and **S** does not precess about the bond axis. **S** remains fixed in space. Λ is added vectorally to **I** giving a resultant angular momentum **N** and

$$\mathbf{N} = \mathbf{R} + \Lambda \qquad (14.19)$$

and then **N** combined with **S** to give total angular momentum **J**, that is,

$$\mathbf{J} = \mathbf{N} + \mathbf{S} \qquad (14.20)$$

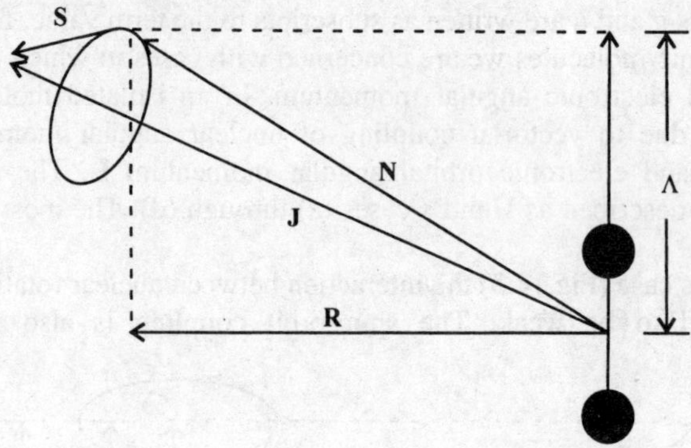

Fig. 14.4 Schematic for angular momentum coupling for Hund's case b

14.3 Vibrational Structure of Electronic Transitions

The total energy E and term value T of the molecule are given by Eqs. (14.11) and (14.12), respectively. In a transition from the upper electronic state T' to lower electronic state T", the wave number of spectral line is given by

$$\bar{\nu} = T' - T'' = (T'_e - T''_e) + [G'(v) - G''(v)] + [F'(J) - F''(J)] \tag{14.21}$$

The vibrational and rotational constants for the primed and double primed states are different because they are associated with different electronic levels.
Putting

$$\bar{\nu}_e = T'_e - T''_e, \quad \bar{\nu}_v = G'(v) - G''(v), \quad \text{and} \quad \bar{\nu}_r = F'(J) - F''(J)$$

in Eq.(14.21)

$$\bar{\nu} = \bar{\nu}_e + \bar{\nu}_v + \bar{\nu}_r \tag{14.22}$$

For a given electronic transition $\bar{\nu}_e$ is constant and corresponds to the energy difference in wavenumbers between the minima of the upper and lower electronic states. Let us consider the case where the rotational structure is ignored. Putting $\bar{\nu}_r = 0$ in Eq. (14.22)

$$\bar{\nu} = \bar{\nu}_e + \bar{\nu}_v = \bar{\nu}_e + G'(v) - G''(v) \tag{14.23}$$

Substituting the values of G' (v) and G"(v) from Eq. (13.40)

$$\bar{\nu} = \bar{\nu}_e + [\bar{\omega}'_e(v'+1/2) - \bar{\omega}'_e x'_e(v'+1/2)^2 + \bar{\omega}'_e y'_e(v'+1/2)^3] - $$
$$- [\bar{\omega}''_e(v''+1/2) - \bar{\omega}''_e x''_e(v''+1/2)^2 + \bar{\omega}''_e y''_e(v''+1/2)^3] \tag{14.24}$$

The spectral location corresponds to $J = 0 \rightarrow J = 0$ rotational transitions. In an electronic transition, there are no strict selection rules on v and it may take any positive or negative integer. For $v = 0 \rightarrow v = 0$ transition, Eq. (14.24) is

$$\bar{v}_{00} = \bar{v}_e + G'(0) - G''(0) \tag{14.25}$$

$$\bar{v}_{00} = \bar{v}_e + \left(\frac{\overline{\omega}'_e}{2} - \frac{\overline{\omega}'_e x'_e}{4} + \frac{\overline{\omega}'_e y'_e}{8} \right) - \left(\frac{\overline{\omega}''_e}{2} - \frac{\overline{\omega}''_e x''_e}{4} + \frac{\overline{\omega}''_e y''_e}{8} \right) \tag{14.26}$$

$\bar{v}_{00}$ is the wavenumber of (v', v''), that is, $(0,0)$ band which is often one of the most intense band in the system. For transition $v' \rightarrow v''$ we have

$$\bar{v} = \bar{v}_e + \left(\frac{\overline{\omega}'_e}{2} - \frac{\overline{\omega}'_e x'_e}{4} + \frac{\overline{\omega}'_e y'_e}{8} \right) - \left(\frac{\overline{\omega}''_e}{2} - \frac{\overline{\omega}''_e x''_e}{4} + \frac{\overline{\omega}''_e y''_e}{8} \right) \; +$$

$$+ \left[\left(\overline{\omega}'_e - \overline{\omega}'_e x'_e + \frac{3}{4}\overline{\omega}'_e y'_e \right) v' - \left(\overline{\omega}'_e x'_e - \frac{3}{2}\overline{\omega}'_e y'_e \right) v'^2 + \overline{\omega}'_e y'_e v'^3 \right] +$$

$$+ \left[\left(\overline{\omega}''_e - \overline{\omega}''_e x''_e + \frac{3}{4}\overline{\omega}''_e y''_e \right) v'' - \left(\overline{\omega}''_e x''_e - \frac{3}{2}\overline{\omega}''_e y''_e \right) v''^2 + \overline{\omega}''_e y''_e v''^3 \right] \tag{14.27}$$

Putting

$$\overline{\omega}_0 = \overline{\omega}_e - \overline{\omega}_e x_e + \frac{3}{4}\overline{\omega}_e y_e$$

$$\overline{\omega}_0 x_0 = \overline{\omega}_e x_e - \frac{3}{2}\overline{\omega}_e y_e \tag{14.28}$$

$$\overline{\omega}_0 y_0 = \overline{\omega}_e y_e$$

The vibrational levels measured relative to the $v = 0$ level in each state, are written by using Eqs.. (14.26) and (14.28) in Eq. (14.27) as

$$\bar{v} = \bar{v}_{00} + [\overline{\omega}'_0 v' - \overline{\omega}'_0 x'_0 v'^2 + \overline{\omega}'_0 y'_0 v'^3] - [\overline{\omega}''_0 v'' - \overline{\omega}''_0 x''_0 v''^2 + \overline{\omega}''_0 y''_0 v''^3]$$

$$\bar{v} = \bar{v}_{00} + G'_0(v') - G''_0(v'') \tag{14.29}$$

where

$$G'_0(v') = \overline{\omega}'_0 v' - \overline{\omega}'_0 x'_0 v'^2 + \overline{\omega}'_0 y'_0 v'^3$$

$$G_0''(v'') = \overline{\omega}_0'' v'' - \overline{\omega}_0'' x_0'' v''^2 + \overline{\omega}_0'' y_0'' v''^3$$

Fig. (14.5) shows a set of vibrational energy levels associated with two electronic states between which electronic transitions are allowed. The vibrational transitions accompanying an electronic transition are called vibronic transitions. The vibronic transitions may be divided into progressions and sequences.

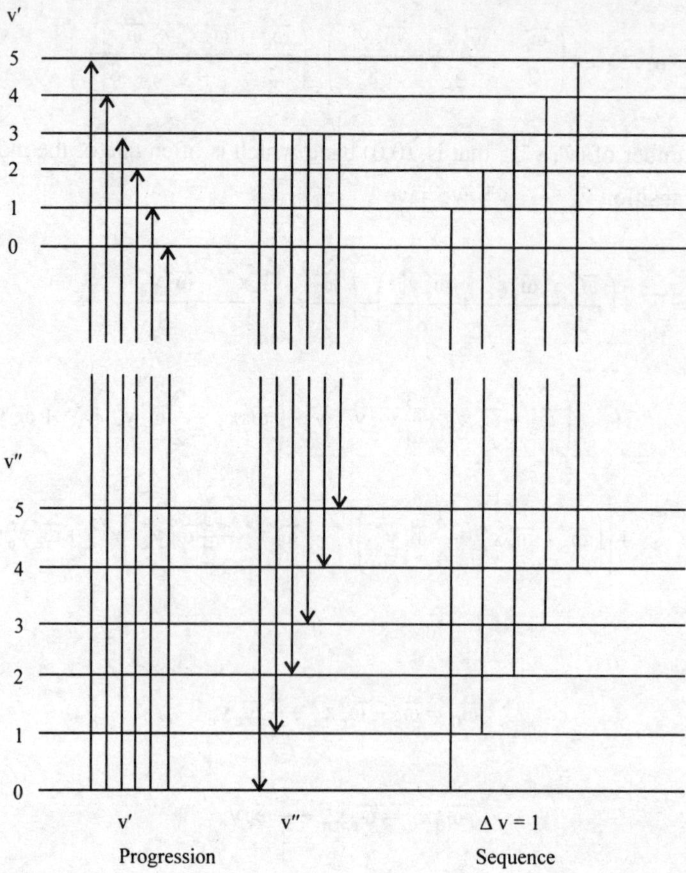

Fig. 14. 5 Vibrational progressions and sequence in the electronic spectrum of a diatomic molecule

A progression involves a series of vibronic transitions with a common lower or upper level. For example, the v'' progression members all have $v' = 3$ level common and $v' = 0$ progression members all have $v'' = 0$ level common ($v' = 0,1,2,3,......$) . A v'' progression extends towards the lower wavenumber as v'' increases and terminates in a continuum where the lower electronic state dissociates. A v' progression extends towards higher wavenumber as v' increases and terminates in a continuum where the upper electronic state dissociates. A group of transitions with the same value of Δv is referred to as a sequence (Fig.14.5). Because of population requirements long sequences are observed mostly in emission. The progressions and sequences are not mutually exclusive. Each member of a sequence is also a member of two progressions. The members of the progressions are generally widely spaced with approximate separation of $\overline{\omega}_e''$ in emission and $\overline{\omega}_e'$ in absorption.

Because $\overline{\omega}_e''$ and $\overline{\omega}_e'$ have slightly different values for the combining states, therefore members of the sequence are closely spaced with approximate separation equal to $\overline{\omega}_e'' - \overline{\omega}_e'$. The bands in each sequence are generally found to be grouped together and overlap each other partially in the spectrum.

In order to analyse the components of vibrational electronic transitions, the wavenumber of each of the (v' , v") band heads rather vibronic band origins are arranged in what is known as Deslandres table (Table 14.1). A band origin corresponds to the position of the hypothetical $J = 0 \rightarrow J = 0$ transition between two states in which there is no rotation. If either a rotational analysis has not been carried out or resolution is too low for rotation fine structure to be resolved, band head measurement has to be used and a constant head-origin separation is assumed for all bands. These are arranged in such a way

Table 14.1 Deslandres table

v" $\rightarrow$				First Diff-erence
v' $\downarrow$	0	1	2	
0	$\overline{v}_{00} + G_0'(0) - G_0''(0)$	$\overline{v}_{00} + G_0'(0) - G_0''(1)$	$\overline{v}_{00} + G_0'(0) - G_0''(2)$	
				$\Delta\,G'(1/2)$
1	$\overline{v}_{00} + G_0'(1) - G_0''(0)$	$\overline{v}_{00} + G_0'(1) - G_0''(1)$	$\overline{v}_{00} + G_0'(1) - G_0''(2)$	
				$\Delta\,G'(3/2)$
2	$\overline{v}_{00} + G_0'(2) - G_0''(0)$	$\overline{v}_{00} + G_0'(2) - G_0''(1)$	$\overline{v}_{00} + G_0'(2) - G_0''(2)$	
	$\Delta G''(1/2)$	$\Delta\,G''(3/2)$	$\uparrow$ $\leftarrow$ First Diff-erence	

that the difference in the frequencies (or wavenumbers) in adjacent columns is approximately constant and varies uniformly and the difference in the wavenumbers in adjacent rows is likewise approximately constant and varies uniformly. The column in the table are labelled with the vibrational quantum number v" of the lower electronic state and rows with the quantum number v' of the upper state. Each wavenumber in the table can then be identified as a transition between v' value of the row it occupies and v" value of the column. The difference between the rows of the Deslandre table gives the spacing of the vibrational levels in the upper electronic state while the difference between the values in the column gives vibrational spacing in the lower electronic state.

The first difference $\Delta G'(1/2)$ value is obtained by subtracting v' = 0 value from v' = 1 value vertically below it in the table. On application of this to table several values of $\Delta G'(1/2)$ would result. An arithmetic mean of the resulting $\Delta G'(1/2)$ value would be regarded as $\Delta G'(1/2)$. By a similar procedure the value of $\Delta G'(3/2)$, and $\Delta G'(5/2)$ etc. would follow. The value of $\Delta G''(1/2)$ would be obtained by subtracting adjacent horizontal numbers in the v"= 1 column from the ones in v" = 0 column. An arithmetic mean of the resulting $\Delta G''(1/2)$ value would then be regarded as $\Delta G''(1/2)$ values. Similarly $\Delta^2 G''(3/2)$, $\Delta^2 G''(5/2)$, etc. are obtained. The arithmetic mean value is obtained from the values which do not deviate much. Eq. (13.48) can be used to determine the values of $\overline{\omega}_e, \overline{\omega}_e x_e$ etc. for both the states.

14.4 Rotational Structure of Electronic Transitions

The possible changes in the rotational state for any given vibrational transition from Eq. (14.22) is

$$\bar{\nu} = \bar{\nu}_0 + \bar{\nu}_r = \bar{\nu}_0 + F'(J') - F''(J'') \tag{14.30}$$

where $\bar{\nu}_0 = \bar{\nu}_e + \bar{\nu}_v$ is constant for a specific vibrational transition and is called the band origin or zero line. $\bar{\nu}_r$ depends on the different values of the rotational quantum number in the upper and lower state, respectively. F' (J') and F''(J'') are the rotational terms of upper and lower state, respectively. Here F' (J') and F''(J'') belong to different electronic states.

   For non rigid vibrating oscillator we have from Eq. (13.74)

$$F_v(J) = B_v J(J+1) - D_v J^2 (J+1)^2$$

The Eq. (14.30) is therefore,

$$\bar{\nu} = \bar{\nu}_0 + [B'_v J'(J'+1) - D'_v J'^2 (J'+1)^2 -$$

$$- [B''_v J''(J''+1) - D''_v J''^2 (J''+1)^2] \tag{14.31}$$

The selection rules for the rotational transitions are

$$\Delta J = J' - J'' = 0, \pm 1 \tag{14.32}$$

$\Delta J = 0$ is forbidden if both the electronic states have $\Lambda = 0$ (for example $^1\Sigma - {}^1\Sigma$ transition) . For allowed transitions we have three series of lines or branches. These branches are designated as R, Q and P corresponding to $\Delta J = +1$, 0 and -1, respectively.

For R branch ($\Delta J = 1$), Eq. (14.31) is

$$\bar{\nu}_R = \bar{\nu}_0 + [B'_v(J+1)(J+2) - D'_v(J+1)^2(J+2)^2] -$$

$$- [B''_v J(J+1) - D''_v J^2 (J+1)^2] \tag{14.33}$$

Neglecting small term in D_v, the Eq.(14.33) becomes

$$\bar{\nu}_R = \bar{\nu}_0 + 2B'_v + (3 B'_v - B''_v)J + (B'_v - B''_v)J^2 = R(J) \tag{14.34}$$

For Q branch ($\Delta J = 0$), Eq. (14.31) is therefore,

$$\bar{\nu}_Q = \bar{\nu}_0 + [B'_v J(J+1) - D'_v J^2 (J+1)^2] -$$

$$-[B_v'' J(J+1) - D_v'' J^2 (J+1)^2] \qquad (14.35)$$

neglecting small term in D_v

$$\bar{\nu}_Q = \bar{\nu}_0 + (B_v' - B_v'')J + (B_v' - B_v'')J^2 = Q(J) \qquad (14.36)$$

For P branch ($\Delta J = -1$), Eq. (14.31) is

$$\bar{\nu}_P = \bar{\nu}_0 + [B_v' J(J-1) - D_v' J^2 (J-1)^2] -$$

$$-[B_v'' J(J+1) - D_v'' J^2 (J+1)^2] \qquad (14.37)$$

Neglecting small term in D_v

$$\bar{\nu}_P = \bar{\nu}_0 - (B_v' + B_v'')J + (B_v' - B_v'')J^2 = P(J) \qquad (14.38)$$

P and R branches can be represented by a single expression

$$\bar{\nu}_{P,R} = \bar{\nu}_0 + (B_v' + B_v'')m + (B_v' - B_v'')m^2 \qquad (14.39)$$

and Q branch can be represented by

$$\bar{\nu}_Q = \bar{\nu}_0 + (B_v' - B_v'')q + (B_v' - B_v'')q^2 \qquad (14.40)$$

where

$$m = -J \text{ for P branch with } J = 1, 2, \ldots\ldots\ldots\ldots \qquad (14.41)$$

$$m = J+1 \text{ for R branch with } J = 0, 1, 2, \ldots\ldots\ldots \qquad (14.42)$$

$$q = J \text{ for } J = 0, 1, 2, \ldots\ldots \qquad (14.43)$$

In the absence of Q branch, there is one simple series of lines in which the separation between lines changes regularly. Fig.14.6 shows all the three branches on an energy level diagram for $^1\Pi - {}^1\Sigma$ transition. For singlet Π state, the possible values of J are Λ, $\Lambda +1$, $\Lambda + 2,\ldots\ldots\ldots$, that is, 1, 2, 3, ... For singlet Σ state, J values are Λ, $\Lambda +1$, $\Lambda + 2,\ldots\ldots\ldots$, that is, 0,1, 2,. As a result of this the lowest level of upper state corresponds to $J = 1$ while for the lower state it is $J = 0$. The first lines in the R, Q and P branches are those having $J = 0$, 1, and 2, respectively. The P branch start from $J = 2$, and there is no transition corresponding to $J = 0$ and 1. Substituting these values of J in Eq. (14.38) shows that there are missing lines at $\bar{\nu} = \bar{\nu}_0$ and $\bar{\nu} = \bar{\nu}_0 - 2B_v''$. The first line of Q branch (corresponding to $J = 1$) lies at [Eq. (14.36)]

$$\bar{\nu} = \bar{\nu}_0 + 2(B_v' - B_v'')$$

Since B_v' and B_v'' are not very different and hence Q branch starts in the neighbourhood of $\bar{\nu}_0$ and therefore gap at $\bar{\nu} = \bar{\nu}_0$ is not so apparent. P and R branches are represented by Eqs. (14.38) and (14.34). Plots of J against frequency of the rotational lines are called Fortrat diagram. A typical example of a Fortrat parabola is shown in Fig.14.7. Due to the quadratic term in the Eq. (14.39) one of the two branches turn back, thus forming a band head.

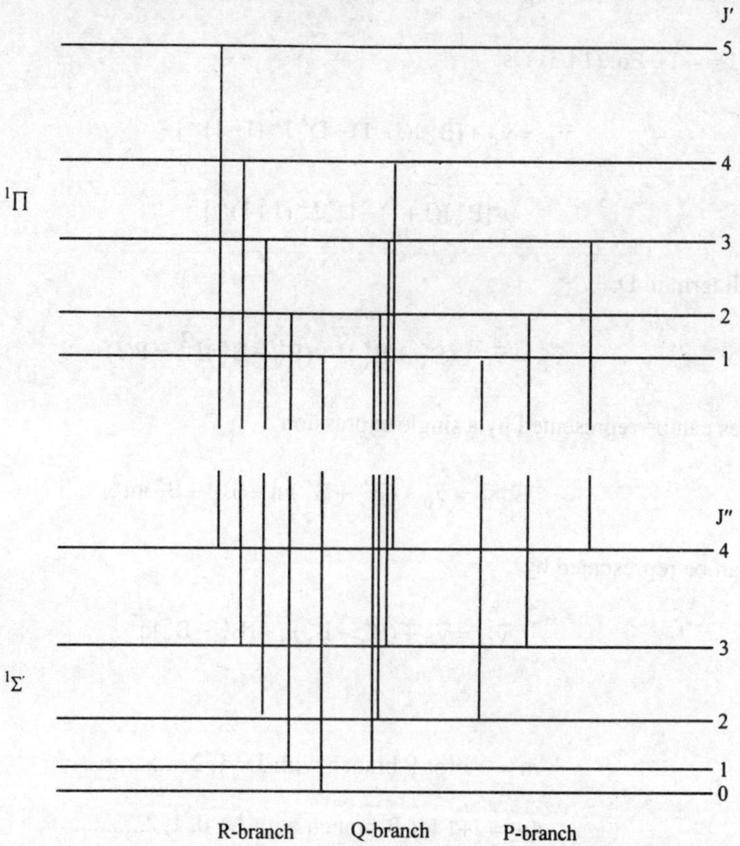

Fig. 14.6 Rotational fine structure of a $^1\Pi$—$^1\Sigma$ vibronic transition in a diatomic molecule for which $r'_e > r''_e$. The Λ - type doubling in $^1\Pi$ state is not shown

Let $B_v' < B_v''$. Since the linear and quadratic term in Eq. (14.39) have opposite sign for R branch $(m = J +1)$ therefore the separation between lines goes on decreasing as J value increases. For a certain J value, the quadratic term contribution exceeds the linear term contribution and the lines of the R branch turn back towards lower value of $\bar{\nu}$ or higher value of wavelength as value of J is further increased. Therefore, in this case the band head lies on the short wavelength side of the zero line and the band is shaded (degraded) towards the red (towards the larger wavelengths). $B_v' < B_v''$ indicates that the internuclear distance in vibrational state of upper electronic state is greater than that of vibrational state of the lower electronic state.

For P branch $(m = -J)$ the linear and quadratic terms in Eq. (14.39) have same negative sign. Therefore, $\bar{\nu}_P$ decreases as the value of J increases. For P branch the spacing between lines increases at longer wavelength side and no band head is formed. For Q branch $(q = J)$ and both linear and

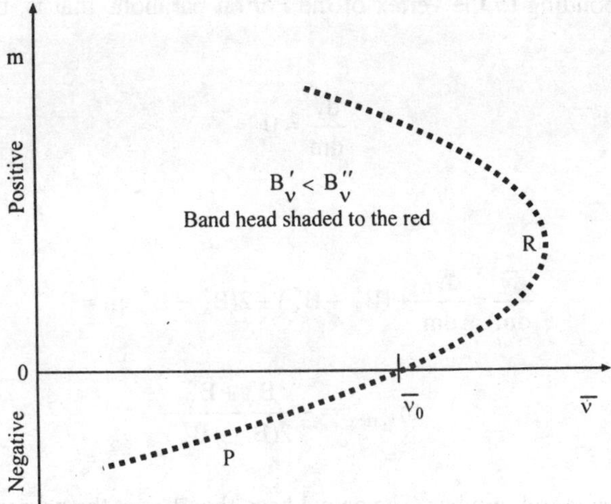

Fig. 14.7 Fortrat parabola

quadratic term of Eq. (14.40) have negative sign. As a result of this, the Q branch extends to lower wavenumber or higher wavelength side of $\overline{v}_0$ as the J value increases. If $B_v' - B_v''$ is very small, lines in Q branch may not be separately resolved for small value of J. Hence a single broad line may appear at $\overline{v}_0$ which is shaded towards the red.

Consider the case $B_v' > B_v''$. This indicates that internuclear distance in upper electronic state is smaller than that of lower electronic state. The coefficient $B_v' - B_v''$ is positive. For R branch the linear and quadratic terms of Eq. (14.39) have positive contribution. Thus $\overline{v}_R$ goes on increasing as the J value increases. The separation between lines goes on increasing on the shorter wavelength side.

For P branch, the linear term gives negative contribution to $\overline{v}_P$ while quadratic term gives positive contribution for a J value. For small J, the linear term is prominent and therefore, $\overline{v}_P$ shifts towards lower wavenumber side. However, for large J the contribution of quadratic term to $\overline{v}_P$ becomes important and as a result of the competition between linear and quadratic terms, the spacing between lines decreases while $\overline{v}_P$ also decreases. After a particular value of J, $\overline{v}_P$ stops decreasing and starts increasing. For this J value the quadratic term contribution exceeds the linear term contribution and lines of P branch turn towards higher value of $\overline{v}_P$ or lower value of wavelengths as J value is further increased. Therefore, in this case the band head lies on the longer wavelength side of the zero line and band is shaded towards violet. In Q branch, the linear and quadratic terms have positive contribution and therefore, the spacing between the lines goes on increasing with J towards higher side of $\overline{v}_0$.

If $B_v' - B_v''$ is small the band head in P and R lies at a great distance from the zero line. If $B_v' \approx B_v''$ the head in P and R branch may lie at such a large distance from the band origin that it is not observed since for the corresponding m values the intensity of lines may have decreased to zero. For Q branch the linear term is small for $B_v' \approx B_v''$ and Q branch parabola intersect the abscissa axis at about right angle. A large difference in B_v' and B_v'' causes the spreading out the Q branch components and thereby prevents the sharp Q branch. The bands having a Q branch often show two heads either in P and Q branch ($B_v' > B_v''$) shaded to violet or in R and Q branch ($B_v' < B_v''$) shaded to red.

The m value corresponding to the vertex of the Fortrat parabola, that is, band head is obtained by setting

$$\frac{d\bar{v}}{dm} = 0 \tag{14.44}$$

or

$$\frac{d\bar{v}}{dm} = \frac{d\bar{v}_0}{dm} + (B'_v + B''_v) + 2(B'_v - B''_v)m = 0$$

$$m_{vertex} = -\frac{B'_v + B''_v}{2(B'_v - B''_v)} \tag{14.45}$$

usually m_{vertex} is not an integral number. The actual head then lies at the nearest whole numbered value of m. For R branch $m = J + 1$ and Eq. (14.45) is

$$(J+1)_{vertex} = -\frac{B'_v + B''_v}{2(B'_v - B''_v)}$$

$$J_{vertex} = \frac{B''_v - 3B'_v}{2(B'_v - B''_v)} \tag{14.46}$$

For P branch, $m = -J$ and Eq. (14.45) gives

$$J_{vertex} = \frac{B'_v + B''_v}{2(B'_v - B''_v)} \tag{14.47}$$

Substituting the value of m_{vertex} from Eq.(14.45) in Eq.(14.39)

$$\bar{v}_{vertex} - \bar{v}_0 = -\frac{(B'_v + B''_v)^2}{4(B'_v - B''_v)} \tag{14.48}$$

If $\bar{v}_{vertex} - \bar{v}_0$ is positive ($B_v' < B_v''$) the band shaded to red. On the other hand if $\bar{v}_{vertex} - \bar{v}_0$ is negative ($B_v' > B_v''$) the band is shaded towards violet.

14.5 Intensity Distribution in the Vibrational Structure

The characteristic time for an electronic transition is $\sim 10^{-16}$ sec. whereas for a nuclear vibration the time has much larger value $\sim 10^{-13}$ sec. Therefore, it can be assumed that during an electronic transition, nuclei remain fixed. Franck in 1925 explained the intensity distribution in vibronic transitions. Franck Condon principle states that " An electronic transition in a molecule takes place so much rapidly than a vibrational transition that, in a vibronic transition, the internuclear distance can be regarded as fixed and velocity is nearly the same before and after the transition".

The electronic states are represented by the potential curves. The shapes of potential curves are different for different electronic states. Let $r_e^{''}$ and $r_e^{'}$ correspond to the minimum of potential curve of ground state and excited state, respectively.

(a) $r_e^{''} < r_e$

Consider the potential curve for ground state and excited state as shown in Fig.14.8. Let us consider a transition CD taking place from the minimum of ground state level to the excited state. Since the time taken by the electronic transition is much smaller than vibrational transition, during electronic

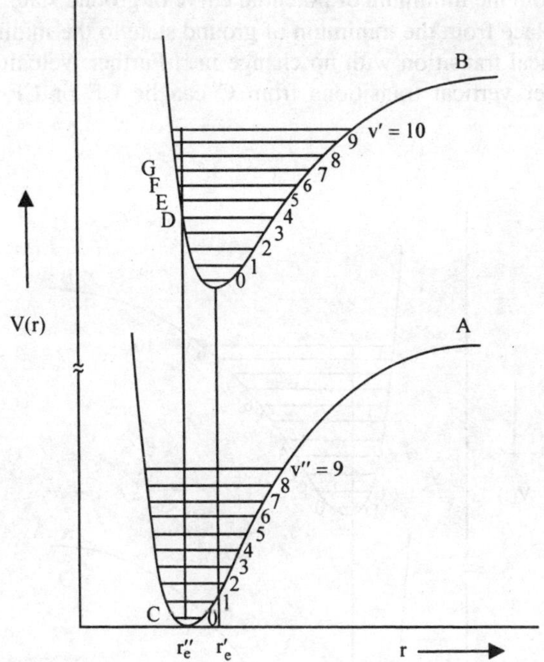

Fig. 14.8 Illustration of the Franck-Condon principle for $r_e^{''} < r_e^{'}$

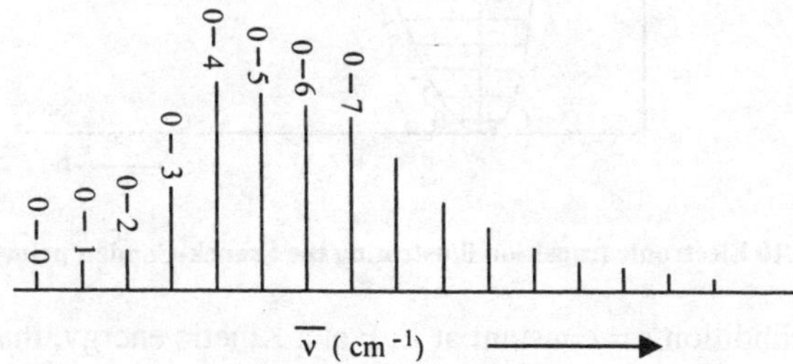

Fig. 14.9 Intensity of vibrational transitions for $r_e^{''} < r_e$

transition nuclei will not change their position. This means r will not change. The transitions are between points, which lie on the same vertical. Such transitions are called vertical transitions. Secondly, the velocity at the minimum of ground state is zero, which means in the excited state it should end up at positions where the velocity is again zero. At the turning points of the potential curve, the energy is totally potential, which implies that at these points velocity is zero. The vertical transitions from ground state may end up in the excited state at D, E, F, G.. etc. However, only D is a turning point while E, F and G are not, that is, they have certain velocities. Transition from C to D which is 0 – 4 (v",v') is the most probable and hence the most intense while other transitions 0 – 1 , 0 – 2, 0 – 3, 0 – 5 etc. are less probable and less intense. In this case, the intensity of vibrational transitions first increases, reaches a maximum and then decreases as shown in the Fig.14.9.

(b) $r_e^{"} = r_e^{'}$

Let the transition be from the minimum of potential curve of ground state. Since $r_e^{"} = r_e^{'}$, therefore, transition CD will take place from the minimum of ground state to the minimum of potential curve of excited state. It is a vertical transition with no change in r. Further; velocity is also same before and after the transition. Other vertical transitions from C can be CE or CF etc. (Fig.14.10) All the

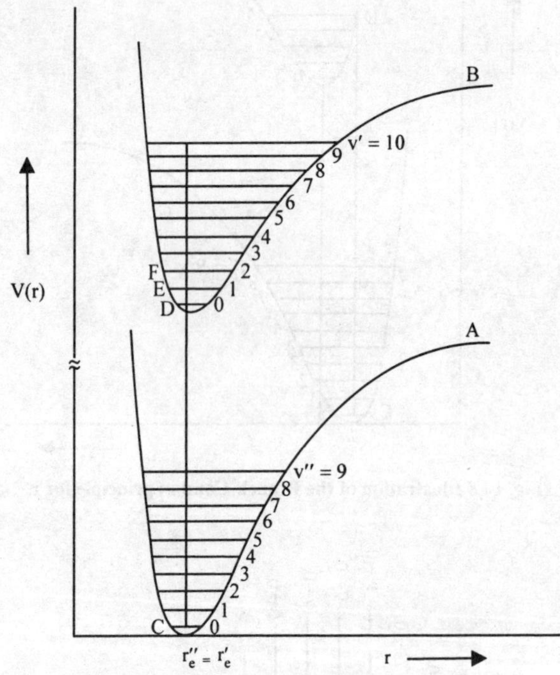

Fig. 14.10 Electronic transition illustrating the Franck-Condon principle for $r_e^{"} = r_e^{'}$

transitions satisfy the condition r = constant at E, F etc. Kinetic energy, that is, velocities are not zero, and in fact velocities go on increasing as we move from E to F etc. The velocity change during transition from C to E, C to F results in the decrease of transition probability. Therefore, the most probable and most intense transition would be 0 – 0. Intensity of 0 – 1, 0 – 2,.. transitions goes on decreasing as shown in the Fig.14.11   .

(c) $r_e^{"} \ll r_e^{'}$

In this case the transition is from the lowest vibrational state of the electronic ground state to higher

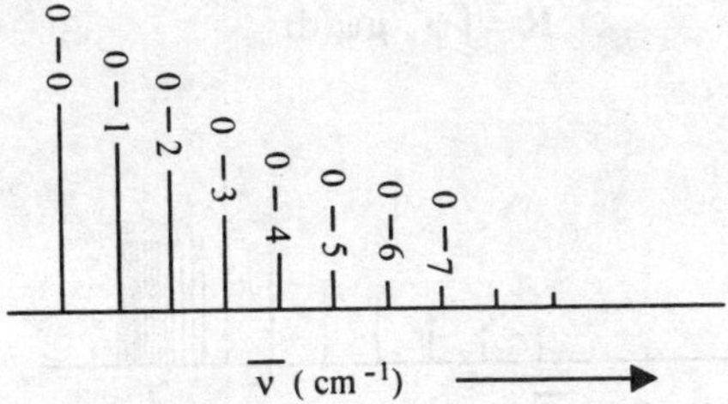

Fig. 14.11 Intensity of vibrational transitions for $r_e'' = r_e'$

vibrational state of excited electronic state (Fig.14.12) . The conditions, that r should be the same and no velocity change, are satisfied for very high value of v'. Vertical transitions to higher v' values also to take place but with velocity change.Therefore, the intensity of the transition decreases. As the value

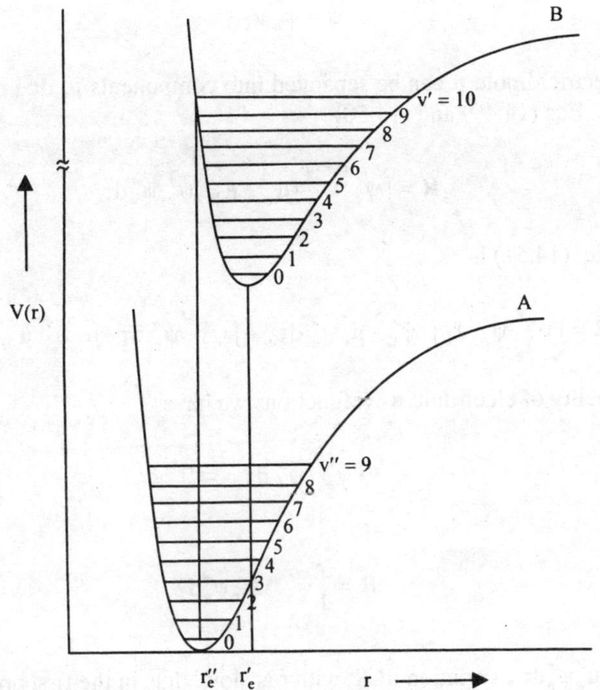

Fig. 14.12 Electronic transition illustrating the Franck-Condon principle for $r_e'' \ll r_e'$

of v' is very high, we are very near to continuum of the upper state. As a result of this we obtain intensity distribution which looks like continuum for higher values of v' as shown in Fig.14.13 .

For $r_e'' > r_e'$ intensity distribution is same as that of case (a) .

In 1928, Condon treated the intensities of vibronic transitions quantum mechanically. The intensity of a vibronic transition between two states is proportional to the square of the transition moment **R** which is given by

$$R = \int \psi'^{*} \mu \psi'' d\tau \tag{14.49}$$

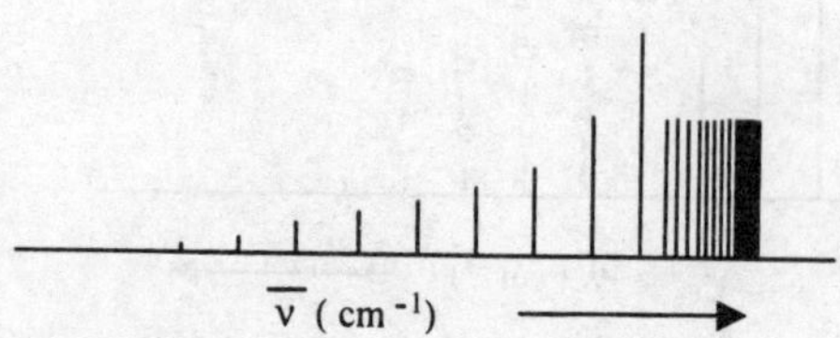

Fig. 14.13 Intensity of vibrational transitions for $r_e'' \ll r_e'$

where μ is the electric dipole moment operator. The wavefunction ψ' and ψ'' characterize the upper and lower electronic states, respectively. Neglecting the rotation and assuming that Born - Oppenheimer approximation holds we have

$$\psi = \psi_e \psi_v \tag{14.50}$$

Assuming that electric dipole μ can be separated into components μ_e due to electron and μ_N due to nuclei, we have from Eqs.(14.49) and (14.50)

$$R = \int \psi_e'^{*} \psi_v'^{*} (\mu_e + \mu_N) \psi_e'' \psi_v'' d\tau \tag{14.51}$$

where $d\tau = d\tau_e \, d\gamma$. Eq. (14.51) is

$$R = \int \psi_v'^{*} \psi_v'' \, d\gamma \int \psi_e'^{*} \mu_e \psi_e'' \, d\tau_e + \int \psi_e'^{*} \psi_e'' \, d\tau_e \int \psi_v'^{*} \mu_N \psi_v'' d\gamma$$

Because of orthogonality of electronic wavefunctions we have

$$\int \psi_e'^{*} \psi_e'' \, d\tau_e = 0$$

Therefore

$$R = \int \psi_v'^{*} R_e \psi_e'' d\gamma$$

where $R_e = \int \psi_e'^{*} \mu_e \psi_e'' d\tau$, variation of R_e with r is slow, thus in the first approximation, R_e can be replaced by its average.

$$R = \overline{R}_e \int \psi_v'^{*} \psi_v'' d\gamma$$

The quantity $\int \psi_v'^{*} \psi_v'' d\gamma$ is called the vibrational overlap integral. Its square is known as Franck-Condon factor. The classical turning point of a vibration, where nuclear velocities are zero, is replaced in quantum mechanics by a maximum or minimum, in ψ_v near the turning point. As shown in

Fig. 14.14 , large the v the closer is maximum, or minimum, in ψ_v to the classical turning points. The intensity is maximum for the transition having maximum overlap.

Let us consider a radiative transition from excited electronic state B to ground electronic state A. The emission may be either stimulated or spontaneous. The conventional emission spectroscopy deals with the spontaneous emission. Spontaneous emission is referred to as fluorescence if there is no change in the spin quantum number and phosphorescence if there is a change in the spin quantum number.

Consider the molecule in the state A . After absorbing energy it is in a vibrational level of state B.

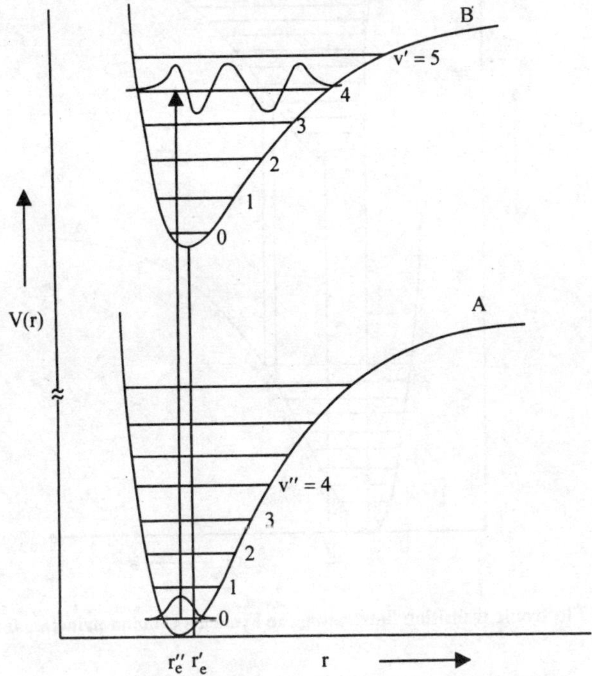

Fig. 14.14 Electronic transition illustrating the Franck - Condon principle for $r_e'' < r_e'$

The energy of this vibrational level relative to the ground vibrational level of state B depends on the separation $r_e' - r_e''$. The energy difference of the vibrational level v' to ground vibrational level v' = 0 is called reorganization energy. The emission spectrum is identical to the absorption spectrum if the downward transition take place before the molecule relax from v state to v = 0 state of B. However, non-radiative processes causes the molecule to relax from v' state to v' = 0 vibrational state of B. Fig. 14.15 shows the vibronic transition from v' = 0 level of electronic state B. The emission electronic transition from v' = 0 vibrational level of B will lead to the point v'' = v_{vert} , and the width of the emission spectrum increases with displacement of the ground and excited electronic state potential.The emission spectrum is at a larger wavelength than absorption. For emission with v' > 0 and $r_e' > r_e''$ ($r_e' < r_e''$) there may be two intensity maxima. When both vibrational levels v' and v'' are excited states, the most favoured transitions will be those that occur at constant r with the nuclei in the electronic state A and B are at the classical end points of their vibrational motion. This is shown in Fig. 14.15 for $v' = 3$ to $v'' = 8$ and $v' = 4$ to $v'' = 9$.

The measured intensities of the bands of an electronic transition can be put in a table form similar to that of Deslandres table as shown in Table 14.2. By connecting the most intense band by a line we obtain a parabola called Condon parabola. The parabola is broad when r_e' and r_e'' are quite different and is narrow when $r_e' \approx r_e''$. For $r_e' = r_e''$, Condon parabola is a line.

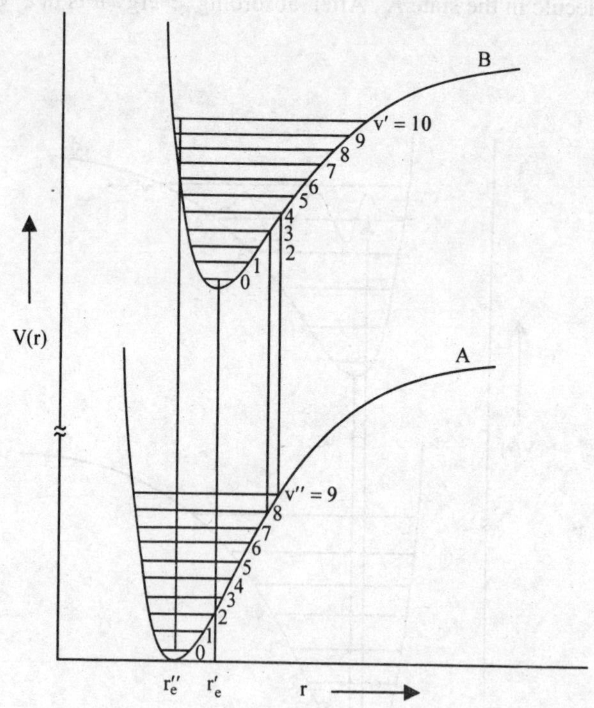

Fig. 14.15 Electronic transition illustrating the Franck - Condon principle for $r_e'' < r_e'$

Table 14.2 Condon parabola

$v' \rightarrow$ / $\downarrow v''$	0	1	2	3	4	5	6
0	20	10	3				
1	3	3	8	4			
2	2	7	4	4	1		
3		3	3		2	1	
4			3		2	2	1
5				2			2
6				1			

Examples

1. For diatomic molecules having the following electronic configurations, determine the electronic state term, its multiplet (assuming weak Λ,S coupling) and the total degeneracy of each component of the multiplet (a) $S = 1/2$, $\Lambda = 1$, (b) $S = 1, \Lambda = 2$.

(a) $\Omega = 1 \pm \dfrac{1}{2} = \dfrac{3}{2}, \dfrac{1}{2}$ the state is $^2\Pi_{1/2,3/2}$

(b) $\Omega = 2 \pm 1 = 3,2,1$ the state is $^3\Delta_{1,2,3}$

2. Wavenumber of the observed lines of 0 - 0 band of Green Be – O bands (in units of cm^{-1}) are as follows :

R(0) = 21199.81	P(1) = 21193.75
R(1) = 21202.88	P(2) = 21189.97
R(2) = 21205.74	P(3) = 21186.41
R(3) = 21208.52	P(4) = 21182.66
R(4) = 21211.12	

Calculate the values of rotational constants B_0'' and B_0' .

From combination relations using Eqs. (14.34) and (14.38)

$$R(J-1) - P(J+1) = 4B_v'' \left(J + \frac{1}{2} \right)$$

$$R(J) - P(J) = 4B_v' \left(J + \frac{1}{2} \right)$$

For $J = 1$

$$R(0) - P(2) = (21199.81 - 21189.97)\, cm^{-1} = 9.84\, cm^{-1} = 6B_v''$$

$$B_v'' = 1.64\, cm^{-1}$$

For $J = 2$

$$R(1) - P(3) = (21202.88 - 21186.41)\, cm^{-1} = 16.47\, cm^{-1} = 10B_v''$$

$$B_v'' = 1.647\, cm^{-1}$$

For $J = 3$

$$R(2) - P(4) = (21205.74 - 21182.66)\ \text{cm}^{-1} = 23.08\ \text{cm}^{-1} = 14B_v''$$

$$B_v'' = 1.6486\ \text{cm}^{-1}$$

The average value is

$$B_v'' \approx 1.645\ \text{cm}^{-1}$$

3   Wavenumber of the various rotational branches of the 4 – 11 band of the fourth positive group of CO molecule (in cm^{-1}) are as follows:

R(0) = 48338.37	Q(1) = 48355.0	P(2) = 48328.58
R(1) = 48340.94	Q(2) = 48333.95	P(3) = 48324.14
R(2) = 48342.87	Q(3) = 48322.64	
R(3) = 48345.16		

Calculate the rotational constant of lower and upper state.

From combination relations obtained from Eqs. (14.34) , (14.36) and (14.38)

$$R(0) - Q(1) = 2B_v'' = (48338.37 - 48335.0)\ \text{cm}^{-1} = 3.37\ \text{cm}^{-1}$$

$$B_v'' = 1.685\ \text{cm}^{-1}$$

From

$$R(1) - Q(2) = (48340.94 - 48333.95)\ \text{cm}^{-1} = 6.99\ \text{cm}^{-1} = 4B_v''$$

$$B_v'' = 1.7475\ \text{cm}^{-1}$$

From

$$R(2) - Q(3) = (48342.87 - 48332.64)\ \text{cm}^{-1} = 10.23\ \text{cm}^{-1} = 6B_v''$$

$$B_v'' = 1.705\ \text{cm}^{-1}$$

From

$$Q(1) - P(2) = (48335.0 - 48328.58)\ \text{cm}^{-1} = 6.42\ \text{cm}^{-1} = 4B_v''$$

$$B_v'' = 1.605\ \text{cm}^{-1}$$

From

$$Q(2) - P(3) = (48333.95 - 48324.14)\ cm^{-1} = 9.81\ cm^{-1} = 6B''_v$$

$$B''_v = 1.671\ cm^{-1}$$

The average value is

$$B''_v \approx 1.671\ cm^{-1}$$

From

$$R(1) - Q(1) = (48340.94 - 48335.0)\ cm^{-1} = 5.94\ cm^{-1} = 4B'_v$$

$$B'_v = 1.485\ cm^{-1}$$

From

$$R(2) - Q(2) = (48342.87 - 48333.95)\ cm^{-1} = 8.92\ cm^{-1} = 6B'_v$$

$$B'_v = 1.487\ cm^{-1}$$

$$R(3) - Q(3) = (48345.16 - 48332.64)\ cm^{-1} = 12.52\ cm^{-1} = 8B'_v$$

$$B'_v = 1.565\ cm^{-1}$$

$$Q(2) - P(2) = (48333.95 - 48328.58)\ cm^{-1} = 5.37\ cm^{-1} = 4B'_v$$

$$B'_v = 1.3425\ cm^{-1}$$

$$Q(3) - P(3) = (48332.64 - 48324.14)\ cm^{-1} = 8.50\ cm^{-1} = 6B'_v$$

$$B'_v = 1.4166\ cm^{-1}$$

The average value is

$$B'_v \approx 1.459\ cm^{-1}$$

4. The rotational constants by analysing the band spectrum of C_2 molecule is obtained as $B'_v = 1.7527$ cm^{-1} and $B''_v = 1.6326$ cm^{-1}. The band origin is found to be at 19378 cm^{-1}. estimate the position of the band head.

From Eq.(14.47) the J value corresponding to the vertex

$$J_{vertex} = \frac{B'_v + B''_v}{2(B'_v - B''_v)} = \frac{1.7527 + 1.6326}{2(1.7527 - 1.6326)} \cong 14$$

The frequency at the vertex from Eq.(14.48) is

$$\bar{v}_{vertex} = \bar{v}_0 - \frac{(B'_v + B''_v)^2}{4(B'_v - B''_v)} = 19378 \text{ cm}^{-1} - \frac{(1.7527 + 1.6326)^2}{4(1.7527 - 1.6326)} \text{ cm}^{-1}$$

$$\bar{v}_{vertex} = (19378 - 23.86) \text{ cm}^{-1} = 19354.1 \text{ cm}^{-1}$$

5. In the N_2 molecule, the ground state $X\,{}^1\Sigma_g^+$ and the excited state b' $\,{}^1\Sigma_u^+$ have the following constants

	$\bar{\omega}_e$ (cm^{-1})	$\bar{\omega}_e x_e$ (cm^{-1})	B (cm^{-1})
${}^1\Sigma_g^+$	2359.6	14.456	2.010
${}^1\Sigma_u^+$	751.7	4.82	1.145

The wavenumber of electronic transition $X\,{}^1\Sigma_g^+ \rightarrow {}^1\Sigma_u^+$ is $\bar{v}_e = 103678.9$ cm^{-1} . Construct a Deslandres table for v' = 0, 1, 2 and v" = 0, 1, 2. Calculate R(1) and P(1) for the transition v' = 0 → v" = 0. Find the position of the band head in cm^{-1}.

From Eq.(14.26) neglecting the $\bar{\omega}_e y_e$ term

$$\bar{v}_{00} = 103678.9 \text{cm}^{-1} + \frac{751.7}{2} \text{ cm}^{-1} - \frac{4.82}{4} \text{ cm}^{-1} - \frac{2359.6}{2} \text{ cm}^{-1} + \frac{14.456}{4} \text{ cm}^{-1}$$

$$\bar{v}_{00} = 102877.359 \text{ cm}^{-1}$$

From Eq.(14.28) neglecting the $\bar{\omega}_e y_e$ term

$$\bar{\omega}'_0 = \bar{\omega}'_e - \bar{\omega}'_e x'_e = (751.7 - 4.82) \text{ cm}^{-1} = 746.88 \text{ cm}^{-1}$$

$$\bar{\omega}'_0 x'_0 = \bar{\omega}'_e x'_e = 4.82 \text{ cm}^{-1}$$

$$\overline{\omega}_0'' = \overline{\omega}_e'' - \overline{\omega}_e''x_e'' = (2359.6 - 14.456)\ \text{cm}^{-1} = 2345.144\ \text{cm}^{-1}$$

$$\overline{\omega}_0''x_0'' = \overline{\omega}_e''x_e'' = 14.456\ \text{cm}^{-1}$$

From Eq.(14.29)

$$\overline{v}_{10} = \overline{v}_{00} + \overline{\omega}_0' - \overline{\omega}_0'x' = (102877.359 + 746.88 - 4.82)\ \text{cm}^{-1} = 103619.419\ \text{cm}^{-1}$$

$$\overline{v}_{11} = \overline{v}_{00} + \overline{\omega}_0' - \overline{\omega}_0'x_0' - \overline{\omega}_0'' + \overline{\omega}_0''x_0'' = (102877.359 + 746.88 - 4.82 - 2345.144 + 14.456)\ \text{cm}^{-1}$$

$$\overline{v}_{11} = 101288.731\ \text{cm}^{-1}$$

$$\overline{v}_{20} = \overline{v}_{00} + 2\overline{\omega}_0' - 4\overline{\omega}_0'x_0' = (102877.359 + 2\times746.88 - 4\times4.82)\ \text{cm}^{-1} = 104351.839\ \text{cm}^{-1}$$

$$\overline{v}_{01} = \overline{v}_{00} - \overline{\omega}_0'' + \overline{\omega}_0''x_0'' = (102877.359 - 2345.44 + 14.456)\ \text{cm}^{-1} = 100546.375\ \text{cm}^{-1}$$

Similarly

$$\overline{v}_{02} = 98244.303\ \text{cm}^{-1},\ \overline{v}_{12} = 98986.353\ \text{cm}^{-1},\ \overline{v}_{21} = 102020.855\ \text{cm}^{-1},\ \overline{v}_{22} = 99718.783\ \text{cm}^{-1}$$

Deslandres table is thus

$v'' \rightarrow$ $\downarrow v'$	0	1	2
0	102877.359 cm^{-1}	100546.375 cm^{-1}	98244.303 cm^{-1}
1	103619.419 cm^{-1}	101288.731 cm^{-1}	98986.353 cm^{-1}
2	104351.839 cm^{-1}	102020.855 cm^{-1}	99718.783 cm^{-1}

From Eq.(14.33)

$$R(1) = \overline{v}_{00} + 6B_v' - 2B_v'' = (102877.359 + 6\times1.145 - 2\times2.010)\ \text{cm}^{-1} = 102880.209\ \text{cm}^{-1}$$

From Eq.(14.38)

$$P(1) = \overline{v}_{00} - 2B_v'' = (102877.359 - 2\times2.010)\ \text{cm}^{-1} = 102873.339\ \text{cm}^{-1}$$

From Eq.(14.48)

$$\overline{v}_{\text{vertex}} = \overline{v}_0 - \frac{(B_v' + B_v'')^2}{4(B_v' - B_v'')} = 10287.359\ \text{cm}^{-1} - \frac{(2.010 + 1.145)^2}{4(1.145 - 2.010)}\ \text{cm}^{-1}$$

$$\overline{\nu}_{vertex} = 102880.266 \, cm^{-1}$$

6. Calculate $\overline{\omega}'_e, \overline{\omega}''_e, \overline{\omega}'_0, \overline{\omega}''_0, \overline{\omega}'_e x'_e, \overline{\omega}''_e x''_e$, $\overline{\nu}_e$ and force constants k'_e , k''_e from the part of the Deslendres table for $C^{12}S^{32}$ as given below (the number in the table are in unts of cm^{-1}):

$v'' \rightarrow$ $\downarrow v'$	0	1	2	3
0	38825.9	37557.3	36301.6	35059.4
1	39883.5	38616	37355.2	36111.5
2	40903.1	39632.2	38374.5	37130.6
3	-	40647.1	39390.2	38144.6

The difference between the rows of the Deslenders table give the spacing of the vibrational levels in the upper electronic state while thwe difference between the values in the column gives the vibrational spacing in the lower electronic state. The first diffetrence $\Delta G'(1/2)$ value is obtained by substracting v' = 0 from v' = 1 value vertically below it. On applying this to the above table, the average value is

$$\frac{1057.6 + 1058.7 + 1053.6 + 1052.1}{4} \, cm^{-1} = 1055.5 \, cm^{-1}$$

By a similar procedure

$$\Delta G'(3/2) = 1018.55 \, cm^{-1}, \Delta G'(5/2) = 1014.866 \, cm^{-1},$$

$$\Delta G''(1/2) = 1269 \, cm^{-1}, \Delta G''(3/2) = 1257.775 \, cm^{-1}, \Delta G''(5/2) = 1243.85 \, cm^{-1}$$

The mean second difference is obtained by taking the average of

$$[G(3/2) - G(1/2)], [G(5/2) - G(3/2)],..etc.$$

Thus

$$\Delta^2 G' = -\frac{36.95 + 3.684}{2} \, cm^{-1} = -20.317 \, cm^{-1}$$

$$\Delta^2 G'' = -\frac{11.225 + 13.925}{2} \, cm^{-1} = -12.575 \, cm^{-1}$$

From

$$\Delta^2 G' = -20.317 \, cm^{-1} = -2\overline{\omega}'_e x'_e = -2\overline{\omega}'_0 x'_0$$

$$\overline{\omega}'_e x'_e = \overline{\omega}'_0 x'_0 = 10.1585 \text{ cm}^{-1}$$

$$\Delta^2 G'' = -12.575 \text{ cm}^{-1} = -2\overline{\omega}''_e x''_e = -2\overline{\omega}''_0 x''_0$$

$$\overline{\omega}''_e x''_e = \overline{\omega}''_0 x''_0 = 6.287 \text{ cm}^{-1}$$

The first vibrational quanta is

$$\Delta G_{1/2} = \overline{\omega}_0 - \overline{\omega}_0 x_0 = \overline{\omega}_e - 2\overline{\omega}_e x_e$$

$$\Delta G'_{1/2} = 1055.5 \text{ cm}^{-1} = \overline{\omega}'_0 - \overline{\omega}'_0 x'_0 = \overline{\omega}'_e - 2\overline{\omega}'_e x'_e = \overline{\omega}'_e - 2 \times 10.1585 \text{ cm}^{-1}$$

$$\overline{\omega}'_e = 1075.817 \text{ cm}^{-1}$$

$$\Delta G''_{1/2} = 1269 \text{ cm}^{-1} = \overline{\omega}''_0 - \overline{\omega}''_0 x''_0 = \overline{\omega}''_e - 2\overline{\omega}''_e x_e = \overline{\omega}''_e - 2 \times 6.287 \text{ cm}^{-1}$$

$$\overline{\omega}''_e = 1281.575 \text{ cm}^{-1}$$

$$\overline{\omega}'_0 = \overline{\omega}'_e - \overline{\omega}'_e - \overline{\omega}'_e x'_e = (1075.817 - 10.1585) \text{ cm}^{-1} = 1065.658 \text{ cm}^{-1}$$

$$\overline{\omega}''_0 = \overline{\omega}''_e - \overline{\omega}''_e x''_e = (1281.575 - 6.287) \text{ cm}^{-1} = 1275.288 \text{ cm}^{-1}$$

The reduced mass of $C^{12}S^{32}$ is

$$\mu = \frac{12 \times 32}{(12 + 32)} \times 1.6606 \times 10^{-27} \text{ Kg} = 14.492 \times 10^{-27} \text{ Kg}$$

We have

$$\overline{\omega}_e = \frac{1}{2\pi c} \sqrt{\frac{k}{\mu}} \quad \text{or} \quad k = (2\pi c \overline{\omega}_e)^2 \mu$$

From this

$$k'_e = (2 \times 3.14 \times 2.9979 \times 10^8 \text{ ms}^{-1} \times 107581.7 \text{m}^{-1})^2 \times 14.492 \times 10^{-27} \text{ Kg} = 594.55 \text{ Nm}^{-1}$$

$$k''_e = (2 \times 3.14 \times 2.9979 \times 10^8 \text{ ms}^{-1} \times 128157.5 \text{m}^{-1})^2 \times 14.492 \times 10^{-27} \text{ Kg} = 843.72 \text{ Nm}^{-1}$$

From Eq.(14.26)

$$\overline{\nu}_e = \overline{\nu}_{00} - \frac{\overline{\omega}'_e}{2} + \frac{\overline{\omega}''_e}{2} + \frac{\overline{\omega}'_e x'_e}{4} - \frac{\overline{\omega}''_e x''_e}{4}$$

$$\overline{\nu}_e = (38825.9 - \frac{1075.817}{2} + \frac{1281.575}{2} + \frac{10.1585}{4} - \frac{6.287}{4})\,cm^{-1} = 38929.75\,cm^{-1}$$

Problems

14.1   For diatomic molecules having the following electronic configurations, determine the electronic state term, its multiplet (assuming weak Λ,S coupling) and the total degeneracy of each component of the multiplet (a) $S = 0$, $\Lambda = 1$, (b) $S = 3/2$, $\Lambda = 2$ (c) $S = 1$, $\Lambda = 1$.

14.2   The part of the Deslenders table for $P^{31}N^{14}$ is as follow (the number in the table are in unts of cm^{-1}):

$v''\rightarrow$ $\downarrow v'$	0	1	2	3
0	39698.8	38376.5	37068.7	-
1	40780.2	39467.2	38155.5	36861.3
2	41859.1	40536.2	-	37932.9
3	-	41597.4	40288.3	-

Calculate $\overline{\omega}'_e, \overline{\omega}''_e, \overline{\omega}'_0, \overline{\omega}''_0, \overline{\omega}'_e x'_e, \overline{\omega}''_e x''_e, \overline{\nu}_e$ and force constants k'_e, k''_e.

14.3   The part of the Deslendres table for the origin of fourth positive group of $C^{12}O^{16}$ is as follow (the number in the table are in unts of cm^{-1}):

$v''\rightarrow$ $\downarrow v'$	0	1	2	3
0	64746.5	62601.8	60484.7	58393.2
1	66231.3	64087.6	-	59881.6
2	67674.8	65533.1	63416.1	61325.2
3	69087.8	66944.3	64828.1	-

Calculate $\overline{\omega}'_e, \overline{\omega}''_e, \overline{\omega}'_0, \overline{\omega}''_0, \overline{\omega}'_e x'_e, \overline{\omega}''_e x''_e, \overline{\nu}_e$ and force constants k'_e, k''_e.

15

Raman Spectra of Diatomic Molecules

15.1 Raman Effect

When a monochromatic light beam is incident on a system such as transparent gases, liquids or solids, most of it is transmitted without change. However, a very small portion of the incident light is scattered. From an analysis of scattered light, it is observed that most part of it has the same frequency as that of incident radiation. However, a small portion of it has different frequencies. The scattering of light at different frequencies is called Raman scattering or Raman effect. The scattering radiation with decreased frequency is called the Stokes Raman scattering and that with increased frequency, the anti-Stokes Raman scattering. The physical origin of the Raman effect lies in inelastic collisions between incident photons and molecules composing the system. Inelastic collision means that there is an exchange of energy between the photon and the molecule with a consequent change in energy, and hence frequency, of the photon. The total energy is conserved during inelastic scattering process. The energy gained or lost by the photon must be equal to the energy change within the molecule. By measuring the energy gained or lost by the photon, change in molecular energy (rotational, vibrational or electronic) can be probed.

15.1.1 Classical theory of Raman effect

A light wave is a travelling wave of electric and magnetic fields, of which only the electric component gives rise to Raman effect. When a light wave meets a molecule, the electric field of light wave exerts the same force on all the electrons of the molecules because the wavelength of light wave is large in comparison with the size of the molecule. The electric field of the light tends to displace electrons in one direction and positively charged nuclei in the opposite direction. This displacement of positive and negative charges result in an induced dipole moment $\mathbf{P}$ in the molecule that is, to a good approximation, proportional to the electric field strength $\mathbf{F}$. Thus

$$\mathbf{P} = \alpha \mathbf{F} \tag{15.1}$$

where α, the proportionality constant is called the electric polarizability of the molecule. In general, the direction of the vector $\mathbf{P}$ does not coincide with the direction of $\mathbf{F}$. Therefore α is not a simple scalar quantity. In fact the magnitudes of the three components of $\mathbf{P}$, namely P_x, P_y and P_z are related to the magnitude of the electric field $\mathbf{F}$ by the relation

$$\begin{pmatrix} P_x \\ P_y \\ P_z \end{pmatrix} = \begin{pmatrix} \alpha_{xx} & \alpha_{xy} & \alpha_{xz} \\ \alpha_{yx} & \alpha_{yy} & \alpha_{yz} \\ \alpha_{zx} & \alpha_{zy} & \alpha_{zz} \end{pmatrix} \begin{pmatrix} F_x \\ F_y \\ F_z \end{pmatrix} \tag{15.2}$$

which is often written in shorthand as $\{P\}=\{\alpha\}\{F\}$. Another way to represent the polarizability is in the form of an ellipsoid defined by

$$\alpha_{xx} x^2 + \alpha_{yy} y^2 + \alpha_{zz} z^2 + 2\alpha_{xy} xy + 2\alpha_{yz} yz + 2\alpha_{zx} zx = 1$$

When represented graphically, the ellipsoid is centred at (0,0,0) but its principal axes are not necessarily aligned along the axes of lab frame because of the cross – terms $(2\alpha_{xy} xy + 2\alpha_{yz} yz + 2\alpha_{zx} zx)$.

The Eq. (15.2) express the fact that F_x, F_y and F_z make contribution to each of the three components of **P**. The nine coefficients α_{ij} are called the components of the polarizability $\boldsymbol{\alpha}$ with the subscript i denotes the component of **P** and subscript j denotes the component of **F** related by the element α_{ij}. α is said to be a tensor.

Let the fixed molecule be irradiated with monochromatic radiation of frequency ν and is plane polarised in z direction. The oscillating electric field is given by

$$F_z(t) = A\cos 2\pi \nu t \qquad (15.3)$$

where A is the maximum value of F and t is the time in seconds from an arbitrary starting time. The z component of P is

$$P_z(t) = \alpha_{zz} A\cos 2\pi \nu t \qquad (15.4)$$

Since P_z depends on α_{zz} as well as on F_z the properties of the molecule change P_z via α.

Let us consider a vibrating diatomic molecule without rotation. The difference $\Delta r(t)$, from equilibrium in the internuclear distance at time t, can be written as

$$\Delta r(t) = \Delta r_{max} \cos(2\pi \nu_{vib} t) \qquad (15.5)$$

where ν_{vib} is the vibrational frequency of the molecule. Δr_{max} is the maximum extension of the distance between the two atoms of diatomic molecule.

Assuming that the polarizability of the diatomic molecule depends linearly on Δr, the α_{zz} component of $\boldsymbol{\alpha}$ can be written using Taylor series as

$$\alpha_{zz}(t) = (\alpha_{zz})_{equil} + \left(\frac{d\alpha_{zz}}{dr}\right)_{equil} \Delta r(t) + \frac{1}{2!}\left(\frac{d^2\alpha_{zz}}{dr^2}\right)_{equil} [\Delta r(t)]^2 + \ldots\ldots$$

Ignoring third and higher terms on the right of the above equation and putting $(d\alpha_{zz}/dr)_{equil} \Delta r_{max} = \alpha_{1v}$ and using Eq.(15.5)

$$\alpha_{zz} = (\alpha_{zz})_{equil} + \alpha_{1v}\cos 2\pi(\nu_{vib} t) \qquad (15.6)$$

The constant $(\alpha_{zz})_{equil}$ is the polarizability element of the nonvibrating molecule and $(d\alpha_{zz}/dr)_{equil}$ is the change of polarizability with separation r between the nuclei of the molecule.

The dependence of P on time t from Eqs. (15.4) and (15.6) is

$$P_z(t) = A(\alpha_{zz})_{equil}\cos 2\pi \nu t + A\alpha_{1v}\cos 2\pi \nu t \cos 2\pi \nu_{vib} t \qquad (15.7)$$

Using $\cos(\theta+\varphi) + \cos(\theta-\varphi) = 2\cos\theta\cos\varphi$

$$P_z(t) = A(\alpha_{zz})_{equil} \cos 2\pi vt + \frac{\alpha_{1v}}{2} A\cos 2\pi(v - v_{vib})t + \frac{\alpha_{1v}}{2} A\cos 2\pi(v + v_{vib})t \qquad (15.8)$$

Therefore, classically, vibration of the molecule causes a variation of induced electric displacement, which can be broken down in the three terms of the Eq. (15.8). The first term on the right of Eq. (15.8) corresponds to the scattered radiation of unchanged frequency with a magnitude determined by $(\alpha_{zz})_{equil}$ and A. According to the classical electromagnetic theory, an oscillating dipole radiates energy in the form of scattered light. Hence, as a result of the first term, light of frequency v will be emitted and will be observed in a direction different from that of the incident light. This is the phenomenon of Rayleigh scattering. The second and third terms of Eq. (15.8) correspond to scattered radiation with a frequency $(v - v_{vib})$ and $(v + v_{vib})$. The second term is known as Stokes Raman scattering and the third term is known as anti-Stokes Raman scattering. Both these components have magnitude that depend on the strength of the electric field component of the light, amplitude of the vibration and the polarizability derivative $(d\alpha_{zz}/dr)_{equil}$. Thus by analyzing the scattered light we can monitor the vibration within the molecule.

Let the molecule be rotating but not vibrating. The Eq. (15.7) takes the form

$$\alpha_{zz} = (\alpha_{zz})_{equil} + \alpha_{1r} \cos 2\pi(2v_{rot}t) \qquad (15.9)$$

where α_{1r} is amplitude of the change in polarizability for rotation about the rotational axis. The frequency with which polarization changes during the rotation is twice the rotational frequency, since the polarizability is same every π radians. The induced dipole moment in the case of rotating molecule from Eqs. (15.4) and (15.9) is

$$P_z(t) = A(\alpha_{zz})_{equil} \cos 2\pi vt + \frac{\alpha_{1r}}{2} A\cos 2\pi(v - 2v_{rot})t + \frac{\alpha_{1r}}{2} A\cos 2\pi(v + 2v_{rot})t \qquad (15.10)$$

The first term of Eq. (15.10) corresponds to scattered radiation of unchanged frequency, the Rayleigh scattering. The second and the third terms correspond to scattered radiation with frequency $(v - 2v_{rot})$ and $(v + 2v_{rot})$ i.e. the Stokes and anti-Stokes Raman scattering, respectively.

15.1.2 Quantum theory of Raman effect

When the incident light of frequency v collides with a molecule, it can either be scattered elastically or inelastically. The light quanta can add or subtract from the system only amounts of the energy equal to the energy differences between the stationary states of the system. Let $\Delta E = E_b - E_a$ be such an energy difference and let the system be initially in the lower state a. The system is brought to the upper state by scattering of light quanta, the energy ΔE being subtracted from the light quantum. This scattering of radiation is two-step process. In step 1, photon of energy hv is absorbed exciting the molecule from a state a to state m. In step 2, the molecule emits a photon of energy $hv - \Delta E$ and is deexcited from the state m to the final state b (Fig.15.1). The two steps can occur in reverse order, i.e. the photon of energy hv is absorbed by the molecule raising it from excited state b to the state n and molecular emits a photon of energy $hv + \Delta E$ and is deexcited from state n to lower state a. If the final state b of the molecule is the same as the initial state a, the emitted radiation has the same frequency as the incident radiation. This process is called Rayleigh scattering. If the final state is different from the initial state, the scattering is inelastic and conservation of energy gives

$$hv' = hv \pm (E_b - E_a) \qquad (15.11)$$

This inelastic scattering process is called Raman scattering or Raman effect. If $v' < v$, the observed line is called Stokes line and for $v' > v$, the observed line is called the anti-Stokes line. Fig.15.1 shows relationship for light scattering in an energy level diagram. The levels indicated by the broken lines do not correspond to any possible energy states of the system but only give the energy of the incident light quanta about the initial state.

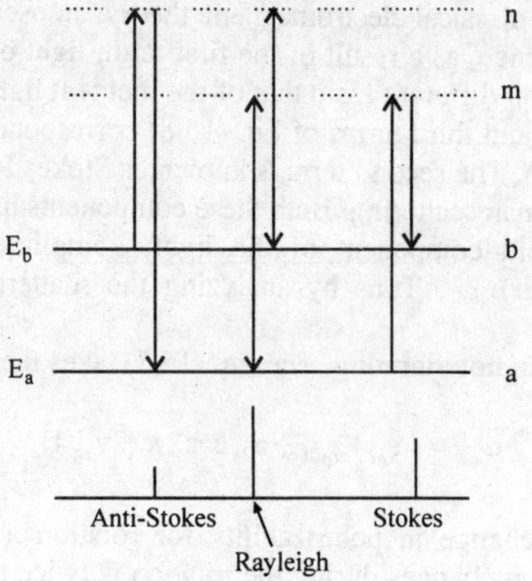

Fig. 15.1 Raman scattering

For the electric dipole transition to take place, the intermediate state m (or n) must have opposite parity to the state a and b. Thus Raman scattering does not change the parity of the molecule. Raman effect can take place for any frequency of the incident light. Raman effect does not require the presence of a permanent electric dipole moment, but rather an induced dipole moment developed under the electric field of incident radiation. For this reason, Raman lines are observed for homonuclear molecules like H_2 , O_2 , etc. which do not exhibit pure rotational or vibrational spectra.

The Raman and Infrared spectra are complementary. The reason lies in the different nature of the processes involved in the two effects. Raman process is a scattering effect involving an induced dipole, which in turn depends on the change of molecular polarizability during vibration. In contrast infrared spectroscopy is an absorption process caused by the change in the permanent molecular dipole.

15.2 Rotational Raman Spectra

The rotational energy levels of diatomic molecules are given by Eq. (12. 67)

$$F(J) = BJ(J+1) - DJ^2(J+1)^2$$

where J= 0, 1, 2, ... But in Raman spectroscopy, the precision of the measurements does not normally require to retain the term involving centrifugal distortion constant D. Thus the rotation energy levels are

$$F(J) = BJ(J+1)$$

(15.12)

The selection rules for the transition between rotational levels in Raman spectroscopy are $\Delta J = 0, \pm 2$. The rotational quantum number change of two units is connected with the symmetry of polarizability ellipsoid. For a diatomic molecule, during end-over-end rotation, the ellipsoid presents the same appearance to an observer after π radians. The magnitude of the frequency shift is

$$\Delta \overline{v}_r = F(J+2) - F(J) = B(J+2)(J+3) - BJ(J+1) = 4B\left(J + \frac{3}{2}\right) \tag{15.13}$$

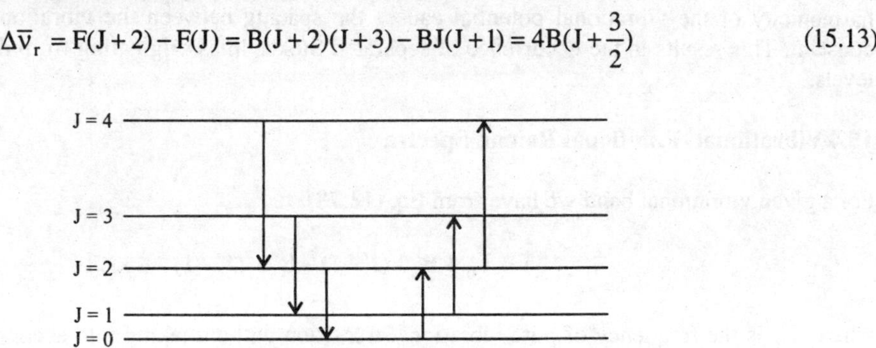

Fig. 15.2 Rotational Raman spectrum

Fig. 15.2 shows the transitions between the rotational energy levels. As a result of the selection rules, a series of equally spaced Raman lines on either side of the undisplaced line is obtained. The two series of lines both having $\Delta J = 2$ are called S branches. The transition $J \to J + 2$ results in a shift to longer wavelengths (Stokes lines) and transition $J + 2 \to J$ results in a shift toward a shorter wavelengths (anti-Stokes lines). The separation between lines of both the series is 4B. The separation between the exciting line (undisplaced line) and first of each of the anti-Stokes and Stokes lines is 6B. The intensities of the Raman lines reflect the population of the rotational levels. Since the initial state of anti-Stokes lines is higher in energy than that of the corresponding Stokes lines, the anti - Stokes lines are correspondingly weaker. However, this effect is small as the rotational energy levels are closely spaced except in light molecules.

15.3 Vibrational Raman Spectra

In simple harmonic oscillator model of diatomic molecule, the energy levels are given by Eq. (13.28)

$$G(v) = \overline{\omega}\left(v + \frac{1}{2}\right)$$

The selection rules for vibrational Raman spectra are $\Delta v = \pm 1$. The vibrational transitions are observed at

$$\overline{v}_{vib} = G'(v) - G''(v) = G(v+1) - G(v) = \overline{\omega}\left(v + \frac{3}{2}\right) - \overline{\omega}\left(v + \frac{1}{2}\right) = \overline{\omega} \tag{15.14}$$

Thus Stokes and anti -Stokes components are obtained shifted by frequency $\Delta E / h$ from Rayleigh line.

At normal temperature, most of the molecules are in the lowest vibrational state and few are in the excited vibrational states. The intensity of the Stokes lines corresponding to the transition $0 \rightarrow 1$ is thus much greater than that of $1 \rightarrow 0$ transition. In most of the diatomic molecules the separation between the lowest vibrational levels is so large that only the Stokes lines can be observed at normal temperature. As described in Sec. 13.2 the potential is not that of simple harmonic oscillator. The non-harmonicity of the vibrational potential causes the spacing between the vibrational levels not to be constant. This results in the occurrence of separated Stokes lines originating from different vibrational levels.

15.4 Vibrational -Rotational Raman Spectra

For a given vibrational band we have from Eq. (13.78)

$$\bar{v} = \bar{v}_0 + B'_v J'(J'+1) - B''_v J''(J''+1) \tag{15.15}$$

where $\bar{v}_0$ is the frequency of pure vibrational transition without taking into account the rotation. The selection rules for vibrating rotator are those of anharmonic oscillator and the rotator, i.e.,

$$\Delta v = \pm 1, \pm 2, \ldots\ldots$$

$$\Delta J = 0, \pm 2$$

For $J' = J'' + 2 = J + 2$ (S-branch), Eq. (15.15) gives

$$\bar{v}_S(J) = \bar{v}_0 + B'_v(J+2)(J+3) - B''_v J(J+1)$$

$$\bar{v}_S(J) = \bar{v}_0 + 6 B'_v + (5B'_v - B''_v)J + (B'_v - B''_v)J^2 \tag{15.16}$$

where $J = 0, 1, 2, \ldots\ldots$
For $J' = J'' - 2 = J - 2$ (O-branch), Eq. (15.15) gives

$$\bar{v}_O(J) = \bar{v}_0 + B'_v(J-2)(J-1) - B''_v J(J+1)$$

$$\bar{v}_O(J) = \bar{v}_0 + 2B'_v - (3B'_v + B''_v)J + (B'_v - B''_v)J^2 \tag{15.17}$$

where $J = 2, 3, \ldots\ldots$
For $J' = J'' = J$ (Q branch)

$$\bar{v}_Q(J) = \bar{v}_0 + B'_v J(J-1) - B''_v J(J+1)$$

$$\bar{v}_Q(J) = \bar{v}_0 + (B'_v - B''_v)J + (B'_v - B''_v)J^2 \tag{15.18}$$

where $J = 0, 1, 2, \ldots\ldots$
Fig. 15.3 shows the rotational transitions accompanying a vibrational transition in a **Raman** spectrum.

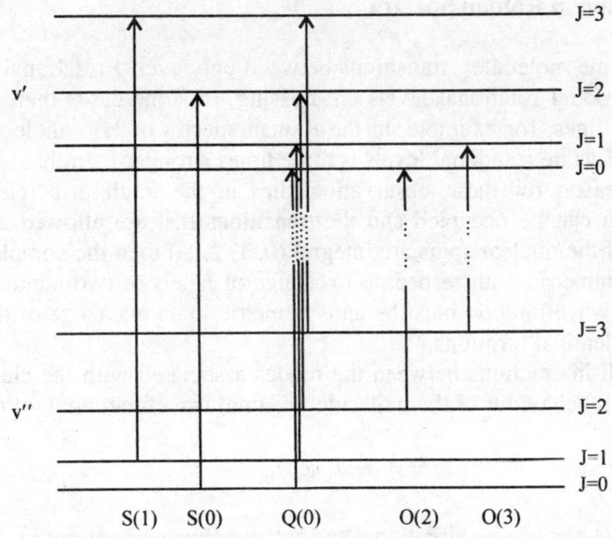

Fig. 15.3 Rotational transitions accompanying a vibrational transition in a Raman spectrum

For vibrational transition $0 \rightarrow 1$, the difference between B_v' and B_v'' is very small, therefore putting $B_v' = B_v'' = B$ in Eqs. (15.16), (15.17) and (15.18)

$$\bar{v}_S(J) = \bar{v}_0 + 4BJ + 6B \tag{15.19}$$

$$\bar{v}_O(J) = \bar{v}_0 - 4BJ + 2B \tag{15.20}$$

$$\bar{v}_Q(J) - \bar{v}_0 \tag{15.21}$$

and hence lines of Q branch are very close to one another and are usually not resolved. This gives rise to an intense line at $\bar{v}_0$ possibly somewhat diffused at shorter wavelength side. The separation of the first S-branch line S(0) and the first O(2) is 12 B. The S branch occurs at higher frequency while O branch is at lower side of the frequency of Q branch. The lines of S and O branches are very much weak since their lines are not superimposed. They form a series of somewhat similar to R and P branches of infrared bands except that the line separation is twice as large.

To obtain the value of B_v' and B_v'' we can use the method of combination differences. From Eqs.(15.16) and (15.17)

$$\Delta_4'' F(J) = \bar{v}_S(J-2) - \bar{v}_O(J+2) = 8B_v''(J+1/2) \tag{15.22}$$

$$\Delta_4' F(J) = \bar{v}_S(J) - \bar{v}_O(J) = 8B'(J+1/2) \tag{15.23}$$

Graphs of $\Delta_4' F(J)$ versus $(J+1/2)$ and $\Delta_4'' F(J)$ versus $(J+1/2)$, gives straight lines with slopes $8B_v'$ and $8B_v''$, respectively.

15.5 Intensity Alternation in Raman Spectra

For Raman spectra of some molecules, transitions between only even J rotational levels are observed and transitions between odd J rotational levels are missing. In some cases there is an alternation in intensity of the adjacent lines, for example, in the Raman spectra of H_2 molecule, the intensity of transitions between odd J value rotational levels is three times stronger than that between even J value rotational levels. The reason for these observations lies in the Pauli principle. For homonuclear molecules, the states that can be occupied and the transitions that are allowed are restricted by the symmetry requirement. If the nuclear spins are integral (0, 1, 2, ...) then the complete wavefunction of the molecule must be symmetric with respect to exchange of labels of two identical nuclei. For half-integral nuclear spin the wavefunction must be antisymmetric in an exchange of the labels of the two nuclei because they are identical fermions.

If we neglect the small interactions between the modes associated with the electronic, vibrational, rotational and nuclear spin behaviour of the molecule, the total wavefunction Ψ of the molecule is

$$\Psi = \psi_e \psi_v \psi_r \Psi_{ns} \qquad (15.24)$$

where ψ_e, ψ_v and ψ_r are the electronic, vibrational and rotational wavefunctions and ψ_{ns} is nuclear spin wavefunction. As described in Sec.12.5, ψ_e and ψ_v are symmetric in an exchange of the labels of the two nuclei. Thus the symmetry of the wavefunction Ψ is governed by the symmetry properties of the product of $\psi_r \psi_{ns}$.

For I having half integral value, Ψ and hence $\psi_r \psi_{ns}$ must be antisymmetric in a nuclear label exchange. Thus if ψ_r is symmetric, ψ_{ns} must be antisymmetric. Let us consider the H_2 molecule . The spin of hydrogen nuclei is 1/2 , which makes it a fermion. Therefore Ψ must be antisymmetric with respect to interchange of nuclei. The ψ_r may be symmetric or antisymmetric depending upon whether J is even or odd. For I = 1/2 , M_I takes the values + 1/2 or – 1/2 . the nuclear spin wavefunction ψ_{ns} is written as α or β corresponding to M_I = +1/2 or – 1/2, respectively. Both H^1 nuclei labelled 1 and 2 can have either α or β nuclear wavefunction. There are four possible forms of ψ_{ns} for the molecule as a whole

$$\psi_{ns} = \alpha(1)\alpha(2); \beta(1)\beta(2); \alpha(1)\beta(2); \beta(1)\alpha(2)$$

$\alpha(1)\alpha(2)$ and $\beta(1)\beta(2)$ are symmetric with respect to interchange of 1 and 2. On the other hand, $\alpha(1)\beta(2)$ and $\beta(1)\alpha(2)$ are neither symmetric nor antisymmetric . The linear combination of $\alpha(1)\beta(2)$ and $\beta(1)\alpha(2)$ can be made symmetric or antisymmetric. The linear combinations, after normalisation, are

$$\frac{1}{\sqrt{2}}\left[\alpha(1)\beta(2) - \beta(1)\alpha(2)\right] \quad \text{and} \quad \frac{1}{\sqrt{2}}\left[\alpha(1)\beta(2) + \beta(1)\alpha(2)\right]$$

are antisymmetric and symmetric, respectively. Thus there are three symmetric and one antisymmetric ψ_{ns} . In order for $\psi_r \psi_{ns}$ to be always antisymmetric for H_2 molecule, the single antisymmetric ψ_{ns} is associated with even J value while three symmetric ψ_{ns} are associated with odd J value. Thus transitions between odd J rotational levels are three times stronger than those between even J rotational levels.

Let us now consider example of CO_2 molecule which is linear and carbon atom is at the centre of gravity of the molecule. Rotation about centre of gravity involves only oxygen atoms and C does not move. The nuclei of O^{16} atom have a nuclear spin $I = 0$. Particle with integral spins obey Bose Einstein statistics. According to the Pauli principle, Ψ must be symmetric with respect to the interchange of the particle labelling (in this case oxygen nuclei). The electronic wavefunction ψ_e is symmetric with respect to interchange of oxygen nuclei since nothing changes in the interchange of the nuclei. The ψ_v is symmetric with respect to the interchange of oxygen nuclei. The symmetry of ψ_r depends on the value of J. Since spin $I = 0$, therefore there is no nuclear spin function and we can ignore it. In order for Ψ to be symmetric with respect to interchange of oxygen atoms, the ψ_r must be symmetric as ψ_e and ψ_v are symmetric. Therefore, only the states for even J exist and odd J states are missing, that is, only even J states are populated. Hence, transitions between even J value rotational levels can take place and transitions between odd J value rotational levels are missing. Thus in Raman spectra of CO_2 every other line is missing.

In general case, the number of antisymmetric spin states to the number of symmetric spin states determine the relative intensity of rotational lines. It is shown that the ratio R of antisymmetric states to symmetric spin states gives the ratio of strong and weak lines if two nuclei of homonuclear molecule each having spin I

$$R = \frac{I+1}{I} \quad \text{for fermion and} \quad R = \frac{I}{I+1} \quad \text{for boson}$$

Examples

1. In the pure rotational Raman spectrum of a diatomic molecule the separation from Rayleigh line of the first Stokes line is different from the subsequent line spacing. Find the ratio of these spacing when only odd levels are populated.

   The spacing between the Rayleigh line and the first Stokes line corresponding to transition $J = 0 \rightarrow J = 2$ is 6B. Separation between adjacent Stokes lines is 4B. Since only odd J levels are populated therefore, only transitions between odd J value levels will be observed. Hence first observed Stokes line would correspond to the transition $J = 1 \rightarrow J = 3$. This line is at a separation of 10B to Rayleigh line. The Stokes lines corresponding to transitions between even J values are missing, hence the separation between Stokes lines corresponding to two consecutive odd J value transitions would be 8B. The required ratio is then $10B/8B = 5/4$.

2. Calculate in units of B, the frequency of the rotational lines of H_2 resulting from the transitions to the excited state characterized by the quantum number $J = 4$. If the bond length of H_2 is 0.07417 nm, determine the spacing between the lines. The mass of hydrogen atom is 1.673×10^{-27} Kg.

   For Stokes lines transition take place from $J = 2 \rightarrow J = 4$ and for anti-Stokes line the transition is from $J = 6 \rightarrow J = 4$. The position of the first Stokes and anti-Stokes line is at 6B to the position of Rayleigh line. The spacing between the adjacent Stokes lines is 4B and same is true for anti-Stokes lines. The first anti-Stokes line corresponds to the transition $J = 2 \rightarrow J = 0$ and therefore anti-Stokes transition $J = 6 \rightarrow J = 4$ transition would be $6B + 16B = 22$ B away from the Rayleigh line. On the other hand first Stokes line corresponds to the transition $J = 0 \rightarrow J = 2$ and $J = 2 \rightarrow J = 4$ transition would be 8B away from the first Stokes line or $6B + 8B = 14$ B away from the Rayleigh line.

   The rotational constant B is given by

$$B = \frac{h}{8\pi^2 Ic}, \quad I = \mu r^2$$

The reduced mass μ of hydrogen molecule is $= (1.673/2) \times 10^{-27}$ Kg, $r = 0.07417$ nm. Hence the rotational constant B is

$$B = \frac{6.6262 \times 10^{-34} \, Js \times 2}{8(3.14)^2 \times 1.673 \times 10^{-27} \, Kg \times (0.07417 \times 10^{-9} \, m)^2 \times 2.9979 \times 10^8 \, ms^{-1}} \approx 60.8 cm^{-1}$$

The spacing between the lines is $4B = 4 \times 60.8 \, cm^{-1} = 243.2 \, cm^{-1}$.

3. Predict the form of rotational Raman spectra of a molecule for which $B = 9.997 \, cm^{-1}$ when it is exposed to monochromatic $29697 \, cm^{-1}$ laser radiation.

From Eq. (15.13) the frequency shift from exciting radiation is given by $4B(J+3/2)$. For Stokes lines the resulting frequency is

$$\overline{v}_{stoke} = \overline{v}_0 - 4B(J+3/2)$$

$$\overline{v}_{anti\text{-}stoke} = \overline{v}_0 + 4B(J+3/2)$$

the position of calculated Stokes and anti - Stokes lines are given in the following table.

J	0	1	2	3
Stokes lines	29637.2	29597.2	29557.2	29517.2
Anti - Stokes lines			29757.2	29797.2

There will be a strong central line at $29697.2 \, cm^{-1}$.

4. The ground electronic state of O_2 is $^3\Sigma_g^-$, and the nuclear spin I of O^{16} is zero. Show that transitions can take place only between odd J values.

For I = 0, the nuclear wavefunction is symmetric but the electronic wavefunction is antisymmetric. In order to make the total wavefunction symmetric the rotational wavefunction must be antisymmetric. Therefore only odd J value levels are populated. As a result of this only transitions between them are observed.

Problems

15.1 In N_2 it is observed that transitions involving even J rotational states yield the most intense lines. Determine the symmetry character of the nuclei in the molecule. The electronic ground state of N_2 is $^1\Sigma_g^+$.

15.2 Show that the ratio of the number of antisymmetric spin states to the number of symmetric spin states is I/(I+1).

16

Lasers

16.1 Spontaneous and Stimulated Emission, Absorption

In an atomic, molecular or solid system there are an infinite set of discrete energy levels. Let us consider two of these energy levels of an atom as shown in Fig.16.1. The atom can make a transition between these two levels by emission or absorption of a photon of energy

$$E = E_n - E_m = h\nu \qquad (16.1)$$

where E_n is the energy of the level n and E_m is the energy of level m such that $E_n > E_m$.

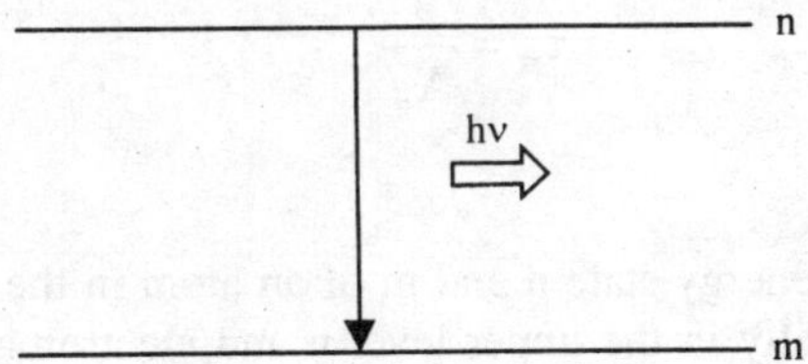

Fig. 16.1 Energy level diagram

16.1.1 Spontaneous emission

Let us assume that the atom is initially at level n. Since $E_n > E_m$, the atom will tend to decay to level m when the corresponding energy difference as given by Eq. (16.1) is emitted. If the energy is emitted in the form of photon of frequency ν, the process is called spontaneous emission. The photon is emitted in a random direction with an arbitrary polarization. The probability of such a spontaneous emission is given by Einstein A coefficient defined as A_{nm} = probability per second of a spontaneous jump from level n to level m. It is also called radiative transition rate or radiation transition probability. A_{nm} has a unit of 1/time.

Let N_n is the number of atoms per unit volume in level n. Then the spontaneous jump from level n to level m is $N_n A_{nm}$ per second. The total rate at which jumps are made from level n to m is

$$\frac{dN_n}{dt} = -N_n A_{nm} \qquad (16.2)$$

There is a negative sign because the population of level n is decreasing due to spontaneous emission.

In case the level n is not the first excited level, the electron can jump to more than one lower level. The total probability that the electron will make a spontaneous jump to any lower level is $A_n \, s^{-1}$ where

$$A_n = \sum_m A_{nm} \qquad (16.3)$$

The summation is over all levels lower in energy than level m. The change of population density N_n of the level n due to spontaneous emission is

$$\frac{dN_n}{dt} = -N_n A_n \tag{16.4}$$

On integration

$$N_n = \text{constant } \exp(-A_n t)$$

If at time $t = 0$, $N_n = N_n^{\,0}$, then

$$N_n = N_n^0 \exp(-A_n t) \tag{16.5}$$

The population of level n decay exponentially with time. The time in which population falls to 1/e of its initial value is called natural life time of level n, τ_n, where

$$\tau_n = \frac{1}{A_n} \tag{16.6}$$

16.1.2 Stimulated emission

Consider transition between energy state n and m of an atom in the presence of an electromagnetic radiation. Let the atom be initially in the upper level n and electromagnetic radiation of frequency ν given by Eq. (16.1) is incident on it. Since this radiation has the same frequency as the frequency of transition between levels n and m, there is a finite probability that the incident radiation will force the atom to undergo transition from n to m. In this case, a photon of frequency ν is emitted preferentially in the direction of the incident electromagnetic beam, which is thereby amplified in intensity (Fig. 16.2).

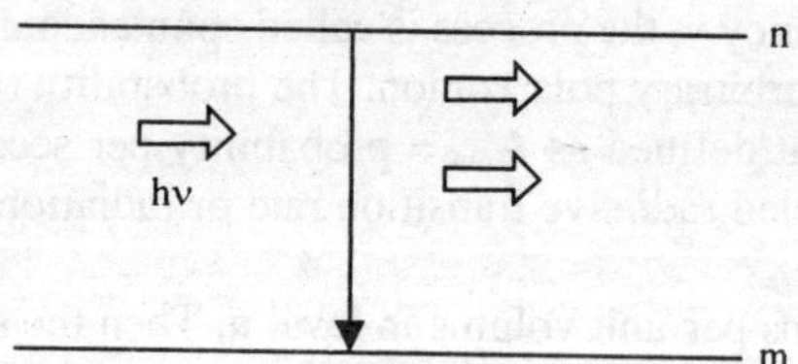

Fig. 16.2 Stimulated emission process for two energy levels of an atom

This behaviour contrasts markedly with the completely random direction over which spontaneous emission occurs. This is the phenomenon of stimulated emission: it is the emission, which is stimulated by other photons of the appropriate frequency.

Let the energy density of externally applied electromagnetic radiation at frequency ν be $\rho(\nu)$ (energy per unit volume per unit frequency interval, that is, $\text{Jm}^{-3}\text{Hz}^{-1}$). The rate of change of population of the upper level due to stimulated emission is proportional to the population of the upper level and the energy density of incident electromagnetic radiation. Therefore,

$$\frac{dN_n}{dt} = -N_n B_{nm}\rho(v)$$ (16.7)

where B_{nm} is Einstein coefficient for stimulated emission. B_{nm} has units of $m^3J^{-1}s^{-2}$.

16.1.3 Absorption

Consider that atom is initially lying in level m. If this is the ground level, the atom will remain in this level. Let an electromagnetic radiation of frequency v given by Eq. (16.1) incident on the atom. There is a finite probability that the atom will be raised to level n (Fig. 16.3). The energy difference h v

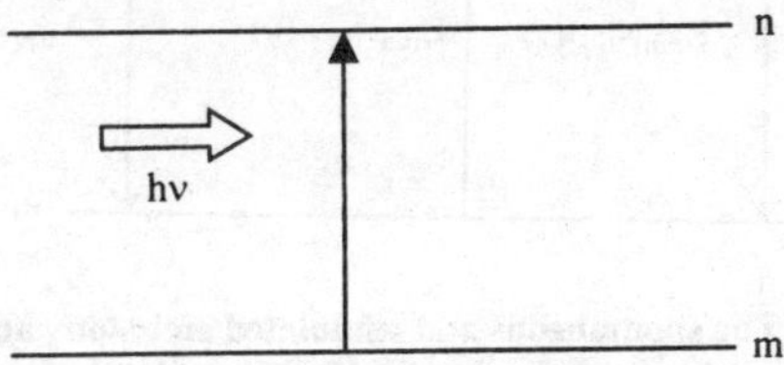

Fig. 16.3 Stimulated absorption between two levels n and m

required by the atom to undergo the transition is obtained from the energy of the incident electromagnetic radiation.

The rate of change of population of the upper level n due to absorption is proportional to the population N_m of the level m and to the incident radiation density $\rho(v)$, that is

$$\frac{dN_n}{dt} = N_m B_{mn}\rho(v)$$ (16.8)

B_{mn} is Einstein coefficient for stimulated absorption. The coefficients B are also referred to as transition cross section.

In stimulated emission and absorption processes, an applied electromagnetic radiation causes the transition between the two energy levels n and m with the emission or absorption of photon, respectively. In both processes, the atom recoils to conserve the linear momentum. Spontaneous emission process takes place without an external electromagnetic radiation. The spontaneous emission process is strictly a quantum mechanical effect. Quantum electrodynamics shows that there are fluctuations in the electromagnetic field. Because of zero point energy of the electromagnetic field, these fluctuations occur even when classically there is no field. It is these fluctuations that induce the spontaneous emission of radiation. Thus spontaneous emission corresponds to stimulated emission resulting from this zero point energy of the radiation field.

16.2 Einstein Coefficients

Consider a collection of atoms inside a cavity at temperature T. The energy density of radiation within a cavity is given by

$$\rho(v) = \frac{8\pi hv^3}{c^3}\frac{1}{\exp(hv/k_BT)-1}$$ (16.9)

Atoms in such a cavity posses many discrete energy levels. Consider two energy levels n and m with energy E_n and E_m, respectively. The energy difference between two levels is hv. Let N_n and N_m be the number of atoms per unit volume in upper level n and lower level m, respectively. The radiation of the cavity may interact with the atoms and a transition between the two states may occur. Spontaneous emission, stimulated emission and absorption can occur between these levels as described in Sec. 16.1. In the Fig. 16.4, the processes are expressed using the Einstein coefficients. Spontaneous emission is independent of energy density of radiation

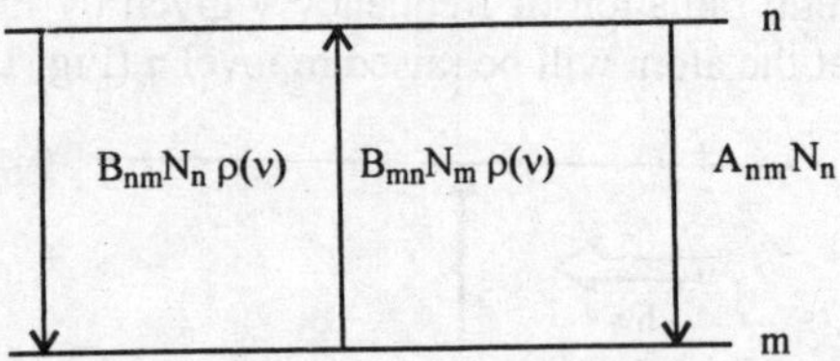

**Fig. 16.4 The spontaneous and stimulated emission , absorption
processes connecting the levels n and m**

The rate of change of population from Eqs. (16.2), (16.7) and (16.8)

$$\frac{dN_n}{dt} = -\frac{dN_m}{dt} = N_m B_{mn}\rho(v) - N_n B_{nm}\rho(v) - N_n A_{nm} \tag{16.10}$$

At equilibrium

$$\frac{dN_n}{dt} = -\frac{dN_m}{dt} = 0$$

$$\rho(v)\left[\frac{N_m}{N_n} B_{mn} - B_{nm}\right] = A_{nm}$$

$$\rho(v) = \frac{A_{nm}}{\dfrac{N_m}{N_n} B_{mn} - B_{nm}} \tag{16.11}$$

For the system in thermal equilibrium, population densities are described by the Boltzmann distribution

$$\frac{N_n}{N_m} = \frac{g_n}{g_m}\exp\left[-\frac{(E_n - E_m)}{k_B T}\right] = \frac{g_n}{g_m}\exp\left[-\frac{hv}{k_B T}\right] \tag{16.12}$$

where g_n and g_m represents the degeneracy of levels n and m , respectively.
From Eqs. (16.11) and (16.12)

$$\rho(\nu) = \cfrac{A_{nm}}{\cfrac{g_m}{g_n}\exp\left(\cfrac{h\nu}{k_B T}\right)B_{mn} - B_{nm}}$$ (16.13)

comparing Eqs.(16.9) and (16.13)

$$\frac{8\pi h\nu^3}{c^3}\cfrac{1}{\exp\left(\cfrac{h\nu}{k_B T}\right)-1} = \cfrac{A_{nm}}{\cfrac{g_m}{g_n}\exp\left(\cfrac{h\nu}{k_B T}\right)B_{mn} - B_{nm}}$$

we obtain

$$g_m B_{mn} = g_n B_{nm}$$ (16.14)

and

$$A_{nm} = \frac{8\pi h\nu^3}{c^3}B_{nm}$$ (16.15)

$$A_{nm} = B_{nm}\frac{8\pi\nu^2}{c^3}h\nu$$

which can be described as

$$A_{nm} = B_{nm} \times \text{No. of modes per unit volume per frequency interval} \times \text{photon energy}$$

For non degenerate levels $g_m = g_n$ and Eq. (16.14) is

$$B_{mn} = B_{nm}$$ (16.16)

Eqs. (16.14) and (16.15) are called Einstein coefficients. Eq. (16.16) indicates that the probability for absorption and stimulated emissions are the same for a transition between states m and n. Thus materials characterized by strong absorptions are also expected to exhibit also a large stimulated emission. Eq. (16.15) shows that a material, in which spontaneous emission does not take place, does not exhibit stimulated emission either. The Einstein relations determine the principal conditions, which should be fulfilled when looking for a material to be used as an active medium in lasers. From Eq. (16.15), the ratio between the number of spontaneous and stimulated emissions increases as ν^3, thus the upper level n will be comparatively rapidly depopulated by spontaneous emission and a very efficient pumping mechanism is necessary to achieve laser action. Therefore, it is very difficult to construct very high frequency lasers like X-ray lasers.

16.3 The Laser Idea

Consider two energy levels n and m of a given atom. Let N_n and N_m be their respective population density. In Sec. 16.1, while describing the three radiative processes, that is spontaneous emission, stimulated emission and absorption, we have assumed that during transition between these two levels a photon of frequency v is emitted or absorbed indicating that energy levels are perfectly sharp. From uncertainity principle it means that if atoms were excited to a perfectly well defined energy levels, it would stay there. Yet we assumed that the excited atom decay by the emission of photon. Therefore, energy level picture is to be slightly modified to take into account that atoms do radiate. Real energy levels are not infinitely sharp, they are broadened. Different levels have different broadening. An atom in a given energy level can actually have energy within a finite range. The frequency spectrum of the emitted or absorbed radiation is described by a line shape function $g(v)$ such that $g(v)\,dv$ is the probability of emission or absorption of a photon with frequency between v and $v + dv$. The function is normalised to unity

$$\int_0^\infty g(v)\,dv = 1 \tag{16.17}$$

In Sec. 16.1 it is assumed that $\rho(v)$ is continuous with bandwidth of $\rho(v)$ much larger than the band of emission or absorption by the atoms. However, if bandwidth of $\rho(v)$ is much smaller than the corresponding spread expressed by the $g(v)$, the Eqs. (16.7) and (16.8) change accordingly to reflect this difference.

The rate of change of population of level n as a result of a monochromatic wave at frequency v with energy density ρ_v (in joules/m^3) for stimulated emission is

$$\frac{dN_n}{dt} = -N_n B_{nm}\,\rho_v\,g(v) \tag{16.18}$$

For absorption

$$\frac{dN_m}{dt} = N_m B_{mn}\,\rho_v\,g(v) \tag{16.19}$$

The energy density of a radiation field ρ_v can simply be related to the intensity of a plane electromagnetic wave. If the intensity of the wave is I_v (watt per unit area per frequency interval) then

$$\rho_v = \frac{I_v}{c} \tag{16.20}$$

where c is the velocity of light in the medium.

Consider an incident beam of light of intensity I_v passing through a medium of thickness Δz and cross sectional area A as shown in Fig.16.5. After emerging from the medium, the output consists of incident intensity plus that added by the radiative processes, that is spontaneous emission, stimulated emission and absorption. Each of these processes contribute (or subtract) a photon of energy hv. However, we neglect the spontaneous emission contribution, since it is radiated in all directions into a solid angle of 4π and this contribution is very little in the direction of the incident beam.

The amount of energy per unit time added is the difference between the number of transitions per unit time m to n and the number of transitions per unit time from n to m within the volume, multiplied by hv per transition.

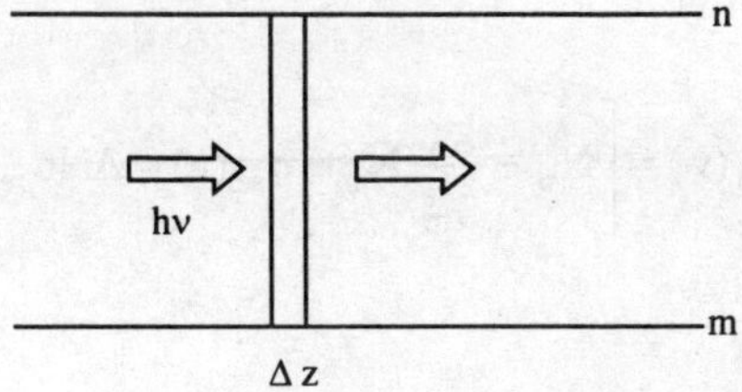

Fig. 16.5 Elemental change in incident photon flux for a plane electromagnetic wave while travelling a distance dz through the material

$$[I_v(z+dz) - I_v(z)]A = [N_n B_{nm} - N_m B_{mn}]\, g(v)\, \frac{I_v}{c}\, hv\, dz$$

or

$$\frac{dI_v}{dz} = (N_n B_{nm} - N_m B_{mn})\, g(v)\, \frac{I_v}{c}\, hv$$

using Eq. (16.14)

$$\frac{dI_v}{dz} = (N_n - \frac{g_n}{g_m} N_m) B_{nm}\, g(v)\, \frac{I_v}{c}\, hv$$

Substituting the value of B_{nm} from Eq. (16.15)

$$\frac{dI_v}{dz} = \frac{A_{nm} c^2}{8\pi v^2}\, g(v) \left[N_n - \frac{g_n}{g_m} N_m \right] I_v \tag{16.21}$$

$$\frac{dI_v}{dz} \cong \gamma_0(v)\, I_v \tag{16.22}$$

where

$$\gamma_0 = \frac{A_{nm} c^2}{8\pi v^2}\, g(v) \left[N_n - \frac{g_n}{g_m} N_m \right] \tag{16.23}$$

is called the gain coefficient (m^{-1}) with the subscript 0 indicating that the incident intensity is sufficiently small to cause negligible perturbation on the population of the levels n and m. Sometimes it is convenient to express $\gamma_0(v)$ in terms of the stimulated emission cross section defined by

$$\sigma_{se} = \frac{A_{nm} c^2}{8\pi v^2}\, g(v) \tag{16.24}$$

and Eq. (16.23) is

$$\gamma_0(v) = \left[N_n - \frac{g_n}{g_m} N_m \right] \sigma_{se}(v) = \Delta N \sigma_{se}(v) \tag{16.25}$$

where

$$\Delta N = N_n - \frac{g_n}{g_m} N_m \tag{16.26}$$

From Eq. (16.22)

$$\frac{dI_v}{I_v} = \gamma_0(v)\, dz$$

On integration

$$I_v(z) = I_v(0)\, \exp[\gamma_0(v)\, z] = G_0(v)\, I_v(0) \tag{16.27}$$

$$I_v(z) = I_v(0)\, \exp\left[\sigma_{se} \left(N_n - \frac{g_n}{g_m} N_m \right) z \right] \tag{16.28}$$

where $I_v(0)$ is the intensity of the beam as it enters the medium. $I_v(z)$ represents the intensity at some distance z. $G_0(v)$ is the power gain of amplifier of length z. If the value of the exponent is positive, the beam will increase in intensity or amplification will occur. If the exponent is negative, the intensity of the beam will decrease and absorption will occur. Since σ_{se} and z are always positive, the amplification will occur if

$$N_n > \frac{g_n}{g_m} N_m \tag{16.29}$$

or

$$\frac{g_m N_n}{g_n N_m} > 1$$

The case of upper level n being more populated (taken into account the statistical weight) than the lower level m is referred to as a population inversion. Population inversion is a necessary condition for amplification. A material having a population inversion is called an active material.

If the frequency of transitions between levels n and m falls in the microwave region then this type of amplifier is called MASER. The word MASER is an acronym for microwave amplification by stimulated emission of radiation. If the transition frequency falls in far or near infrared, in the visible, ultraviolet or even in X-ray region, the amplifier is called LASER. The word LASER is again an acronym for light amplification by stimulated emission of radiation.

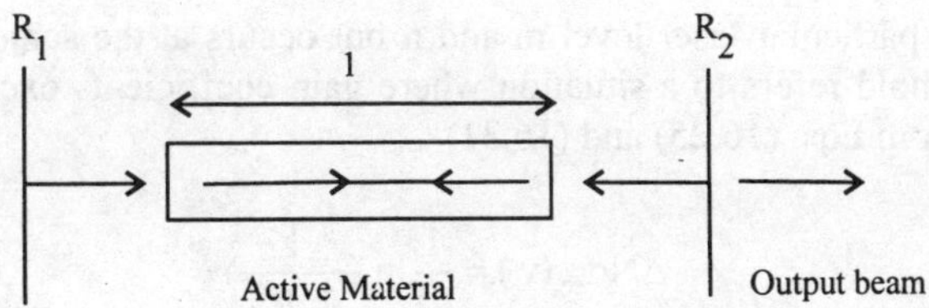

Fig. 16.6 Scheme of Laser

To make an oscillator from an amplifier, it is necessary to introduce suitable positive feedback. Fig. 16.6 shows the basic elements of a laser oscillator. It is composed of two mirrors (e.g. plane mirrors) having reflectivity R_1 and R_2 and an active material of length l. The region bounded by the mirrors is called the laser cavity. One of the two mirrors is made partially transparent to obtain output beam. For gain

$$N_n > \frac{g_n}{g_m} N_m .$$

Since the length of the active material is l, the small signal power gain is

$$G = \exp[\gamma_0(v)l] \tag{16.30}$$

In each pass through the material, the intensity increases by a factor of exp $[\gamma_0(v)l]$. At each reflection a fraction $(1 - R_1)$ or $(1 - R_2)$ of the energy is lost. Starting at one point, the electromagnetic radiation will suffer two reflections before it can pass the same point in the original direction. When the electromagnetic radiation crosses length l it is amplified by a factor of exp $[\gamma_0(v)l]$. After reflection from mirror R_1, a part of it is lost because of mirror losses and R_1 exp $[\gamma_0(v)l]$ is reflected back. This radiation is again amplified by a factor of exp $[\gamma_0(v)l]$. as it crosses length l. The radiation reaching the mirror R_2 is R_1 exp $[2\gamma_0(v)l]$. After reflection from mirror R_2, a part of this is lost and R_2 times R_1 exp $[2\gamma_0(v)$ l] is reflected back in the original direction. The threshold condition is established by requiring that photon density after reflection from R_1 and R_2 be equal to the initial photon density, that is,

$$R_1 R_2 \exp[2\gamma_0(v)l] \geq 1$$

or

$$\gamma_0(v) \geq \frac{1}{2l} \ln \left(\frac{1}{R_1 R_2}\right) \tag{16.31}$$

If we include possible distributed losses α per unit length within the gain medium then

$$R_1 R_2 \exp[\gamma_0(v_0) - \alpha]2l = 1$$

$$\gamma_0(v) = \frac{1}{2l} \ln \frac{1}{R_1 R_2} + \alpha = \alpha - \frac{1}{2l} \ln(R_1 R_2) \tag{16.32}$$

α does not involve the particular laser level m and n but occurs at the same frequency as that of the laser beam. Thus threshold refers to a situation where gain coefficients exceed the loss over a small band of frequencies. From Eqs. (16.25) and (16.31)

$$\Delta N \sigma_{se}(\nu) = \frac{1}{2l} \ln \left(\frac{1}{R_1 R_2}\right)$$

$$\Delta N = \frac{1}{2l\sigma_{se}} \ln \left(\frac{1}{R_1 R_2}\right)$$

From Eq. (16.26)

$$N_n - \frac{g_n}{g_m} N_m = \frac{1}{2l\sigma_{se}} \ln \left(\frac{1}{R_1 R_2}\right)$$

If the levels are non degenerate, that is, $g_n = g_m = 1$

$$N_n - N_m = \frac{1}{2l\sigma_{se}} \ln \left(\frac{1}{R_1 R_2}\right) \tag{16.33}$$

Thus threshold is reached when population inversion is given by Eq. (16.33). $N_n - N_m$ of Eq. (16.33) is known as critical inversion. Once the critical inversion is achieved, the oscillations will build up from the spontaneous emissions. The photons, which are spontaneously emitted in a direction along the axis of the laser cavity, initiate the amplification process.

16.4 Properties of Laser Beams

Laser radiation is characterized by an extremely high degree of monochromaticity, coherence, directionality and brightness.

16.4.1 Monochromaticity

Monochromaticity of laser radiation results from the fact that all the photons are emitted because of a transition between the same two atomic or molecular energy levels, and hence have almost exactly the same frequency. Further, the laser cavity is resonant only for the frequencies of resonant cavity limits the frequency range. However, as the energy levels are not sharp, there is always a small spread to the frequency distribution, which may cover several discrete frequencies. The result is that a small number of closely spaced frequencies may appear in a laser action i.e. the light is not monochromatic. In order to achieve the optimum monochromaticity, generally an etalon is placed within the laser cavity and arranged so that only well defined wavelength can travel back and forth between end mirrors.

One of the important factors which characterizes laser is the quality factor Q defined as the ratio of emission frequency ν to the linewidth, that is,

$$Q = \frac{\nu}{\Delta\nu} = \frac{\lambda}{\Delta\lambda} \tag{16.34}$$

where ν is the central frequency of the laser beam. The degree of non - monochromaticity ξ of laser radiation is the reciprocal of quality factor i.e.

$$\xi = \frac{1}{Q} = \frac{\Delta\nu}{\nu} = \frac{\Delta\lambda}{\lambda} \qquad (16.35)$$

16.4.2 Coherence

Coherence is the property, which results from the nature of the stimulated emission process. There are two types of coherence: temporal coherence and spatial coherence. Consider two points P and Q a distance L apart in the direction of propagation of laser beam as shown in Fig. 16.7. If a definite and fixed phase relationship exists between the wave amplitudes at P and Q at time t and t + τ, then wave

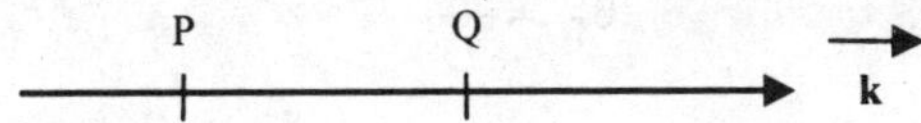

Fig. 16.7 Diagram for explaining temporal coherence

shows temporal coherence for time τ. The maximum separation L at which the fixed relationship is retained is called the coherence length L_c. The coherence length is related to the coherence time τ_c by

$$L_c = c\,\tau_c \qquad (16.36)$$

The coherence time is related to monochromaticity of the laser by

$$\tau_c \cong \frac{1}{\Delta\nu_L} \qquad (16.37)$$

and

$$L_c \approx \frac{c}{\Delta\nu_L} \approx \frac{\lambda^2}{\Delta\lambda} \qquad (16.38)$$

Spatial coherence in the simplest case tells us about the phase relationship between the field amplitudes at two points P and Q in a plane normal to the wave vector $\mathbf{k}$ as shown in Fig.16.8. Light

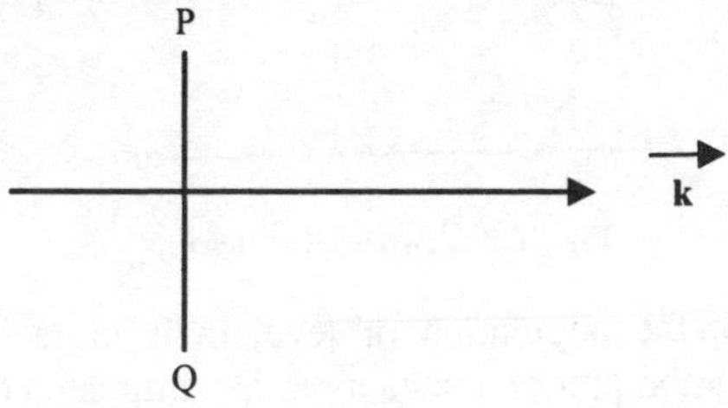

Fig. 16.8 Diagram for explaining spatial coherence

that is emitted from different parts of the conventional light source is not phase- related. Therefore, the light from an extended light source will not be spatially coherent.

A good measure of coherence of a light source is its ability to produce stable interference fringes. Temporal coherence can be studied with a Michelson interferometer. The existence of spatial coherence between the fields at two points can be demonstrated in a Young's slits experiment.

16.4.3 Directionality

Stimulated emission produces photons with almost precisely identical direction of propagation. The mirrors selectively amplify axial beam. The laser thus emits a narrow parallel beam from its output mirror. The extent of beam divergence is essentially determined by the diffraction limit of output aperture. From diffraction theory, the divergence angle θ_D is

$$\theta_D = \frac{\beta \lambda}{D} \tag{16.39}$$

where λ and D are wavelength and diameter of the laser beam, respectively. B is a coefficient whose value is around unity.

If the partial spatial coherent beam has a given intensity distribution over the diameter D and a given coherence area A_{sc}, its divergence is bigger than the diffraction limited divergence, then it can be shown

$$\theta_d = \frac{\beta \lambda}{(A_{sc})^{1/2}} \tag{16.40}$$

16.4.4 Brightness

It is defined, as the power emitted from unit area of the output mirror per unit solid angle. It is extremely high. The reason for this is that although the power may be small but it is distributed over solid angle which is small.

16.5 Rate Equations

Consider a two level system as shown in the Fig. 16.9. At thermal equilibrium the population of level m is more than that of level n . An incoming radiation of appropriate frequency will cause stimulated

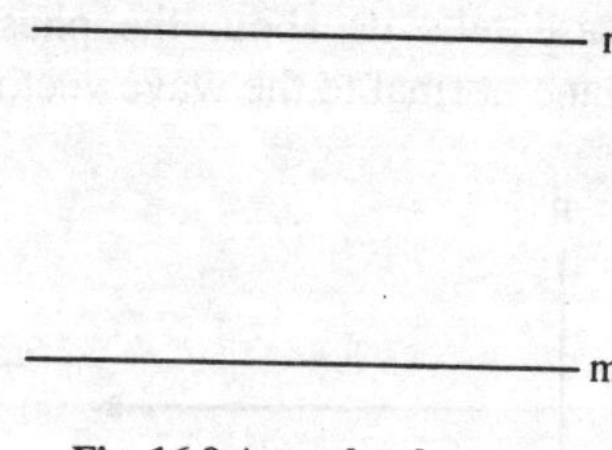

Fig. 16.9 A two level system

emission and absorption. Since the population of level m is large in comparison with that of n, therefore, the absorption of radiation predominates over the stimulated emission. As a result of this the population of level n increases and a condition is reached when the population of level m and n becomes equal. The absorption and stimulated processes will compensate each other and there will be no population inversion.

Now consider a three level system as shown in Fig.16.10. The population of various levels is governed by the Boltzmann distribution. The atoms are in some way raised from level 1 to level 3. Let

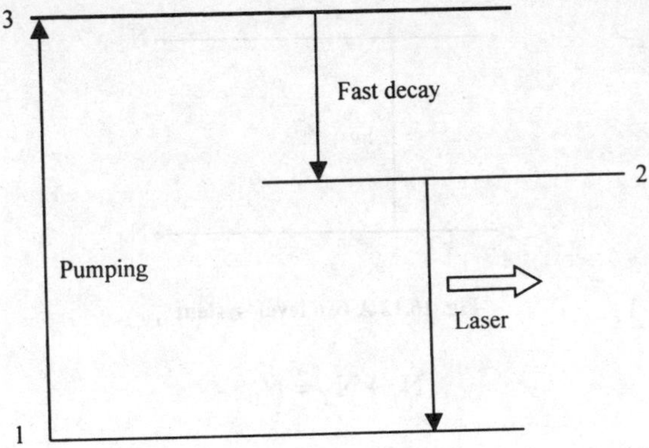

Fig. 16.10 A three level laser system

the atom raised to level 3 decay very rapidly to level 2 and lifetime of level 2 is very large. Thus the atoms rises from level 1 to 3 decay rapidly to level 2. Because the lifetime of level 2 is very large, therefore the atoms start accumulating in level 2 and a population inversion can be achieved between levels 2 and 1.

Fig.16.11 shows a four level laser system. The atoms are raised from level 0 to level 3. The lifetime of level 3 is very small; therefore the atoms, which are transferred to level 3, decay rapidly to level 2. If the lifetime of level 2 is large, a population can be built up in level 2. Since the population of various levels are governed by Boltzmann distribution, the level 1 is nearly empty. Hence a population inversion between levels 2 and 1 is achieved. For maintaining population inversion it is necessary that atoms reaching level 1 should quickly decay to level 0 so that level 1 remains more or less empty. The process by which atoms are raised from level 1 to 3 in three level laser system and from level 0 to level 3 in a four level laser system is known as pumping.

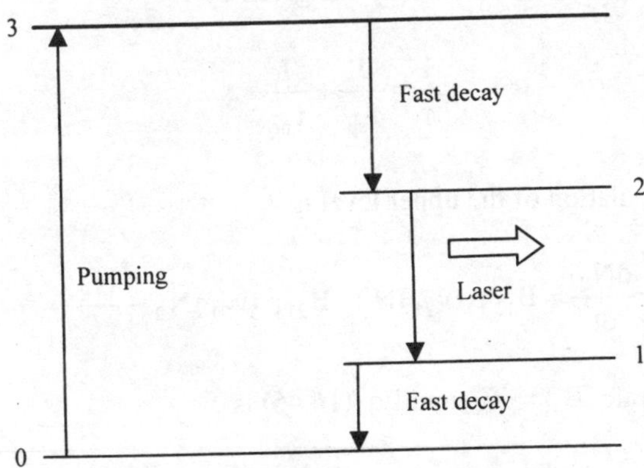

Fig. 16.11 A four level laser system

16.5.1 A two level system

Consider a hypothetical atom having just two levels (Fig. 16.12). Let the population of two levels be N_1 and N_2 per unit volume. We assume that total population of two levels is constant, that is,

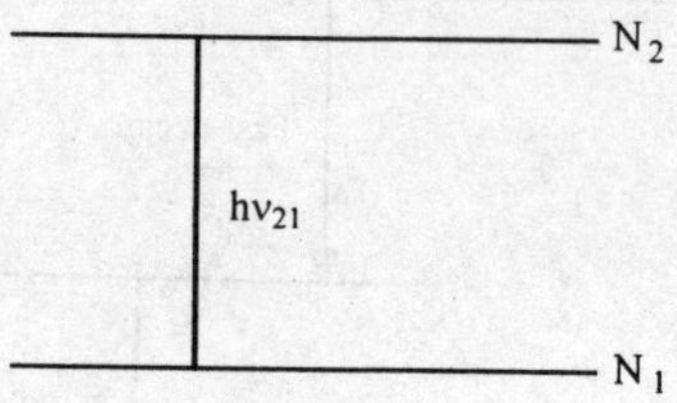

Fig. 16.12 A two level system

$$N_1 + N_2 = N_0 \tag{16.41}$$

$$\Delta N = N_1 - N_2 \tag{16.42}$$

combining Eqs. (16.41) and (16.42)

$$N_1 = \frac{N_0 + \Delta N}{2} \tag{16.43}$$

$$N_2 = \frac{N_0 - \Delta N}{2} \tag{16.44}$$

Consider a monochromatic electromagnetic wave of frequency ν_{21}. This electromagnetic radiation will be absorbed and some of the atoms will go from level 1 to level 2. Excited atoms lose their energy by radiative and non-radiative transitions. The probability for spontaneous emission is $A_{21} = 1/\tau_{sp}$ and for non-radiative transition it is $1/\tau_{nr}$ where τ_{sp} and τ_{nr} are spontaneous emission decay time and non-radiative process lifetime. The overall time decay τ is given by

$$\frac{1}{\tau} = \frac{1}{\tau_{sp}} + \frac{1}{\tau_{nr}} \tag{16.45}$$

The rate of change of population of the upper level is

$$\frac{dN_2}{dt} = B_{12}\,\rho\,(\nu_{21})N_1 - B_{21}\rho\,(\nu_{21})N_2 - \frac{N_2}{\tau} \tag{16.46}$$

If the levels are non degenerate, $B_{21} = B_{12}$ and Eq. (16.46) is

$$\frac{dN_2}{dt} = B_{12}\rho\,(\nu_{21})\Delta N - \frac{N_2}{\tau} \tag{16.47}$$

In the steady state $\dfrac{dN_2}{dt} = 0$. From Eqs. (16.44) and (16.47)

$$\Delta N = \frac{N_0}{1 + 2B_{12}\tau\rho(\nu_{21})} \tag{16.48}$$

Thus the population difference between the two levels in a steady state depends on the decay time of the upper level and on the density of incident radiation. $B_{12}\rho(\nu_{21})$ is the probability per unit time that the atoms are excited to the upper level and is called the pumping rate denoted by W_p. Eq. (16.48) becomes

$$\Delta N = \frac{N_0}{1 + 2W_p\tau} \tag{16.49}$$

whatever the value of W_p is, ΔN is always positive and hence population inversion is not possible. A two level system is not suitable for lasing action.

16.5.2 Three level laser

Consider a three level laser system as shown in Fig. 16.13. Assume that all the levels are non-degenerate. energy levels 1 and 2 are the levels involved in the actual laser transition. The third level is required to achieve a population inversion. The pump lifts atoms from the level 1 to level 3. The level 3 is never populated because there is nearly instantaneous, non-radiative transition from level 3 to level 2. The level 2 is required to be metastable. Thus pump effectively transfers atoms from level 1 to level 2 through level 3.

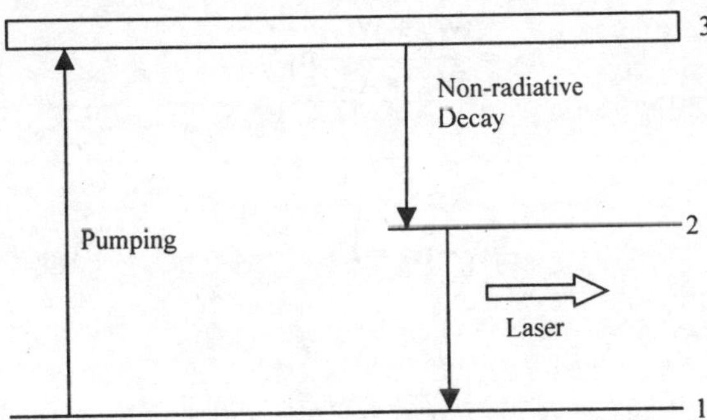

Fig. 16.13 A three level laser system

The rate equation for the time development of N_2 is

$$\frac{dN_2}{dt} = W_p N_1 + W_{12}N_1 - W_{21}N_2 - A_{21}N_2 \tag{16.50}$$

where W_p is pumping rate per atom, W_{21} and A_{21} the stimulated and spontaneous rates from level 2. Because $N_3 \approx 0$, all the atoms exist in level 1 or 2. The first term in Eq. (16.50) represents the number of atoms arriving at level 2 from level 1 via level 3 by the pumping. It is assumed that all the atoms arriving at level 3 from level 1 are instantaneously transferred to level 2 by non-radiative transitions and level 3 remains empty. The second term represents the induced absorptions from level 1 to level 2. The third term represents the number of stimulated emissions from level 2. The last term represent the spontaneous emission. Because stimulated emission and absorption cross-section are equal, hence $W_{12} = W_{21}$. In steady state

$$\frac{dN_2}{dt} = 0 \tag{16.51}$$

and from which

$$(W_{21} + A_{21})N_2 = (W_p + W_{12})N_1 \tag{16.52}$$

Dividing both sides by $N = N_1 + N_2$

$$\frac{N_2}{N}(W_{21} + A_{21}) = \frac{N_1}{N}(W_p + W_{12}) \tag{16.53}$$

or

$$n_2(W_{21} + A_{21}) = n_1(W_p + W_{12}) \tag{16.54}$$

where n_1 and n_2 are normalised populations. From Eq. (16.54)

$$n_2 = \frac{W_p + W_{12}}{W_{21} + A_{21}}n_1 \tag{16.55}$$

Since

$$n_1 + n_2 = 1 \tag{16.56}$$

From Eqs. (16.55) and (16.56)

$$n_2 = \frac{W_p + W_{12}}{W_p + A_{21} + 2W_{12}} \tag{16.57}$$

$$n_1 = \frac{W_{12} + A_{21}}{W_p + A_{21} + 2W_{12}} \tag{16.58}$$

$$n_2 - n_1 = n = \frac{W_p - A_{21}}{W_p + 2W_{12} + A_{21}} \tag{16.59}$$

For amplification

$$W_p > A_{21} \tag{16.60}$$

Thus we must pump to level 2 faster than n_2 is depleted by spontaneous emission. For Eq. (16.60) to be satisfied it is desirable that the lifetime of level 2 should be large, pumping source should be intense and material should strongly absorb the pump energy.

Suppose a parallel beam of light at frequency v_p of pumping light is incident on the surface of an optically thin sample of the material. The power of the pumping light is

$$P = n_i h v_p$$

where n_i is the number of incident photons. The flux density F_p

$$F_p = \frac{P}{A} = \frac{n_i h v_p}{A} \tag{16.61}$$

where A is the area of the sample perpendicular to the incident beam. The pumping rate is by definition

$$W_p = \frac{n_i}{A} \sigma_p \tag{16.62}$$

where σ_p is the absorption cross section of pump light. W_p is the pump power per unit area. For the threshold condition from Eqs. (16.61) and (16.62) we get

$$F_p = \frac{W_p h v_p}{\sigma_p} \tag{16.63}$$

The threshold condition for population inversion is obtained by putting $W_p = A_{21}$

$$F_p = \frac{A_{21} h v_p}{\sigma_p} = \frac{h v_p}{\sigma_p \tau} \tag{16.64}$$

Eq. (16.64) is the required power density of the laser.

Let us estimate the total power. Suppose the pump power absorbed at v_p is equal to P. The total output power is

$$P_0 \approx P \frac{v_{21}}{v_p}$$

because in the steady state, each excitation to the pumping level results in single emission from upper laser level 2. Let us assume that laser action starts when $n_2 - n_1$ is slightly greater than zero. Then nearly half of the atoms are in the upper level and this condition will persist as long as pumping is continued. Let $N_1 + N_2 = N_0$ and $N_1 \approx N_0/2$. The total pumping power absorbed is

$$P = \frac{N_0 V}{2} W_p h\nu_p$$

where V is the volume of the active material. For threshold condition $W_p = A_{21}$, therefore

$$P = \frac{N_0 V}{2} A_{21} h\nu_p = \frac{N_0 V h\nu_p}{2\tau}$$

16.5.3 Four level laser

Let us consider a laser operating on four level scheme as shown in Fig. 16.14. There is only one broad level (level 3). Level 0 is the ground level and levels 1,2 and 3 are excited levels of the system. Atoms from level 0 are excited by a pump to level 3 from which the atoms decay rapidly through some non-radiative transitions to level 2. Level 2 is a metastable level having a long lifetime ($\sim 10^{-3}$s). The level 2 forms the upper laser level and level 1 forms the lower laser level. The lifetime of the level 1 is very short. Therefore, the incoming atoms from level 2 relax down immediately from level 1 to level 0 ready for being pumped to level 3. If the rate of relaxation of atoms from level 1 to level 0 is faster than the rate of arrival of atoms into level 2, one can obtain population inversion between levels 2 and 1 even for very small pump power.

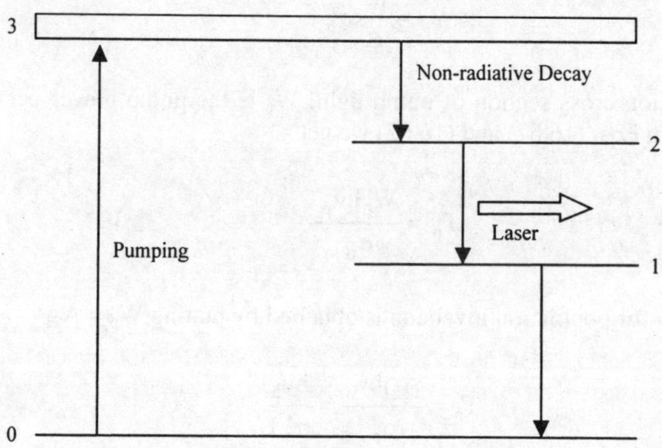

Fig. 16.14 A four level laser system

Let N_0, N_1, N_2 and N_3 represent the population per unit volume of the levels 0, 1, 2 and 3, respectively. The rate of change of population of level 3 is

$$\frac{dN_3}{dt} = B_{03}\rho\,(\nu_{03})N_0 - B_{30}\rho\,(\nu_{03})N_3 - N_3 T_{32} \tag{16.65}$$

$$\frac{dN_3}{dt} = W_p(N_0 - N_3) - N_3 T_{32} \tag{16.66}$$

Here we have ignored the relaxation from level 3 to 1 and 0. $W_p = B_{03}\, \rho\,(\nu_{03})$ and $B_{03} = B_{30}$. The first term represents the net rate of stimulated transitions between levels 0 and 3 caused by the pump. T_{32} is the relaxation rate (both radiative and non radiative) from level 3 to 2.

The rate equation for N_2 would be

$$\frac{dN_2}{dt} = N_3 T_{32} + W_1(N_1 - N_2) - T_{21}N_2 \tag{16.67}$$

The first term represents the rate at which atoms are transferred from level 3 to 2, the second term represents the rate of stimulated transitions from level 1 to 2 because of the presence of laser radiation. Third term represents the rate of loss of atoms from level 2 to level 1, through spontaneous emission. Transitions from level 2 to level 0 are ignored.

The rate equation for N_1 would be

$$\frac{dN_1}{dt} = W_1(N_2 - N_1) - T_{10}N_1 + T_{21}N_2 \tag{16.68}$$

Similarly, the rate equation for N_0 would be

$$\frac{dN_0}{dt} = T_{10}N_1 + W_p(N_3 - N_0) \tag{16.69}$$

Let

$$N = N_0 + N_1 + N_2 + N_3 \tag{16.70}$$

From $\dfrac{dN}{dt} = 0$ we have

$$\frac{dN_0}{dt} + \frac{dN_1}{dt} + \frac{dN_2}{dt} + \frac{dN_3}{dt} = 0 \tag{16.71}$$

At steady state

$$\frac{dN_0}{dt} = \frac{dN_1}{dt} = \frac{dN_2}{dt} = \frac{dN_3}{dt} = 0 \tag{16.72}$$

From $\dfrac{dN_3}{dt} = 0$

$$N_3 = \frac{W_p N_0}{W_p + T_{32}} \tag{16.73}$$

From $\dfrac{dN_2}{dt} = 0$ we have

$$N_2 = \frac{1}{T_{21} + W_1}\left[W_1 N_1 + \frac{W_p N_0}{W_p + T_{32}}T_{32}\right]$$

(16.74)

Similarly, from $\dfrac{dN_1}{dt} = 0$ we get

$$N_2 = \frac{W_1 + T_{10}}{W_1 + T_{21}}N_1$$

(16.75)

From Eqs. (16.74) and (16.75)

$$N_1 = \frac{W_p T_{32}}{(W_p + T_{32})T_{10}}N_0$$

(16.76)

$$N_2 = \frac{(W_1 + T_{10})W_p T_{32}}{(W_1 + T_{21})(W_p + T_{32})T_{10}}N_0$$

(16.77)

$$\frac{N_2}{N_1} = \frac{W_1 + T_{10}}{W_1 + T_{21}}$$

(16.78)

In order to obtain a population inversion, that is, $N_2 > N_1$ we must have

$$T_{10} > T_{21}$$

(16.79)

that is, the rate at which atoms relax from level 1 to level 0 must be greater than the rate at which atoms relax from level 2 to level 1. Under such a condition, the creation of population inversion between level 2 and level 1 is independent of the pumping power, but the magnitude of population inversion depends on W_p.

At and below threshold of oscillations $W_1 \approx 0$ and population difference between levels 2 and 1 is

$$N_2 - N_1 = \frac{W_p T_{32} N_0}{(W_p + T_{32})T_{10}}\left(\frac{T_{10}}{T_{21}} - 1\right)$$

(16.80)

From Eqs. (16.70), (16.73), (16.76) and (16.77)

$$N_0 = \frac{N[W_p + T_{32}]T_{10}T_{21}}{2W_p T_{21}T_{10} + W_p T_{32}T_{10} + W_p T_{32}T_{21} + T_{32}T_{21}T_{10}}$$

(16.81)

From Eqs. (16.80) and (16.81)

$$N_2 - N_1 = \frac{W_p T_{32}(T_{10} - T_{21})}{2W_p T_{21} T_{10} + W_p T_{32} T_{10} + W_p T_{32} T_{21} + T_{32} T_{21} T_{10}} \qquad (16.82)$$

In the approximation $T_{10} > T_{21}; \; T_{32} > T_{21}$

$$\frac{N_2 - N_1}{N} = \frac{\dfrac{W_p}{T_{21}}}{\dfrac{2W_p}{T_{32}} + \dfrac{W_p}{T_{21}} + \dfrac{W_p}{T_{10}} + 1} \cong \frac{\dfrac{W_p}{T_{21}}}{\dfrac{W_p}{T_{21}} + 1} \qquad (16.83)$$

Since $N \gg N_2 - N_1$, we have for threshold pumping

$$\left(\frac{W_{pt}}{T_{21}} + 1\right)\left(\frac{N_2 - N_1}{N}\right) = \frac{W_{pt}}{T_{21}}$$

$$W_{pt} \cong \frac{N_2 - N_1}{N} T_{21} \qquad (16.84)$$

16.6 Methods of Obtaining Population Inversion

Since all laser emission involves radiation from excited states, the energy must be supplied to these atoms to produce the excited states. Methods of pumping, irrespective of the system of levels (three level or four level) involved and of whether lasing is to be pulsed or continuous wave (CW) falls into following categories:

   (a) Optical
   (b) Electrical
   (c) A thermal oven
   (d) Chemical reaction
   (e) Heavy particles
   (f) Ionization radiation

In view of the wide variety of techniques we will discuss only electrical pumping and optical pumping. Electrical pumping is accomplished by means of a sufficiently intense electrical discharge. In an electric discharge electrons are produced by ionization in strong electric field. The electrons achieve very high speed of 10^6 to 10^7 m/s. These electrons on collision transferred energy to atom or molecules of gas and raised them to excited states

$$M + e \rightarrow M^* + e$$

where M and M* represent the atom in the ground and excited states, respectively. For a gas consisting of mixture of gases say X and Y, X serve only to transfer energy from electrons to Y. X is excited to X*, a long lived metastable state, by electron impact. The X is in excited state and Y is in ground state as shown in Fig. 16.15. If the level Y to be pumped is of similar energy to that of X*, then there is an appreciable probability that after collision X* will transfer its energy to Y to raise it to excited state. The process is denoted by

$$X^* + Y \rightarrow X + Y^* + \Delta E$$

where the energy difference ΔE between X^* and Y^* will be added or subtracted.

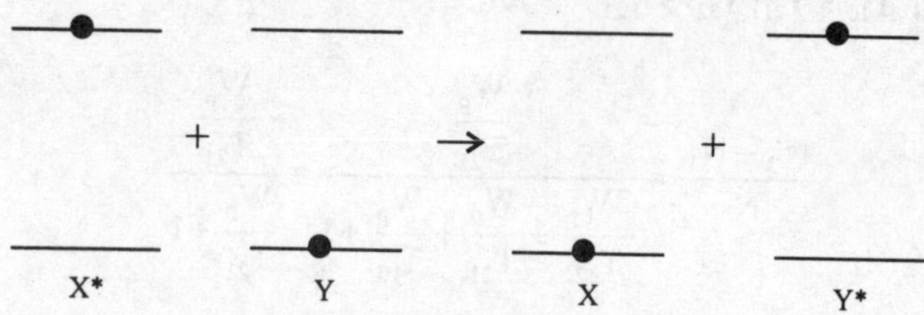

Fig. 16.15 Energy transfer between X and Y

For optical pumping, the excitation involves the transfer of the energy to the system from a high intensity light source. Optical pumping can be accomplished by many different light sources including inert gas flash lamp usually xenon at a pressure of about 100 torr. The result is a pulsed laser, the repetition rate being that of pumping source. CW optical pumping may be achieved by a continuously acting krypton or high pressure mercury lamp.

Electrical pumping is typically used in most gases and semiconductor lasers. On the other hand, optical pumping is often used in liquids (dye) lasers and crystalline solid state lasers. In solids and in liquids, electrons can not easily be accelerated by electric field to excite the laser energy levels of impurities species. The line broadening mechanisms in solids and liquids produces an appreciable broadening. Therefore, one is usually dealing with pump band rather than levels. These bands can therefore absorb a sizeable fraction of light emitted by flash lamps whose energy occurs over a broad wavelength region. However, non laser optical pumping is not feasible in gaseous systems. The gaseous system does not have in general broad absorption bands.

16.7 Laser Resonators

A population inverted medium will amplify an incoming wave of correct frequency through stimulated emission. A laser in the visible region is very noisy because of increasing importance of spontaneous emissions. According to Eq. (16.15) the signal to background ratio will be reduced by cube of frequency. Therefore, lasers are rarely used for light amplification. Lasers are normally used as light oscillator. An amplifier becomes an oscillator if feedback is introduced. This can be achieved by placing the laser medium in a resonator consisting of two mirrors (Fig. 16.6). A pair of resonator facing each other can in general form a resonant cavity. The requirements for the resonator are: (i) it should contain a sufficiently large amount of active material and (ii) it should permit amplification at only a narrow band of frequencies. Laser resonators are usually open, that is, no lateral surface is used. The dimensions of resonators are much larger than the wavelength of the laser because the resonators are open means that there would be some losses due to diffraction losses. In this section we will limit ourselves to passive resonators (which have no gain or loss).

The most widely used laser resonators have either plane or spherical mirrors of rectangular or circular shape separated by some distance.

16.7.1 Plane - parallel (or Fabry – Perot) resonator
Plane parallel resonator consists of two plane mirrors set parallel to each other as shown in Fig.16.16. To a first approximation the modes (the possible standing waves in the cavity) of this

resonator can be thought of as the superposition of two plane electromagnetic waves propagating in opposite direction along the cavity axis. For electric field of Electromagnetic wave to be zero on the two mirrors, the

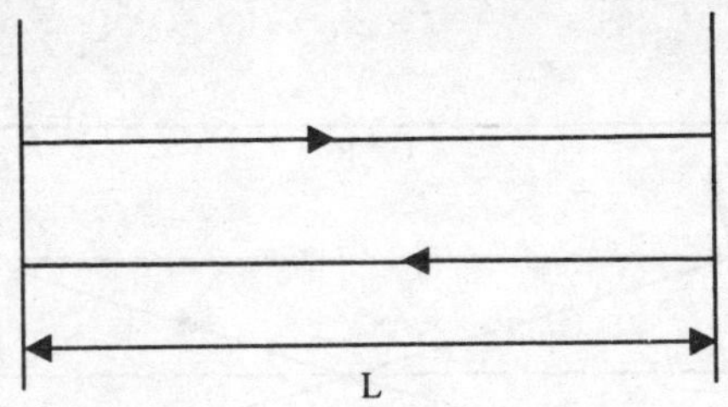

Fig. 16.16 Plane parallel resonator

cavity length must be integral number of half wavelength, that is,

$$L = \frac{n\lambda}{2}$$

$$v = \frac{nc}{2L}$$

(16.85)

The frequency difference between two consecutive modes is from Eq. (16.85)

$$v = \frac{c}{2L}$$

(16.86)

This difference is referred to as the frequency difference between two consecutive longitudinal modes. The word longitudinal is used because the n indicates the number of half wavelengths of the modes along the resonator.

16.7.2 Concentric (or spherical) resonator

This consists of two spherical mirrors with the same radius R separated by a distance L = 2R so that the centres are coincident (Fig. 16.17). In this case the resonance frequencies are given by Eq. (16.85).

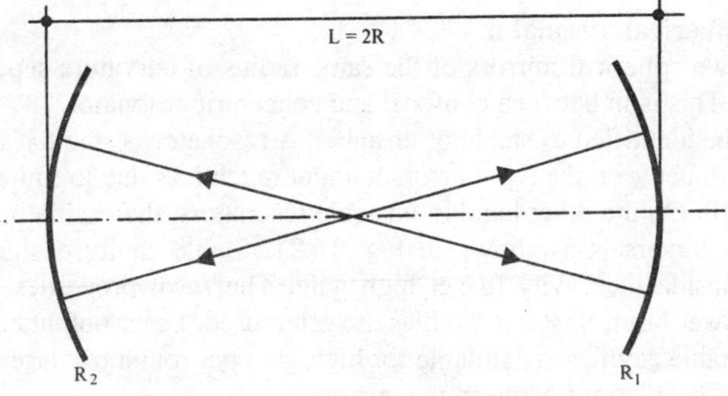

Fig. 16.17 Concentric (spherical) resonator

16.7.3 Confocal resonator

This consists of two spherical mirrors with the same radius of curvature R separated by a distance L as shown in Fig. 16.18. In this case foci of two mirrors coincide and centre of curvature of one mirror lie on the surface of another mirror. The resonant frequency cannot be readily obtained from geometrical optic consideration.

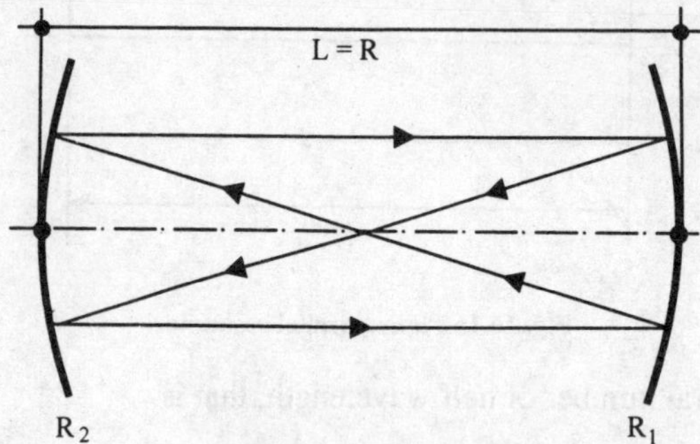

Fig. 16.18 Confocal resonator

16.7.4 Resonator using a combination of plane and spherical mirrors

Examples of these resonators are shown in Figs. (16.19) (hemiconfocal) and (16.20) (hemispherical)

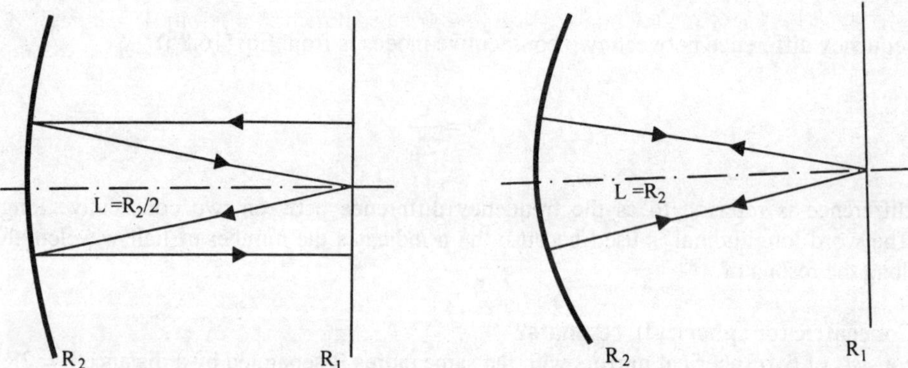

Fig. 16.19 Hemiconfocal resonator   **Fig. 16.20 Hemispherical resonator**

16.7.5 Generalized spherical resonator

This is formed by two spherical mirrors of the same radius of curvature separated by a distance L such that $R < L < 2R$. This is in between confocal and concentric resonator.

The resonators can be identified as stable or unstable. A resonator is stable if the oscillatory beam is bouncing back and forth between the two mirrors without much loss due to finite size of the mirrors as shown e.g. in Fig. 16.18. On the other hand in unstable resonators, the oscillating beam spreads out of the cavity after a few traversals as shown in Fig. 16.21. Stable cavity design allows the beam to oscillate many times inside the cavity to get high gain. The focal properties and directionality are important. For high power laser, unstable cavities are often used. Laser output comes from the edge of the output mirror. Unstable cavities are suitable for high gain per round trip laser system, which do not require large number of oscillations between the mirrors.

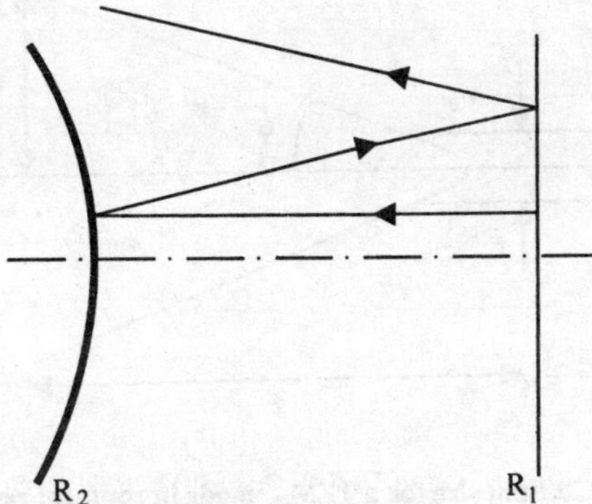

Fig. 16.21 Unstable resonator

The simplest mode in a confocal cavity is described by a ray that travels back and forth along the axis. This is 00 mode. The output of a laser oscillating in this mode is a spherical wave with a Gaussian intensity distribution. The beam width is usually expressed as the radius w at which the beam intensity falls to $1/e^2$ of its maximum value.

It is shown that the field distribution in any curved mirror cavity is characterized by a beam waist, which for a symmetrical cavity, is located in the centre of the cavity. The intensity distribution in the plane of the waist for 00 mode is

$$I(r) = \exp\left(-\frac{2r^2}{w_0^2}\right) \tag{16.87}$$

where

$$w_0 = \left(\frac{L\lambda}{2\pi}\right)^{1/2} \tag{16.88}$$

and r is the distance from the centre of the beam. For convenience, the intensity is normalized to 1 at the centre of the beam. Fig. 16.22 shows the propagation of the beam, which is Gaussian both inside and outside the cavity. It is shown that at a distance z from the beam waist

$$w(z) = w_0\left[1+\left(\frac{\lambda z}{\pi w_0^2}\right)^2\right]^{1/2} = w_0\left[1+\left(\frac{2z}{L}\right)^2\right]^{1/2} \tag{16.89}$$

When the cavity is not confocal, it is customary to define dimensionless stability parameters g_1 and g_2 by the equations

$$g_1 = 1 - \frac{1}{R_1} \tag{16.90}$$

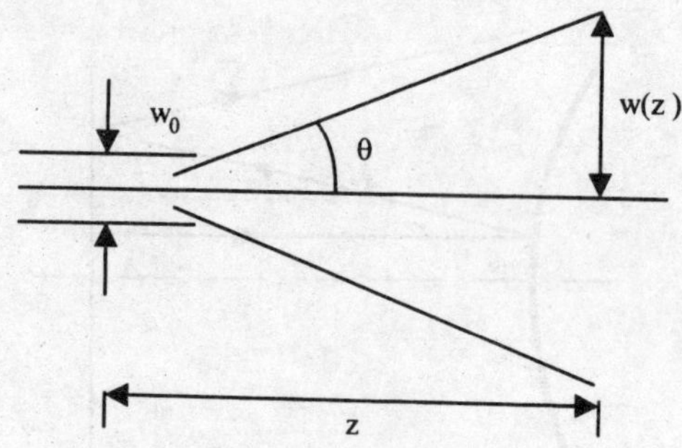

Fig. 16.22 Spot size for a TEM_{00} mode in confocal resonator

$$g_2 = 1 - \frac{1}{R_2} \qquad (16.91)$$

where R_1 and R_2 are the radii of the curvature of the mirrors separated by a distance L. The sign of radius of curvature is taken to be positive for concave mirrors and negative for convex mirrors.

For stable electric field distribution, that is, in which the beam retraces its path and beam resonates inside the resonator, the radius of curvature of the beam wavefront at the positions of the mirror must be exactly equal to the radii of curvature of the wavefront at two locations z_1 and z_2. Distances z_1 and z_2 from waist to mirror R_1 and R_2 are known. It is shown that

$$w_0 = \left(\frac{\lambda L}{\pi}\right)^{1/2} \left[\frac{g_1 g_2 (1 - g_1 g_2)}{g_1 + g_2 - 2g_1 g_2}\right]^{1/4} \qquad (16.92)$$

At mirror R_1

$$w(z_1) = \left(\frac{\lambda L}{\pi}\right)^{1/2} \left[\frac{g_2}{g_1(1 - g_1 g_2)}\right]^{1/4} \qquad (16.93)$$

At mirror R_2

$$w(z_2) = \left(\frac{g_1}{g_2}\right)^{1/2} w(z_1) \qquad (16.94)$$

The expressions for $w(z_1)$ and $w(z_2)$ contain $(1 - g_1 g_2)$. As $g_1 g_2 \to 0$, $w(z)$ at either or both the mirrors become infinite. Laser cavities, for which $g_1 g_2 > 1$ are unstable and for which the product is just less than 1, are on the border of stability, because the spot size may exceed the mirror size and bring about great loss. Thus stability criterion for lasers is

$$0 \leq g_1 g_2 = \left(1 - \frac{L}{R_1}\right)\left(1 - \frac{L}{R_2}\right) \leq 1 \qquad (16.95)$$

This condition can be expressed in the form of stability diagram as shown in Fig. 16.23. The clear

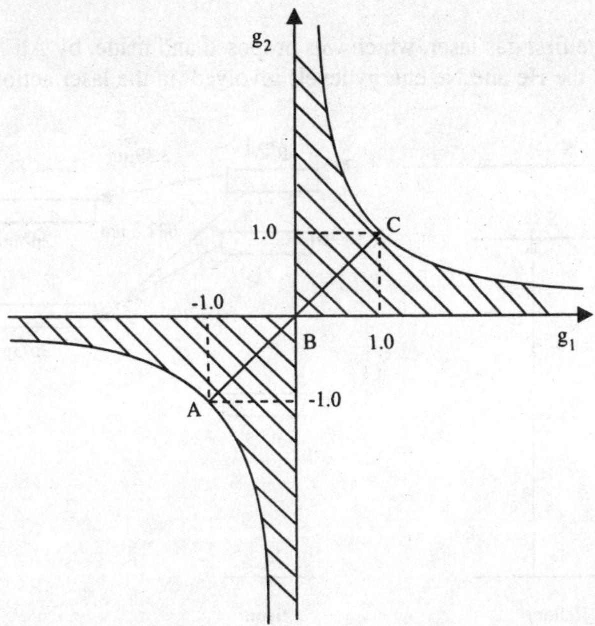

Fig. 16.23 Stability diagram for two mirrors with radii of curvature R_1 and R_2

regions are the regions for which $g_1 g_2 > 1$ and cavity is unstable. For the shaded region $g_1 g_2 < 1$ and the cavity is stable. The limiting case $g_1 g_2 = 1$ is a hyperbola. Stable resonator lies between the two branches of the hyperbola and the axes; unstable resonator lies outside the two branches. For shaded region, Eq. (16.95) is satisfied and the cavity is stable. Three particular points are of interest. These points represent cavities on the verge of instability:

$$R_1 = R_2 = \frac{L}{2} \quad \text{(symmetric concentric)}$$

$$R_1 = R_2 = L \quad \text{(confocal)}$$

$$R_1 = R_2 = \infty \quad \text{(plane parallel)}$$

All three of these are on the edge of stability in the diagram and can become extremely loosy for slight deviation from the shaded region.

The plane parallel and concentric resonators are sensitive to mirror misalignment. The concentric resonators produce a very small spot on the resonator centre. Confocal resonators typically give a spot size that is too small for effective use of all available cross section of the laser medium. The most commonly used laser resonator uses two concave mirrors of large radius of curvature or a plane mirror and a concave mirror of large radius. The resonator gives a spot size somewhat larger than the confocal resonator and a reasonable stability against misalignment.

16.8 Specific Laser Systems

16.8.1 He-Ne laser

The He-Ne laser is the first gas laser, which was proposed and made, by Ali Javan and coworkers in 1959. Fig. 16.24 shows the He and Ne energy levels involved in the laser action. The energy levels of

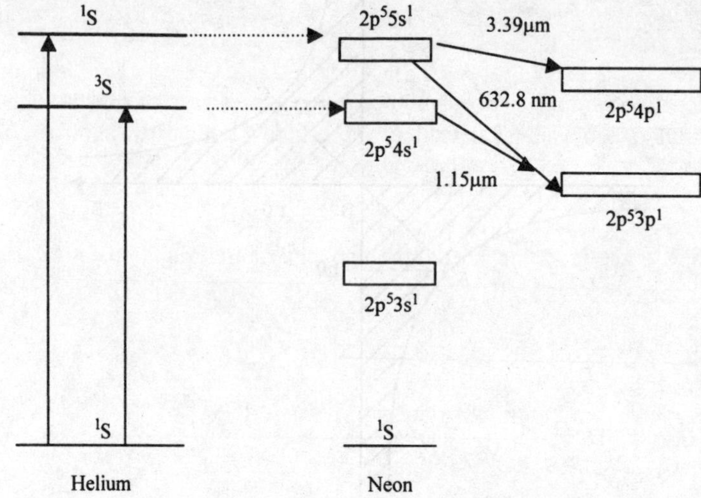

Fig. 16.24 Helium and neon energy levels and the He-Ne laser transitions

He atom are described in Sec.4.3 . The ground state is 1^1S. The excited ^{1}S and ^{3}S are metastable. The normal electronic configuration of Ne (Z=10) is $1s^22s^22p^6$. The ground state is ^{1}S. The excited configuration gives rise to states for which the L - S coupling approximation does not apply. The configurations giving rise to excited states of Ne are of the type $2p^5ns$ and $2p^5np$ (n > 2). These configurations give rise to four (and 12 states) and ten energy levels (and 36 states), respectively. The lifetime of the states corresponding to $2p^5ns$ configurations are somewhat longer (100 ns) than those obtained from the configurations $2p^5np^1$ (~10 ns).

The states arising from the $2p^55s^1$ configuration of Ne have very similar energy to the 2^1S state of He. Similarly the states arising from the $2p^54s^1$ configuration of Ne lie just below 2^3S state of He. An electric discharge is created in He-Ne gas mixture by applying either a DC voltage or microwave radiation. In such a discharge, some fraction of the gas atoms are ionized, creating a mixture of neutral atoms, positively charged ions, the free electrons, and the plasma. Applied electric field accelerates the lighter electrons more effectively than ions or atoms. Accelerated electrons strike the He atoms and promote them to excited states, 2^1S and 2^3S. However, they are forbidden to return to the ground state 1^1S by electric dipole transition and also they are spin forbidden. The Ne atoms are not very strongly excited by electron impact . The lifetime of the He ^{1}S state is 10^{-4} s. The ^{1}S and ^{3}S excited states of He are very close in energy to excited electronic states of Ne (energy difference $2^1S_0 - 3s$ and $2^3S_1 - 2s \approx$ 400 cm^{-1} $\approx k_BT$). Thus once a He atom has been lifted to the 2^3S and 2^1S levels, there is a fair chance that before it decays downward by other means, it will collide with an unexcited Ne atom. If this occurs, there is finite probability an energy transfer between He and Ne atoms may take place, that is, He atom may drop back to the $^1S_0$ state, giving up its energy of excitation, while Ne atom simultaneously absorbs the energy and is excited upwards. Thus electrons pump the He atoms and the He atoms pump the Ne atoms. This mechanism in effect provides a continuous supply of Ne atoms preferentially into the Ne $2s(2p^54s^1)$ and 3s $(2p^55s^1)$ levels. As a consequence, these levels acquire a steady state population, which, though small, can be larger than the population of several lower levels. A population inversion is thus achieved between 2s and 2p $(2p^53p^1)$, 3s and $3p(2p^54p^1)$ levels of Ne. A large number of laser transitions take place between these levels and prominently studied among them are, 1.1523 μm $(2s_2 - 2p_4)$, 632.8 nm $(3s_2\text{-}2p_4)$ and 3.3913 μm $(3s_2\text{-}3p_4)$. The lifetime of the p states of Ne is very small, therefore, atoms in these states decay to 1s $(2p^53s^1)$ state by spontaneous emission. The lifetime of 1s state is large as the transition between 1s and $^1S_0$ of Ne is forbidden. The population of Ne tends to build up in the 1s level and this increases the probability of the 2p – 1s radiation being reabsorbed, a process called radiation trapping, thereby increasing the 2p population and decreasing the laser efficiencies at 632.8 nm and 1.1523 μm. There are other reasons by which the population of 2p is increased. The $2p_4$ can also be excited by the discharge. For instance, the He 2^3S state transfers its energy to 2s manifold in Ne, which in turn radiates to 2p level. Further, 2p state can be excited from Ne ground state or 1s manifold by electron collision in a discharge as is evident from the characteristic colour of Ne sign depends on the 2p – 1s transition. Therefore, to maintain population inversion it is necessary that atoms in 2p levels should transfer their energy quickly to 1s level. Thus it is essential that 1s level should also be empty. Depopulation of the 1s states is achieved by collision with the walls of the discharge tube. For this reason narrow tubes of a few mm diameter are used.

The upper laser levels of 3.3913 μm and 632.8 nm transitions are the same. The former being of larger wavelengths, and thus having a smaller Doppler width, has much higher stimulated emission cross section than the 632.8 nm transition. Thus 3.3913 μm transition will lase and deplete the population inversion of 632.8 nm transition. Hence for 632.8 nm laser beam, it is necessary to reduce the gain of 3.3913 μm transition. For this, a methanol cell is introduced which absorbs 3.3913 μm or a magnetic field is applied which causes Zeeman splitting. The magnetic field is different at different points along the direction of the tube, thus effectively increasing the linewidth of laser transition and reduceing its gain. The effect of magnetic field is much smaller for 632.8 nm because its Doppler broadend linewidth is already large. The suppression of 3.3913 μm can also be achieved by using multilayer cavity mirrors designed specifically for 632.8 nm wavelength.

The He-Ne laser is normally a CW laser with power up to about 100 mW but with very low efficiency. The laser may also be pulsed by pulsing the input power.

Fig.16.25 shows the block diagram of a typical He-Ne laser. A glass or quartz tube of length 30-100 cm and diameter 1-10 mm is filled with He and Ne in the ratio (i) 5:1 at λ = 632.8 nm and (ii) 9:1 at λ = 1.150 μm. Pressure of He ~ 1 torr while that of Ne is 0.1 torr. Pressure of Ne is kept small so that the probability of transfer from Ne to He is decreased. Usually the diameter and pressure is ~ 2.9 – 3.6 torr mm. The gas mixture is contained between a set of mirrors. This forms a resonant cavity. The

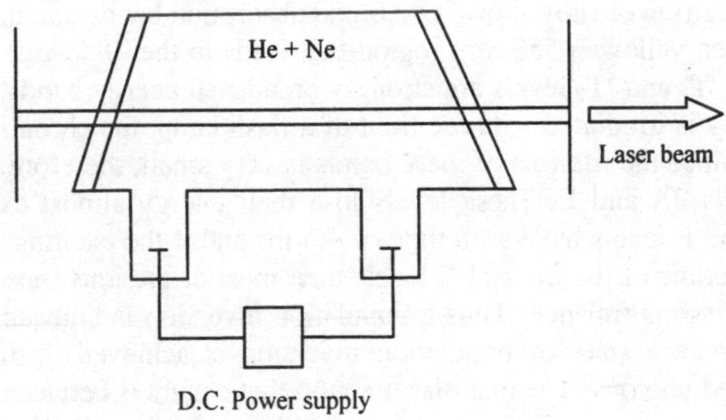

Fig. 16.25 He-Ne laser

reflectivity of one of the mirrors is slightly less than 100%. The mirrors permit most of the photons to bounce back and forth within the cavity , while a small fraction of the total number are transmitted by partially reflecting mirror. The windows of the laser are at Brewster angle with respect to the long axis of the tube. The light, which strikes at surface of the window at Brewster angle, is partially transmitted and partially reflected. The light, which is reflected, has a component of polarization perpendicular to the plane of incidence. This means that horizontally polarised light is rejected by reflection at the tube window. The vertically polarised light is transmitted and can undergo repeated reflections between the mirrors. The light emerging from the laser is thus vertically polarised. The He-Ne laser is electrically pumped, usually with a dc power supply (a typical value 1400 V and discharge current is limited to 5 mA) or by means of radiofrequency (~300 MHz) radiation emanating from metal sleeves around discharge tube.

Applications
   1. Interferometry
   2. Laser printing
   3. Bar code reading
   4. As pointing and directional reference beam
   5. 1.15 µm laser line is used for measurement of optical fibre line, which has a minimum loss in that wavelength region.

16.8.2 Ruby laser

The ruby laser was first demonstrated in 1960 by Maiman. It is a three level solid state laser. The ruby is pale pink crystal of Al_2O_3 doped with about 0.05% of Cr_2O_3 by weight. The concentration of Cr^{3+} is ~10^{25} ions/m^3. At such a low concentration the Cr^{3+} are so far apart from one another that their mutual interaction may be neglected. The ground configuration of Cr^{3+} is $1s^2 2s^2 2p^6 3s^2 3p^6 3d^3$. Terms arising from this configuration are ^2P, ^4P, $^2D(2)$, ^2F, ^4F, ^2G, ^2H. According to Hund's rule, the ground state is ^4F. Of the others ^2G is the lowest excited state. Cr^{3+} ions substitute for Al^{3+} in Al_2O_3 lattice and are in crystalline field of approximately octahedral symmetry. In this crystalline field, ^4F ground term splits into $^4A_2$, $^4T_1$ ($^4F_1$) and $^4T_2$ ($^4F_2$) states while ^2G excited term gives $^2A_1$, ^2E and $^2T_1$ ($^2F_1$) and $^2T_2$ ($^2F_2$) states. The energy levels of Cr^{3+} in ruby that are relevant to laser action are shown in Fig. 16.26. T_1 and T_2 are three-fold degenerate while ^2E splits into two sublevels with a separation of 29 cm^{-1}. The upper one is ^2A and the lower one is ^2E sublevel. These sublevels rapidly assume Boltzmann equilibrium due to thermal relaxation with population ratio of exp (hcΔE/ k_BT). A is non-degenerate. The symbols A, E, T, F etc are from group theory and sometimes F is used in place of T. F used here is different from F of atomic states.

The absorption spectrum of ruby shows two broad absorption bands one in the violet blue (~400 nm) and the other in green yellow (~550 nm) region that leads to the ^4T levels. Due to slight variation of crystalline field, the $^4T_1$ and $^4T_2$ levels are strongly broadened energy bands. The width of these bands ~ 0.1 µm. When ruby is irradiated with the light of a flash lamp, the chromium ions are excited to the $^4T_1$ and $^4T_2$ bands. Since the lifetime of these bands is very small, therefore the ions are transferred to the upper laser levels 2A and E. These levels lose their energy almost exclusively by spontaneous emission. The 2A and E levels have a lifetime of ~ 3 ms and if the exciting flash of lamp is of shorter duration than the lifetime of the 2A and E levels then most of the ions transferred to these levels will remain there when flash is finished. Thus a population inversion is built up between these levels and the ground level. Once a state of population inversion is achieved, lasing action is triggered by spontaneously emitted photons. The main laser action that occurs is between E and the ground state .It is denoted by R_1 and occurs at 694.3 nm. Laser oscillation is also obtained between 2A and ground state. This occurs at 692.8 nm and is denoted by R_2. The radiative decays from $^4T_1$ and $^4T_2$ to the ground state have a lifetime of about 3 μs against a value of about 50 ns into the 2E levels.

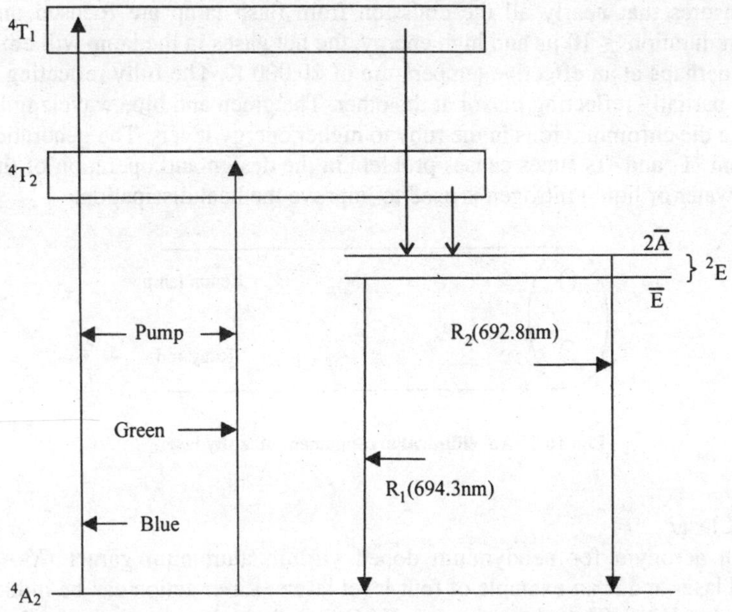

Fig. 16.26 Energy level diagram of ruby. Broken line indicates fast non-radiative decay

As a result of laser action, the population of the upper laser level decreases below threshold condition and laser action stops. However, the flash lamp is still on. Therefore, the population of the upper laser levels increases and again population inversion is achieved. The laser action again starts and the process repeats itself several times during the period of flash of the lamp. The net result is a series of pulses of laser action, which is referred to as spiking.

Very high peak power pulses from the ruby laser can be achieved by Q switching or Q spoiling. The quality factor Q is defined as the ratio of the energy stored in the cavity to the energy loss per cycle. In the technique of Q switching, energy is stored in the amplifying medium by pumping while the cavity Q is lowered so that laser action does not take place. Thus the population inversion is built up above the thermal value without laser action. When a high cavity Q is restored, the stored energy is suddenly released in the form of very short pulses of light. The peak power of the resulting pulse exceeds than that obtained from an ordinary pulse by several order of magnitude. Because of high power of the pulse, it is also called a giant pulse.

The first method used for Q switching is the method of rotating mirrors. In this method one of the two end mirrors of the laser is rotated very rapidly (~ 30,000 revolutions per minute) about an axis perpendicular to the resonator axis. During most of the rotation cycle the two mirrors remain tilted at an angle. The ruby is pumped very strongly with the flash lamp. Near the end of the pump pulse a very large population inversion is set up without laser action because the mirrors are not parallel to each other. At this point if rotating mirror and fixed mirror become parallel to each other, the laser action is initiated. As a consequence, the laser oscillations build up much more rapidly than in the normal operation and will rise rapidly to a peak power much higher than normal. Several other methods are used for obtaining Q switching. The methods, which are widely used, are (i) electro-optical shutters (Pockel cell, Kerr cell), (ii) shutter using saturable absorbers and (iii) accousto- optic Q switching.

A ruby rod of about 5-20 cm long of diameter 5-10 mm and linear xenon lamp are placed in an elliptical cavity (Fig. 16.27). The rod and the lamp are oriented at its foci of highly polished elliptical cylinder. This ensures that nearly all the emission from flash lamp are focused on the rod. For a discharge of short duration ≤ 10 μs and high energy, the hot gases in the lamp will emit approximately as a black body, perhaps at an effective temperature of 20,000 K. The fully reflecting mirror is placed on one end and a partially reflecting mirror at the other. The green and blue wavelengths of the flash of xenon lamp excite the chromium ions in the ruby to higher energy levels. The generation of heat by the energy losses from $^4T_1$ and $^4T_2$ states causes problem in the design and operation of the laser. Cooling of the crystal by water or liquid nitrogen is used to improve the heat dissipation.

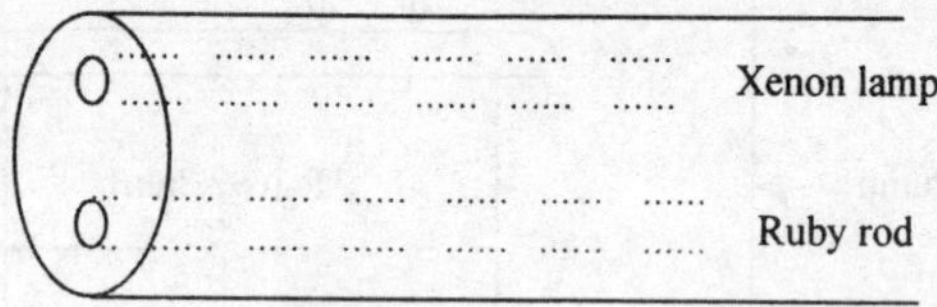

Fig. 16.27 An elliptical arrangement of ruby laser

16.8.3 Nd: YAG laser

Nd:YAG is an acronym for neodymium doped yttrium aluminum garnet ($Y_3Al_5O_{12}$). It is an optically pumped laser and is an example of four level lasers. Laser action can be induced in Nd^{3+} ions incorporated in cubic YAG. The advantages of YAG include high thermal conductivity which allows it to disperse the waste heat from the optical pumping, high mechanical strength and the fact that it can be grown as crystals of large size with good optical quality.

The electronic configuration of Nd (Z = 60) is $KLM4s^24p^64d^{10}4f^45s^25p^6\ 6s^2$ and a ground state $^5I_4$. The ground state configuration of Nd^{3+} is $KLM4s^24p^64d^{10}4f^35s^25p^6$. Of the terms arising from this configuration ^4I and ^4F are important for laser action. The ground state term is ^4I (L = 6, S = 3/2). In L-S coupling approximation, the ground state term is split into states with J = 15/2, 13/2, 11/2 and 9/2. The multiplet is normal with J = 9/2 lying lowest. The ^4F (L = 3, S = 3/2) splits into states with J = 9/2, 7/2, 5/2 and 3/2 with J = 3/2 lying lowest in this multiplet. The Nd^{3+} ions substitutes for Y^{3+} in YAG and are surrounded by several oxygens. Typical doping density of Nd is ~ 1% by weight. The amount of Nd^{3+} doped in the material varies according to its use. For CW output the doping is significantly lower than that of the pulsed laser. The lightly doped CW YAG rods can be optically distinguished by being less coloured at most white, while higher doped YAG rods are pink-purplish. The degeneracy of the energy levels of Nd^{3+} is partially removed in the crystalline field of YAG. The $^4F_{3/2}$ level splits into two components referred to as R_1 and R_2. The R_2 is approximately 84 cm^{-1} above the R_1 level. The $^4I_{11/2}$ level splits into six levels. Fig. 16.28 shows the energy levels of Nd^{3+} ion in the crystalline host lattice YAG. The $^4F_{3/2}$ level is ~11507 cm^{-1} above the ground level while $^4I_{11/2}$ is ~2110 cm^{-1} above the ground level $^4I_{9/2}$. The lifetime of $^4F_{3/2}$ is ~ 230 μs while that of $^4I_{11/2}$ is 30 ns.

Nd: YAG has strong broad absorption bands around 730 - 760 nm and 790 - 820 nm. Therefore, krypton flash lamps, with high output at these bands, are used for pumping. Nd^{3+} ions are pumped to $^4F_{7/2}$, $^4F_{5/2}$ and $^4F_{3/2}$ levels. The ions, which are pumped to $^4F_{7/2}$ and $^4F_{5/2}$, fall to the $^4F_{3/2}$ level, which has a comparative long lifetime. Therefore, population is quickly built up in the $^4F_{3/2}$ level. Population inversion with regards to $^4I_{11/2}$, $^4I_{13/2}$ and $^4I_{9/2}$ is most easily achieved. Laser action involves the transition $^4F_{3/2} \rightarrow {}^4I_{13/2}$ (λ = 1.35 μm), $^4F_{3/2} \rightarrow {}^4I_{11/2}$ (λ = 1.06 μm) and $^4F_{3/2} \rightarrow {}^4I_{9/2}$ (λ = 0.914 μm) but the $^4F_{3/2} \rightarrow {}^4I_{11/2}$ transition is the most easily pumped. Therefore, Nd:YAG laser usually operates at λ = 1.06 μm.

For the transition $^4F_{3/2} \rightarrow \, ^4I_{11/2}$, the lower laser level which is nearly empty at 300 K decays with a lifetime of $\sim$ 30 ns to $^4I_{9/2}$, but the radiation is strongly absorbed by the YAG and thus energy shows

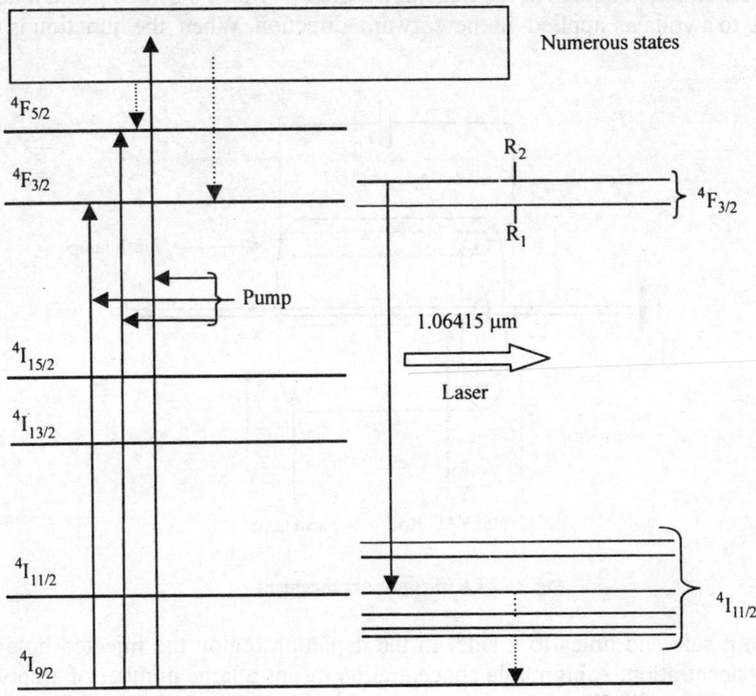

Fig. 16.28 Energy levels of Nd: YAG. Broken lines show non - radiative transitions

up as heat. Since YAG has high thermal conductivity therefore the heat energy is removed by the conduction.

In a typical arrangement of Nd:YAG laser, light from a cylindrical lamp centred along one of the focus of the elliptical cylinder is concentrated at the laser rod which is centred along the conjugate focus. Circulating distilled water cools the laser rod. The size of the rod is about 1-15 cm in length and diameter is about 6 mm. In an improved arrangement particularly for large diameter rods bielliptic pumping cavity for efficient pumping is used as shown in Fig. 16.29. A bielliptic configuration consists of two interconnected incomplete elliptic cylinders and share a common focus where the rod is positioned.

Applications
  1. In material processing, for example, drilling, spot welding and resistor trimming etc.
  2. Medical applications: in many types of surgery such as membrane cutting, gall-bladder surgery
     etc.
  3. Military applications: range finding and target designation.
  4. X-ray production by focusing the laser on to a solid target.
  5. In scientific and general laboratory use.

16.8.4 Semiconductor lasers

Also known as diode laser, junction laser or injection laser, the semiconductor laser is the most widely used of all lasers. The first useful semiconductor laser was made of GaAs, which had been strongly doped. Laser action occurs in the transition region (~ 1 μm) between p and n doped material in a diode subject to a voltage applied in the forward direction. When the junction is forward biased

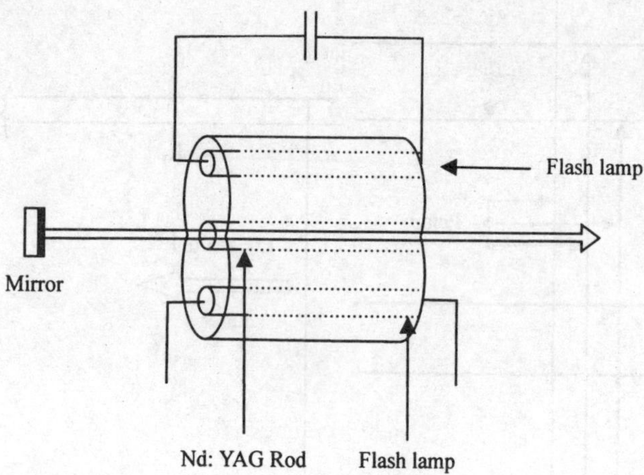

Fig. 16.29 A bielliptic arrangement

electrons flow to p side and holes to n side. In the depletion region the injected holes and electrons appear in high concentration. A high hole concentration means a large number of empty sites are there into which electrons can fall. Thus a population inversion is created between the filled level near the bottom of conduction band and empty level near the top of the valence band. The recombination of electron and holes releases energy. If this energy is in the form of heat (as in the case of Si or Ge) the material is of no use for laser action. In GaAs most of the energy emitted appear as light of wavelength ~ 840 nm. If the photon so emitted is travelling exactly in the plane of the junction there may be amplification and if the photon continues to travel near the junction, the amplification grows. In order to make the amplification radiation travel to and fro as to increase the gain it is necessary to have an optical resonator containing p-n junction. This is done by cutting the crystal so that the two end faces are exactly perpendicular to junction and parallel to each other. Since most of the semiconductors have a high refractive index, reflection at an air- semiconductor interface is high and thus special coating is not necessary. The laser emission is restricted to a very thin region ~ 1μm, therefore, the beam emerges over a wide range of angles $(20^0$-$30^0)$. Because the laser transition occurs between the bands of energies, hence the emission is not as monochromatic as from a gas laser.

A typical diode laser is physically small ~ 1 mm long and with an effective thickness of ~ 1 μm. The lifetime of an electron-hole pair is ~ 1.0-10.0 ns. Thus, very high current densities are required to obtain population inversion. The high current densities raise the temperature of the diode, necessitating even higher current densities for laser action. As diode heats up wavelength changes by 0.3 nm/^{0}C because of change in band gap. Thus by changing the temperature light can be tuned between 800-900 nm. Electrical pumping to laser diode provided by current controlled DC power supply (a typical voltage ~ 2.4 V, operating current 45 mA) is used. The threshold current is ~ 40 mA. A schematic diagram of the diode laser is shown in Fig. 16.30.

Applications
1. In communication systems.
2. In compact disc (CD) players.
3. High speed printing.
4. Pump source of solid state lasers.
5. Laser pointers.
6. In medicine.

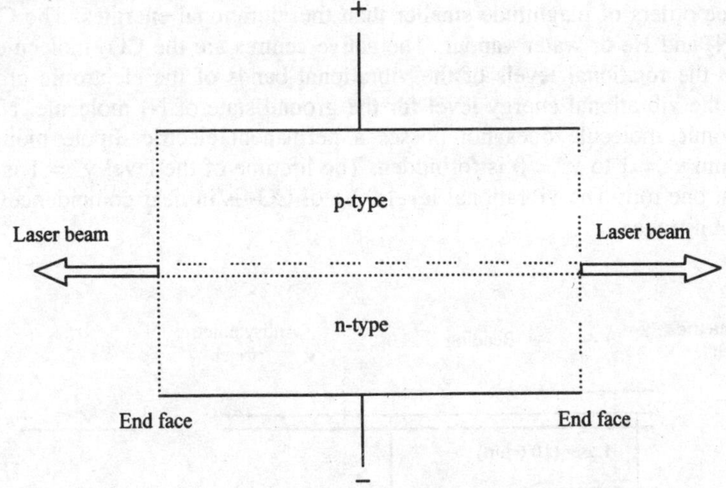

Fig. 16.30 Semiconductor laser

16.8.5 CO_2 laser

CO_2 is a linear and symmetric molecule and its ground electronic state is a $^1\sum$. Since CO_2 is a linear molecule, therefore it has $3N - 5 = 4$ internal vibrational degree of freedom where $N = 3$ is the number of nuclei in the molecule. These four vibrations correspond to (i) symmetric stretching (ii) doubly degenerate bending mode and (iii) asymmetric stretching mode. In the symmetric stretching mode, the oxygen atoms oscillate along the axis of the molecule simultaneously departing or approaching the carbon atom which is stationary [Fig. 16.31(a)]. In the bending mode, which is doubly degenerate, the molecule ceases to be exactly linear. It occurs both in the plane of the figure and the plane perpendicular to it. In this mode all the three atoms in the molecule undergo vibrational motion perpendicular to the molecular axis [Fig.16.31(b)]. In antisymmetric stretching all the three atoms oscillate but both the oxygen atoms move in one direction and carbon atom moves in the opposite direction.

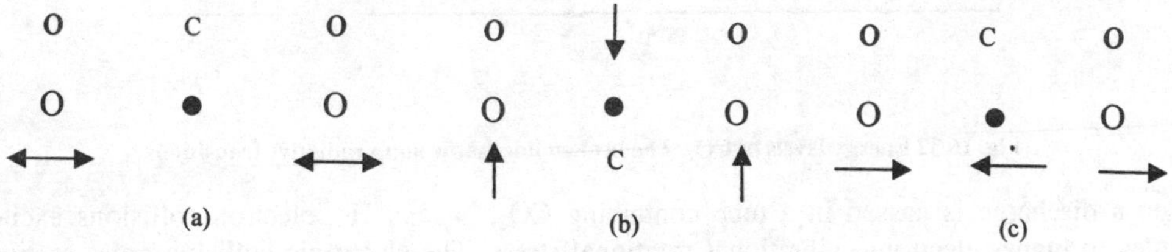

Fig. 16.31 (a) symmetric stretching (b) bending mode (c) antisymmetric stretching

In the first approximation these three modes are described as independent harmonic oscillators, and hence the excited vibrational states are completely described by the number of excited quanta in each mode. These states are represented by a set of three vibrational quantum numbers v_1, v_2^l, v_3 where subscript 1, 2 and 3 refer to the symmetric bending and antisymmetric modes, respectively. The superscript l gives the angular momentum of this vibration about the axis of molecule in units of $\hbar$. For example, 02^00 indicates that the two vibrations combine to give an angular momentum $l = 0$.

The energy level diagram for the lowest levels is shown in Fig. 16.32. Each of the vibrational levels have superimposed upon it a set of rotational levels. The rotational levels are spaced by energies that are more than three orders of magnitude smaller than the vibrational energies. The CO_2 laser uses a mixture of CO_2, N_2 and He or water vapour. The active centres are the CO_2 molecules lasing on the transition between the rotational levels of the vibrational bands of the electronic ground state. Fig. 16.32 also shows the vibrational energy level for the ground state of N_2 molecule. Nitrogen being a homonuclear diatomic molecule does not posses a permanent electric dipole moment and hence radiation decay from $v'' = 1$ to $v'' = 0$ is forbidden. The lifetime of the level $v'' = 1$ is therefore quite long and ~ 0.1 s at one torr. The vibrational level 00^01 of CO_2 is in near coincidence ($\Delta E = 18$ cm^{-1}) with $v'' = 1$ level of nitrogen.

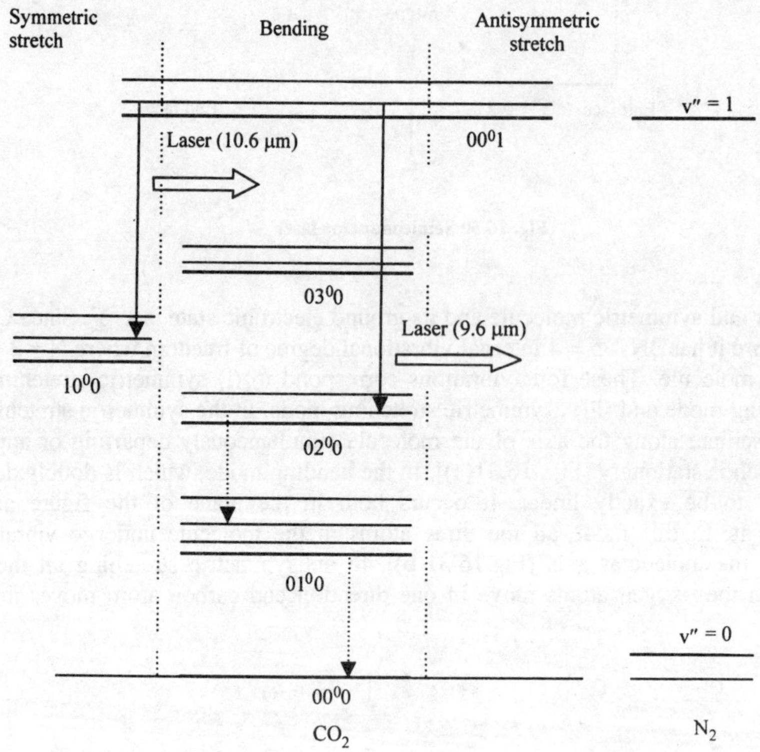

Fig. 16.32 Energy levels of CO_2. The broken lines show non - radiative transitions

When a discharge is passed in a tube containing CO_2, N_2 and He, electron collisions excite the molecules to higher electronic vibrational rotational states. The electronic collision cross section for the excitation to the level 00^01 is very large. The electron collision excite the N_2 molecules by

$$N_2 + e^- \rightarrow N_2^- \rightarrow N_2^* + e^-$$

The cross section for this process is quite large and produce copious amount of metastable N_2^* in the discharge. These nitrogen molecules transfer their internal energy to CO_2 asymmetric stretching modes through near resonance collision.

$$CO_2 + N_2^* \rightarrow CO_2^* + N_2 + \Delta E$$

The cross section for this process is also large therefore, the excitation of CO_2 00^01 level is unusually efficient. Further, the discharge containing the CO_2 produces dissociation of CO_2 into CO and oxygen. Since CO vibrational frequency is not grossly different from that of the CO_2 asymmmetric mode, it will rapidly transfer its internal energy to the CO_2 system through near resonant collision in the same manner as N_2.

The high density of population arises principally from three factors (i) the long vibrational lifetime of the asymmetric mode of CO_2 , (ii) the near- resonant N_2 - CO_2 energy transfer, (iii) the large cross section for the production of the vibrationally excited nitrogen through inelastic collision with electrons (iv) population gain from other levels of the asymmetric mode.

The 10^00 and 02^00 levels are in resonance. The coupling between these levels is sufficiently strong so that 10^00 and 02^00 level system behaves like a single level. The separation between these levels ~ 100 cm^{-1} which is significantly less than the thermal energy of motion k_BT ~ 210 cm^{-1} at room temperature. The similarity of vibrational motion in combination with the near degeneracy of the levels causes the rate of transfer between them by collision process, corresponding to

$$10^00 + 00^00 \rightarrow 02^00 + 00^00 + \Delta E$$

to be extremely rapid. Thus 10^00 and 02^00 levels of CO_2 reach thermal equilibrium in a very short time. For population inversion, between upper 00^01 level and lower 10^00 and 02^00 levels, it is necessary that population of lower levels should decay very fast. The presence of He has a considerable influence on the population decrease of lower levels. The He atoms collide with CO_2 and increase the rate of relaxation of 01^00 levels. In collision of the type

$$01^00 + He \rightarrow 00^00 + He + \Delta E$$

this rate is approximately twenty times that for the corresponding

$$01^00 + 00^00 \rightarrow 00^00 + 00^00 + \Delta E$$

process where two CO_2 molecules collide. This reduces the bending mode population, which in turn reflects itself in reduction of lower level population, and hence increasing the gain. Another function of He is to keep CO_2 cold. The excited CO_2 can decay spontaneously and directly to the ground state, liberating energy as heat. To avoid population of lower level by thermal excitation, it is necessary that the temperature of CO_2 is low. Helium has high thermal conductivity and hence helps to conduct heat away to the walls keeping CO_2 cold. Thus while N_2 helps to increase the population of the upper level, helium helps to depopulate the lower level.

The laser oscillation is at a wavelength corresponding to the transition from the upper vibrational level 00^01 to the lower vibrational level 10^00. With each vibrational level, rotational levels are also associated. Thus a number of laser lines are possible. The population of J' = 21 of 00^01 has the maximum population and transition corresponding to J' = 21 to J" = 22 or P(22) is strongest and this corresponds to the wavelength of 10.6 μm. Another laser line of λ = 9.6 μm from transition between 00^01 and 02^00 is also prominent.

A typical CO_2 laser is shown in the Fig.16.33. The discharge is produced in a tube having diameter 2.5 cm and length 5 m. Brewster windows were used at the end. These windows are made of NaCl,

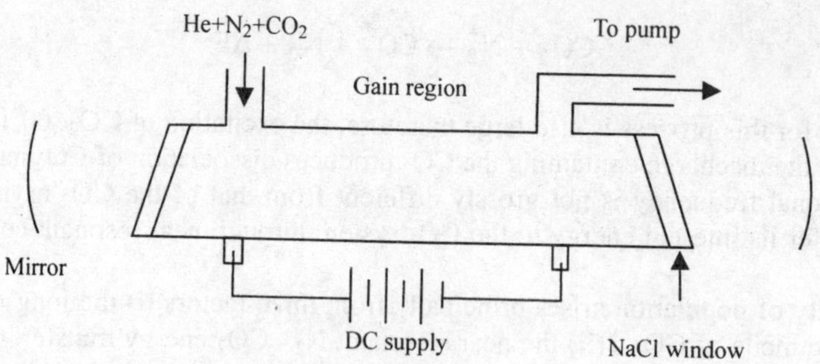

Fig. 16.33 CO₂ laser

CaCl, BaFCl, ZnSe or KCl as these materials are transparent to 10.6 μm radiation. Since glass strongly absorbs most of the frequencies in infrared region therefore, near confocal Si mirrors coated with Al are used. He, N_2 and CO_2 are filled in the ratio 8:1:1 with pressure of He, N_2 and CO_2 as 0.7 torr, 1.2 torr and 0.33 torr, respectively. To remove the dissociation products such as CO and O, which would contaminate the laser, the continuous flow of the gas mixture is maintained in the tube. Grating can be used to separate out different laser lines. CO_2 can be operated in pulsed or CW mode. Pulse energy ~ 2kJ and pulse width is of few μs. For CW operation, power > 50 W is easily available.

Applications

The advantage of using this laser is based on the fact that it provides very intense heating source over a very small area. It has applications in
1. Material processing: cutting, drilling, material removal, welding, cladding, alloying, hardening, melting etc.
2. In medicine it is used for cutting and cauterizing.

16.8.6 Nitrogen laser

N_2 is a homonuclear molecule. The electronic ground state of N_2 is X $^1\Sigma_g^+$. Out of the many electronic states possible, three triplet excited states, that is, A $^3\Sigma_u^-$, B $^3\Pi_g$ and C' $^3\Pi_u$ along with ground state are shown in Fig.16.34. The equilibrium internuclear distances r_e for various states are given in Table 16.1

The values of r_e indicate that the minimum of potential for C state lies almost vertically above that of the X state. On the other hand the minima of A and B are shifted to high r. When the electric discharge is passed through the gas, the electron collides with the molecules and excites them to various levels. According to Franck Condon principle the probability of transition from X to C is much larger than that of X to A or X to B. Thus, population inversion occurs at level C (v = 0 level) relative to level B (v = 0 level).A lasing action is observed between C and B levels (corresponding to 0-0 and 0-1 vibrational levels). Since the lifetime of state C is about 40 ns and that of B is about 100 μs, therefore, the molecule remains in C state for a short duration and after transition accumulates in level B. Hence the

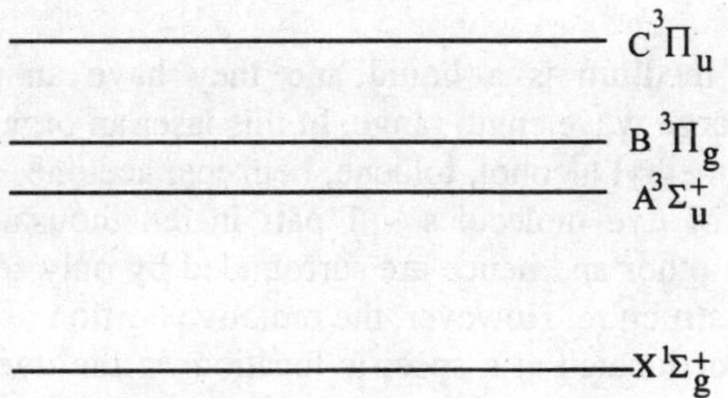

Fig. 16.34 Energy states of N_2 molecule

Table 16.1 The value of r_e for X, A, B and C states of N_2

State	r_e (nm)
$X\,^1\Sigma_g^+$	0.10977
$A^3\Sigma_u^+$	0.12866
$B\,^3\Pi_g$	0.12126
$C\,^3\Pi_u$	0.11487

population difference between B and C decreases very rapidly and this renders laser action impossible. The population inversion can therefore be achieved only for a short time and after this the laser action is self-terminating. Due to high gain of this self-terminating transition, oscillations take place in the form of amplified spontaneous emissions. Thus the laser can be operated without mirror. The mirror placed at one end of the tube reduces the threshold power and also provides unidirectional output. This also reduces the beam divergence. The pulse width for nitrogen laser is normally 10 ns and emission wavelength is at 337.1 nm. Peak power up to 1 MW can be achieved.

A basic arrangement for N_2 laser is shown in the Fig. 16.35. A pulsed high voltage of about 20 kV, triggered by a thyratron or spark gap is applied across the tube through which N_2, at a pressure of about 100 torr, is flowing. On one end of the tube a mirror is used while a window used on the other end has a very high transmission. The length of the tube is usually between 0.1 – 1.0 m.

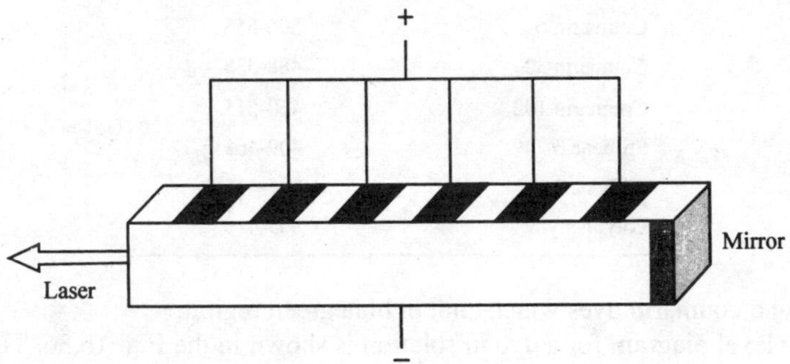

Fig. 16.35 N_2 laser cavity

Nitrogen lasers are used primarily for pumping tunable dye lasers

16.8.7 Dye laser

In dye lasers the active medium is a liquid and they have an important property of being continuously tunable over a large wavelength range. In this laser an organic dye dissolved in a suitable liquid such as ethyl alcohol, methyl alcohol, toluene, benzene, acetone, water etc is used as the active medium. The concentration of dye molecules ~ 1 part in ten thousand, so the dye molecules are somewhat isolated from each other and hence are surrounded by only solvent molecules. The organic dye molecules have complex structure. However, the radiative portion of the dye is relatively simple, it is referred to as a chromophor located at a specific location on the dye molecule. The energy levels associated with the chromophor are similar to those of simpler molecules. Different solvents cause slight shift in the energy levels of dye molecules leading to slight varaiation in the radiative spectrum of these molecules. Conjugate double bonds are a common property of all the organic dyes that can be used as an active medium in a liquid laser. The dyes that are most effective are classified into eight groups:

1. Xanthenes
2. Polymethines
3. Oxaines
4. Caumarins
5. Anthracenes
6. Acridines
7. Azines
8. Phthaloziamins

Laser dyes have strong absorption region in the ultraviolet and visible spectral regions. If they are irradiated or pumped with light at these wavelengths then the dye molecules radiate very effectively at somewhat longer wavelength than the pump wavelength. The lasing range covered by various dyes are given in Table 16.2 . The most widely used dye is rhodamine 6G (Xanthenes type) which emits in

Table 16.2 Dyes and their spectral range

Dye	Wavelength range (nm)
Dicyanomethylene	610-705
Rhodamine 6G	570-640
Rhodamine 110	528-580
Coumarin 6	506-558
Coumarin 30	488-535
Coumarin 102	460-515
Stilbene 3	409-465
Stilbene 1	391-435
Polyphenyl 2	363-410

yellow-red region and coumarin dyes which emit in blue green region.

A typical energy level diagram for a dye in solution is shown in the Fig. 16.36. The molecules have singlet as well as triplet electronic states. Each electronic state contains a series of vibrational levels and each of these, in turn, contains several rotational levels. The separation of the vibrational levels is 1400-1700 cm^{-1} whereas that of rotational levels is ~ 150 cm^{-1}. Since these molecules are in liquid solution, the rotational and vibrational levels smear together owing to collision interaction with solvent to form a continuous spectrum of energy for each electronic state. The population within each electronic state is distributed according to the Boltzmann distribution. The wavelength (T_1-T_2) absorption is same as that of emission of (S_1-S_0).

In the singlet state the spins of the active electron and that of the remainder of the molecule are antiparallel, while in the triplet states the spins are parallel. The allowed selection rules is $\Delta S = 0$

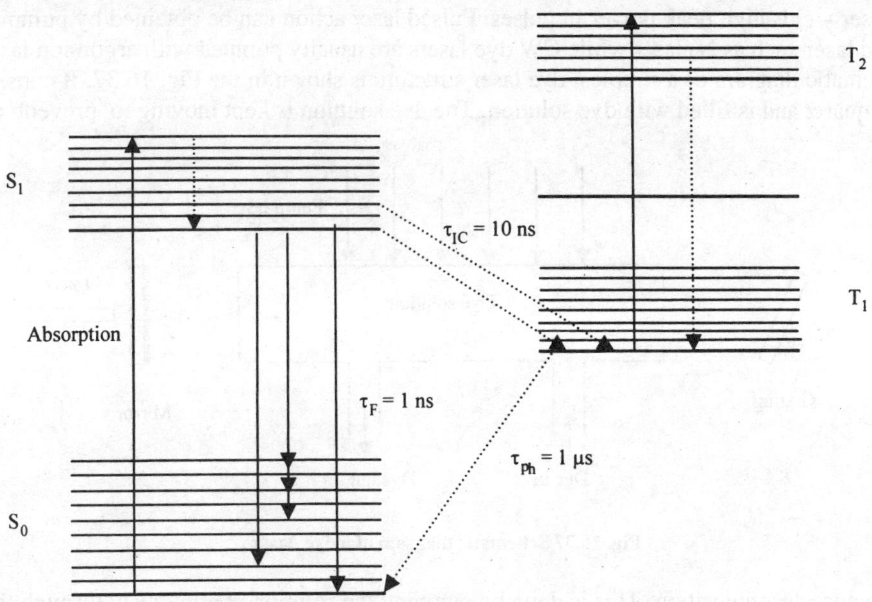

Fig. 16.36 Energy level diagram of a typical dye molecule with radiation (solid lines) and non- radiative (broken lines). τ_F is fluorescence lifetime; τ_{ph} is phosphorescence lifetime

while singlet to triplet transitions are not allowed. The optical pumping raises the molecule from the lowest vibronic level of the ground state S_0 to one of the upper vibronic levels of the excited state S_1. A rapid relaxation, by a non radiative process with a decay time of about $\tau_{nr} \sim 10^{-12}$ s to the lower vibronic level of S_1 state then follows. The lifetime of the lower vibronic level of S_1 is 1 ns. The laser action occurs between this level and full manifold of the levels S_0 , the resulting emission is a continuous spectrum of radiation. This includes transitions originiating from bottom of S_1 terminating at various levels of S_0. The number of possible transitions are restricted by Franck Condon principle. From various levels of S_0 molecule return to the ground state by non radiative decays.

The ground state molecules are not excited to triplet states by optical pumping as the singlet –triplet transitions are not allowed. However, when a molecule is in S_1 state, it can decay to triplet state by collision. This is known as intersystem crossing (IC). The time for this intersystem crossing is $\sim 1\mu s$ while the lifetime of T_1 is 10^{-3} s. Thus molecules accumlate in the state T_1 and can make transition to T_2 state. The energy difference between T_1 and T_2 states is almost equal to the corresponding fluorescence transition S_1-S_0 and therefore the intersystem crossing reduces the number of molecules in the upper laser state. This reduces the gain. The laser action stops when a significant fraction of the molecules is in the triplet state. Further, the molecule in T_1 tends to react and convert to other molecular species, causing a gradual degradation of the dye.

To minimize the effect of triplet state, oxygen is added to the solution which acts as a triplet state quenching additive. The lifetime of T_1 is about 10^{-7} s in oxygen saturated solution in comparison with 10^{-3} s for deoxygenated solution. The lifetime of T_1 is thus reduced. The radiative lifetime of S_1-S_0

transition is of the order of few nanoseconds while lifetime of S_1 is much longer ~ 100 ns. Hence most of the molecules decay from level S_1 by fluorescence, thus contributeing to the laser action. The triplet level problem is also overcome by flowing the dye solution through the laser cavity. If the active volume is small enough and the rate of flow is great enough , molecules are physically removed from the cavity before an appreciable fraction of molecules is lost to the triplet state.

Dye laser yields high peak power in pulses. Pulsed laser action can be obtained by pumping the laser by pulsed laser such as N_2 laser while CW dye lasers are usually pumped with argon ion laser.

A schematic diagram of a simplest dye laser structure is shown in the Fig. 16.37. It consists of a cell made of quartz and is filled with dye solution. The dye solution is kept moving to prevent overheating

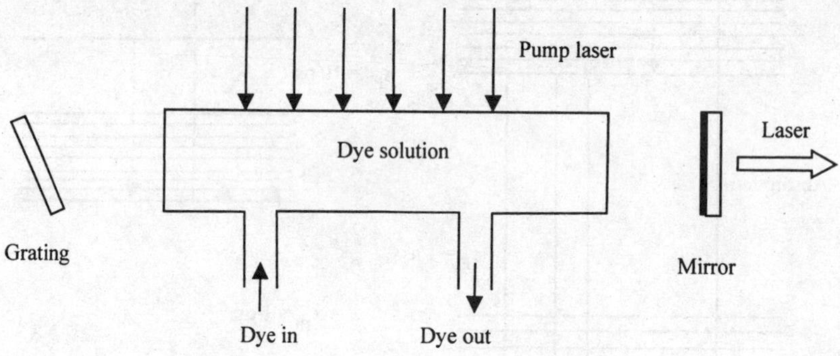

Fig. 16.37 Schematic diagram of a dye laser

and premature decomposition. This is done by pumping the solution continuously through the cell. The dye is pumped using a flash lamp or a laser. Usually N_2 laser (337 nm, as ultraviolet output is suitable for pumping many dyes in the visible region), Nd:YAG laser (1.06 μm and its harmonics 532 nm, 354 nm and 266 nm), ruby laser (694 nm and its harmonic 347 nm) and Argon ion laser (has about 10 lines in the range 454 -529 nm).For tuning of laser radiation wavelength, one mirror of laser cavity is replaced by grating or a dispersive prism. In order that the radiation be reflected back along the cavity axis, the angle that the normal to the grating makes with the cavity axis, must satisfy the condition

$$2d\,\sin\theta = n\lambda , \qquad n = 1,2,......$$

where d is the grating spacing and λ is the radiation wavelength. The radiation will propagate to and fro in the cavity for a particular wavelength of the radiation. The wavelength tuning is achieved by rotating the grating which changes the angle θ and hence output radiation wavelength.

Applications
1. Important tool for spectroscopists as it is continuously tunable over a broad wavelength.
2. In isotope separation.
3. Ultrafast pulses for study of dynamics of excited states of semiconductors and other solids.
4. In medicine to remove birth mark, shatter kidney stones or gall stones etc.
5. For slowing down of atoms to very low speeds.

16.9 NH_3 Maser

The structure of ammonia molecule is similar to a pyramid having a triangular base. Hydrogen atoms form the base and nitrogen atom is at the apex. NH_3 molecule has an electric dipole moment pointing in the direction of the line connecting the apex of the pyramid with the centre of its base. In the presence of an electric field a torque acts on the dipole, which results in the alignment of the dipole in the direction of the field. The molecule has two isomers. One form has atom of nitrogen above the plane of the hydrogen atoms and in the other form it is below the plane of hydrogen atoms. The pyramid can stand on its base or it can be upside down. The difference between the two isomers of ammonia molecule can be seen in the presence of an electric field.

Each isomer of ammonia molecule has a quantum state associated with it. The configuration having nitrogen atom above the hydrogen plane is called up or $|u>$ or ψ_1 and in the other case the configuration is called down or $|d>$ or ψ_2. The two quantum states ψ_1 and ψ_2 do not form a good basis of a two-dimensional Hilbert space. This is because of the fact that molecular fluctuations are present between them all the time, even in the absence of applied force on it.

The general wavefunction is

$$\psi = a_1 \psi_1 + a_2 \psi_2$$

where a_1 and a_2 are constants. $|a_i|^2$ is the probability that the molecule is in a state i . Since molecule is equally likely to be in either state

$$|a_1|^2 = |a_2|^2 = \frac{1}{2}$$

$$\psi_\pm = \frac{1}{\sqrt{2}} \left(\psi_1 \pm \psi_2 \right)$$

The antisymmetric state ψ_- has slightly higher energy than the symmetric state ψ_+ . Thus the ammonia molecule has a level scheme (Fig.16.38) consisting of a pair of energy levels with fairly small separation compared to the separation of one pair from another.

Fig. 16.38 Vibrational energy levels of ammonia molecule

Ammonia maser uses the vibrational levels of the lowest pair. The energy difference between these levels corresponds to a microwave frequency of 23870 MHz (or wavelength of 1.25 cm). The procedure of designing a maser involves (i) obtaining inverted population and (ii) allowing maser action to occur. To obtain $N_2 > N_1$, differing property of the molecule in two states is used. In the presence of electric field E, charge distribution of the molecule is distorted. This induces a dipole moment in the molecule given by

$$P = P_0 + \alpha E$$

where α is polarizability. $P_0 = 0$ for the lowest pair of energy levels and α is of opposite sign in the two states.

Ammonia molecules emerge from the source and are passed through a region of highly non-uniform electric field (Fig.16.39). The field is produced by applying 30 kV to two opposite electrodes and earthing the other two. The inner faces of these electrodes are shaped as segment of hyperbolas. The spacing between the rods is ~ 0.2 cm. Near the electrodes the field is strong.

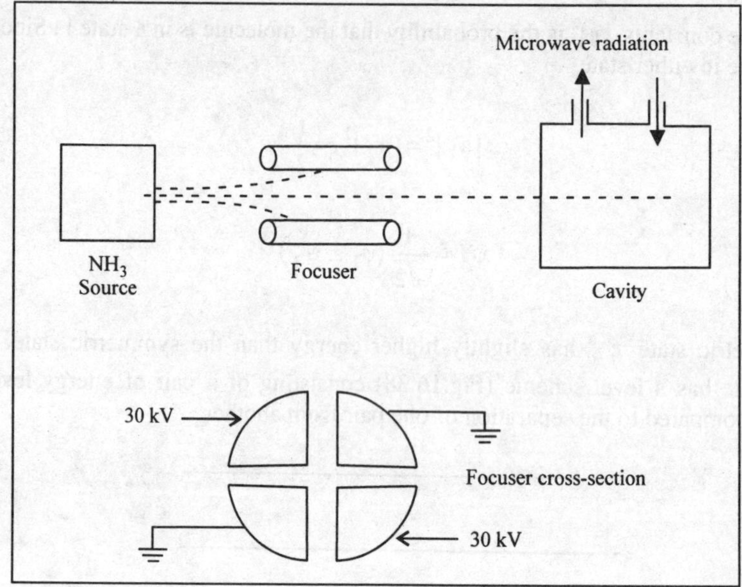

Fig. 16.39 Ammonia maser

When a mixture of upper and lower state molecules are passed through the electrodes the induced dipole interacts with the electrostatic field. Because of this, the molecules in the upper state are pushed towards the axis while molecules in the lower state are pulled towards the electrodes. Thus upper state molecules are being brought to a focus while lower state molecules are diverged by the field and lost from the beam. Hence a population inversion is achieved not by pumping but by depletion of the population of the molecules in the lower state.

The beam is focused to a point in a resonant cavity with highly reflecting walls. In the cavity the upper state ammonia molecules are subjected to incoming microwave radiation which induces v = 1 to v = 0 transitions. Because of the population inversion, the radiation emerging from the cavity has been amplified.

The power output of the NH_3 maser is of the order of 10^{-3} W. As an amplifier the NH_3 maser has a narrow bandwidth and therefore can not amplify waves which depart from the central frequency by more than 3000 to 5000 Hz. It has noise frequency oscillation stability of few parts in 10^{10}.

16.10 Applications of Lasers

16.10.1 Laser isotope separation

The different isotopes of an element are chemically virtually indistinguishable and physically different to only a small extent. Consequently, it has always been difficult to remove, separate pure isotope from the mixture of isotopes found for an element. The use of laser for isotope separation is based on the fact that different isotopes of the same element while chemically identical have different electronic energies and therefore absorb different frequencies of laser light. Since the light emerging from laser is extremely monochromatic, one may shine laser light on a mixture of isotopes and excite the atoms of only one of the isotopes, then separate the excited atoms.

(a) Separation Using Radiation Pressure

A photon of energy hv carries with it a momentum hv/c. When an atom absorbs a photon impinging perpendicular to the direction of flight of the atom, the photon momentum is transferred to the atom. Thus the absorption tends to push the atom in the direction of the travel of the incident photon. For a sodium atom absorbing a D-line quantum, the transverse velocity change is about 3 cm/s. The atom is deexcited by spontaneous emission, which occurs in random directions. Therefore, the direction of recoil is also in random directions. Thus on the average a transverse momentum is transferred to the atomic beam during many absorption and emission processes (Fig. 16.40). This pushes the atom along the direction of the laser beam. The deflection of the laser beam depends on the intensity of the interaction region. Deflection of the atomic beam from the incident direction is normally small ($\sim 1^0$).

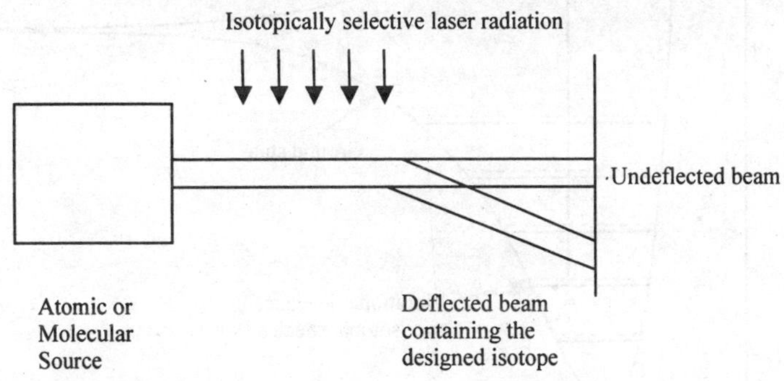

Fig. 16.40 Separation of isotope by deflection caused by selective absorption

(b) Separation by Selective Photoionization or Photodissociation

The basic principle of this method for an atom having two isotopes A and B is shown in the Fig. 16.41.A narrow band tunable laser selectively transfers the isotope A to an excited state. The excited atom is subsequently photoionized using a second laser pulse. The photon energy of the second laser is not sufficient to photoionize the ground state atom. The separation of the ion can be carried out directly by means of an electric field.

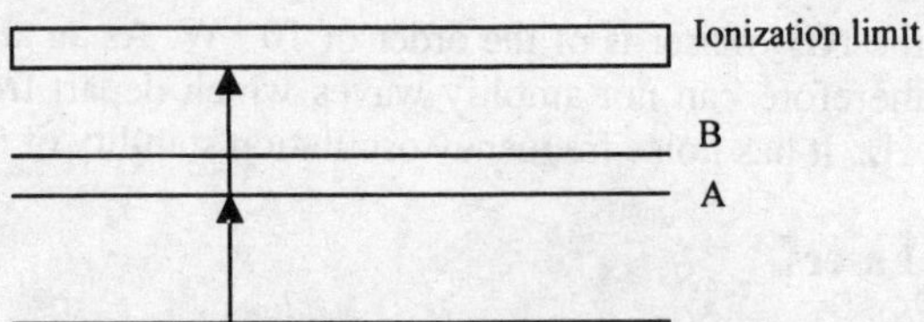

Fig. 16.41 Photoionization of atoms: a two-step process

In photodissociation process an infrared photon selectively excites a vibrational state of one of the isotopes (Fig.16.42). This vibrationally excited isotope is then dissociated with a second energetic photon . The dissociation products are separated by chemical reaction

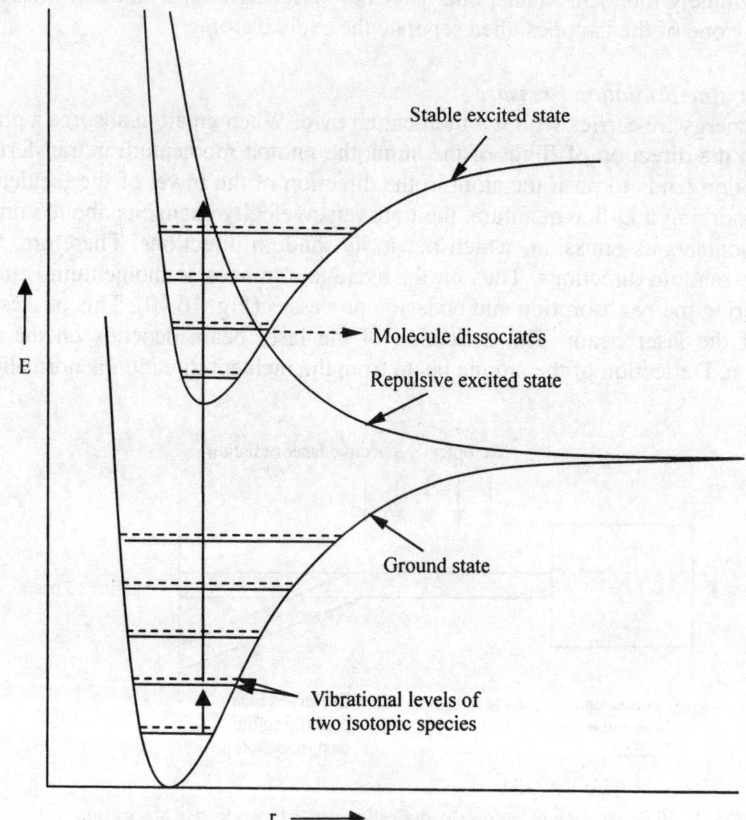

Fig. 16.42 Two- step photodissociation of molecules

(c) Photochemical Separation

Another way of separating selectively excited atoms or molecules from those in the ground states is by means of chemical reactions. This method relies on the fact that excited states of atoms or molecules are frequently more reactive than those in the ground state. For example the isotope of chlorine can be separated by irradiating the mixture of iodine chloride and bromobenzene with red light (of wavelength 605 nm) from a dye laser.

$$ICl^{37} + h\nu \rightarrow (ICl^{37})^*$$

$$(ICl^{37})^* + C_6H_5Br \rightarrow C_6H_5Cl^{37} + I + Br$$

The resultant chlorobenzene $(C_6H_5Cl^{37})$ is enriched in Cl^{37}.

Let the atom or molecule A is in its ground state and does not react with another atom or molecule B. However, when A is excited , it reacts with B to form AB (Fig. 16.43)

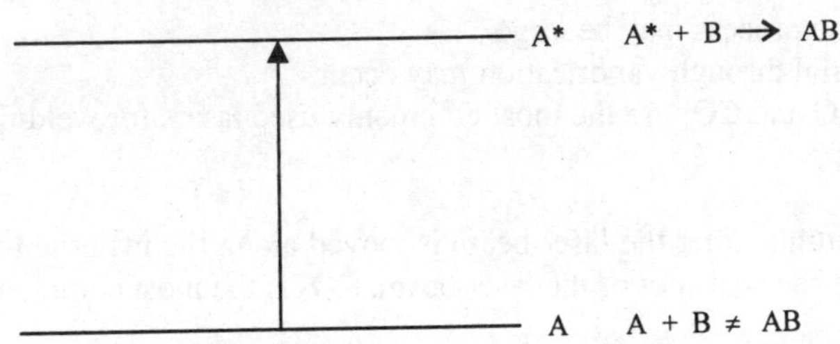

Fig. 16.43 Laser enhanced chemical reaction

16.10.2 Material processing

It is a process in which a work piece is either melted or some material is removed from it. Interest in the laser as processing tool is a result of its unique characteristic as a source of heat. A few of these characteristics are:

    (a) The high spatial coherence of laser beams allows focusing the beam to a small spot of high optical intensity.

    (b) A laser wavelength usually can be found for which the optical absorption depth in the work piece is such that the optical energy is converted into the heat energy.

Material processing with laser beams has the advantages

1. This is a non-contact process in which no tools come into contact with the work piece, thus eliminating the problem of contamination of the work piece by the tool material.
2. The energy beam can be steered through any transparent materials, thus inaccessible parts may be welded or machined.
3. Laser beams are easily modulated electronically. This allows interfacing the laser into complex computer controlled processes.
4. The working atmosphere may be controlled to suit the particular work.

The lasers, which are mainly used in material processing, are:

1. Ruby $(\lambda = 694.3$ nm)
2. Nd:glass $(\lambda = 1.06$ μm)
3. Nd:YAG $(\lambda = 1.06$ μm)
4. CO_2 $(\lambda = 10.6$ μm)

CO_2 laser dominates laser material processing.

(a) Laser Welding

In the laser welding process, a laser beam is incident upon the surface to be welded. A part of the beam is absorbed and part is reflected. The absorbed energy heats the surface and melts it. Laser welding is usually a fusion welding process with some of its characteristics resulting from the short heating and cooling time. Laser fusion welds are easiest to make when the melting point of the materials to be joined are similar. The advantages of laser welding are:

1. It is possible to start or stop the welding very rapidly.
2. Difficult to weld materials like quartz, titanium etc. can be welded.
3. Very narrow and accurate welds can be made.
4. The absence of physical contact with an electrode.
5. Localized heating and rapid cooling due to the high heat flux and small laser spots.
6. The ability to weld components in a controlled atmosphere or sealed within optically transparent materials.

Disadvantages of laser welding are:

1. The cost of the process may be large.
2. Loss of material through vaporization may occur.

Nd:glass, Nd:YAG and CO_2 are the most commonly used lasers for welding.

(b) Laser Cutting

For continuous cutting either the laser beam is moved along the material to be cut or the material is moved relative to the focused spot of the laser beam. CO_2 is the most common laser used for cutting.

(c) Laser Drilling

The ability of the laser beam to concentrate large power in a small volume is utilised for the drilling of small holes in the material. Pulsed ruby laser, CO_2 and Nd:YAG lasers are used for drilling. The advantages are:

1. Very fine holes with dimensions less than 0.001 cm can be made.
2. Precise, close patterns of adjacent areas and no contamination.
3. Hard, brittle materials such as ceramics and gemstones can be drilled.
4. Holes with large depth to diameter can be produced.

(d) Laser Hardening

The laser irradiation on a surface causes very rapid heating of the thin layer of the material of the surface. On switching off the laser, the heated area cools off rapidly due to conducted heat transfer. This rapid cooling of the surface results in an increase in the hardening of the material. The laser hardening is very useful for uneven work piece geometries like corners, gear teeth etc.

16.10.3 Lidar

Lidar, which is an acronym for light detection and ranging is an active technique, similar to radar. In this technique a laser pulse is generated and transmitted into the atmosphere. For this a Nd:YAG laser is used either directly or after frequency conversion in a dye laser. The light is scattered back from a distance R and arrives at the lidar receiver at a time $t = (2R/c)$ after the transmission of the laser pulse. From this delay, range information is obtained. The range resolution ΔR is given by the duration of laser pulse t_p given by $\Delta R = (t_p c/2)$. Light travels about 30 cm in 1ns, thus a segment of the atmosphere about 1 m in depth can be probed in this way. The use of several wavelengths allows a limited amount of constituent identification.

16.10.4 Holography

Holography is the lensless three-dimensional imaging technique. It was devised by D.Gabor in 1948 to solve the problem of aberrations in electron microscopy. A conventional photograph represents a two dimensional object. It records light intensity information. There are two steps to use conventional photography. Taking (developing) the photograph and viewing the photograph. In holography, information of intensity and phase is recorded and the record is called hologram. The holography is also a two-step process, that is, recording (and developing) the hologram and reconstructing the hologram image.

For recording the hologram a laser beam is split into two beams by a beam splitter (Fig. 16.44).The transmitted beam goes directly to the object whose hologram is to be taken. A diffuser such as ground glass plate or grating is introduced between transmitted light and object. Any part of the object surface then receives light from many directions. Light after scattering and reflection from the object falls on the photographic film. The reflected beam B goes directly to the photographic film and this beam is known as reference beam. The superposition of these two beams, that is, reference beam and beam scattered and reflected from the object, produces an interference pattern which is recorded on the photographic film. The developed hologram does not contain a recognizable image of the object as a conventional photograph does. The hologram generally appears as a collection of bright and dark lines with bright and dark pattern of concentric circles.

To reconstruct the image, the object is removed and is placed back to its original position. The laser is turned on. Light falling on the developed hologram reconstructs a virtual image of the object that can be viewed by looking through a hologram toward the original position of the object (Fig. 16.45). A real image is also formed between the observer and the plate, which can be photographed.

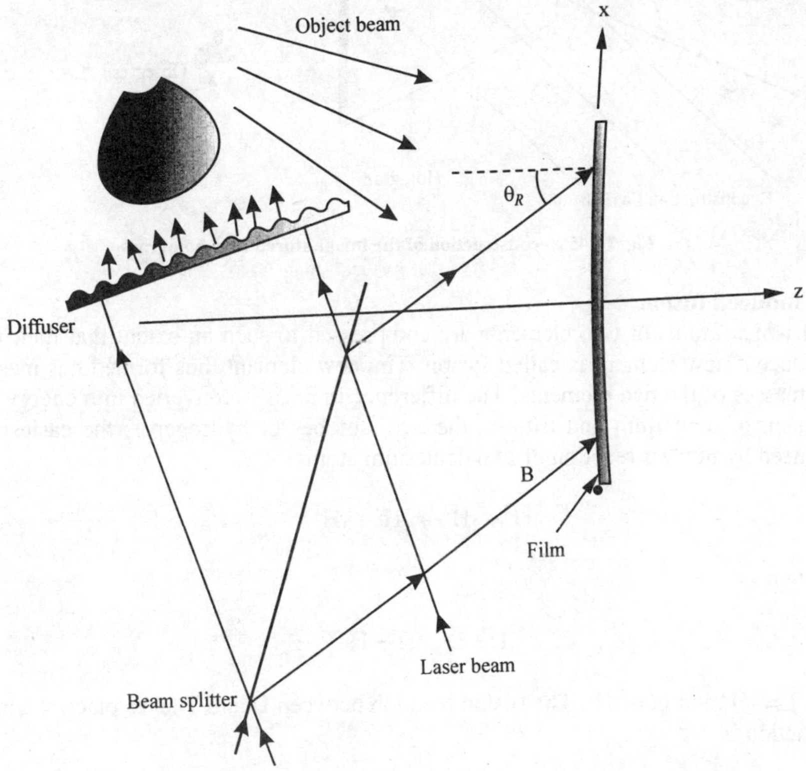

Fig. 16.44 A simple set up for making a hologram

Requirements for Holography
1. The site must be quiet, that is, the object, mirror, reference beam etc. much be motionless relative to each other.

2. Place of recording should be dark.

3. The resolution of film used for recording the hologram ~ 1.222 x 10^{-3} mm (for the He-Ne laser).

4. Since holography is an interference phenomenon. Therefore, the light source should be coherent for the formation of stable fringes. The path difference between the object wave and reference wave should not exceed the coherent length L_c which is defined as the distance travelled by light in the time during which sinusoidal emission exists (coherent time τ_c).

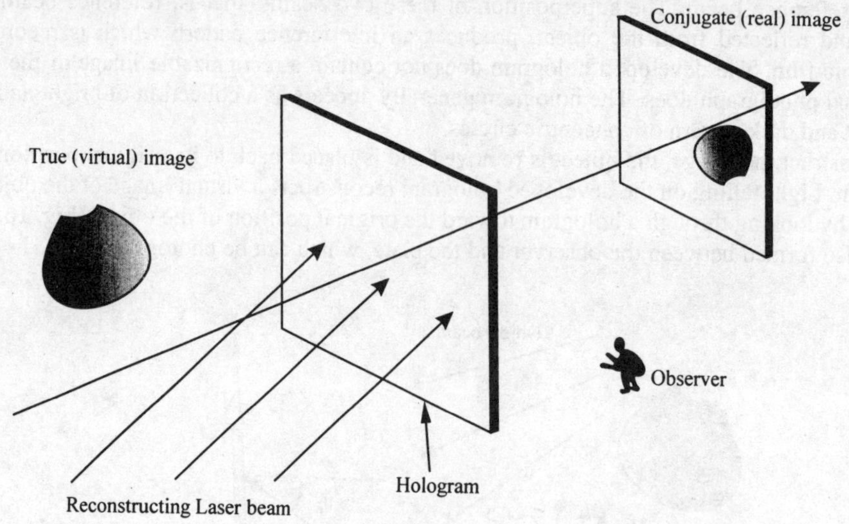

Fig. 16.45 Reconstruction of the image stored in a hologram

16.10.5 Laser induced fusion

A process in which atoms of two elements are compressed to such an extent that their nuclei fuse together to produce a new element is called fusion. The new element thus formed has mass less than the sum of the masses of the two elements. The difference in mass is converted into energy $E = \Delta Mc^2$. The nuclear fusion of deuterium and tritium, the two isotopes of hydrogen is the easiest to realise. Tritium is produced by nuclear reaction of two deuterium atoms

$$_1H^2 + {_1H^2} \rightarrow {_1H^3} + {_1H^1}$$

This is also written as

$$D + D \rightarrow T + H$$

where $D = {_1H^2}$, $T = {_1H^3}$ and H is $_1H^1$. The fusion reaction between D and T takes place at a temperature ~ 10^8 K. The reaction is

$$D + T \rightarrow {_2He^4} + {_0n^1} + 17.6 \text{ MeV}$$

Thus the fusion of deuterium and tritium releases 17.6 MeV of energy.

There is mutual electrostatic repulsion or Coulomb repulsive barrier between the proton clouds of two nuclei. Thermal velocities are imparted to the nuclei to overcome the mutual electrostatic repulsion so that the nuclei can fuse together. This can be done by heating the matter to a high temperature (~100 million K). At such a high temperature matter consists of a mixture of electrons and ions- a plasma. The energy released due to fusion at such high temperatures is much greater than the energy lost by radiation. It is necessary to confine the hot plasma so as to sustain sufficient amount of thermal nuclear energy. Fig. 16.46 shows some of the processes that occur due to laser fusion process.

A pea sized spherical shaped target pellet, a fraction of a millimeter in diameter containing fusion fuel, usually a mixture of D and T, is projected into a reaction chamber. The irradiation of the fuel pellet with high energy laser beam vaporizes its outer surface to form a plasma and thus allowing the laser beam to penetrate a certain distance. The hot plasma (T ~ 100 million K) conducts thermal

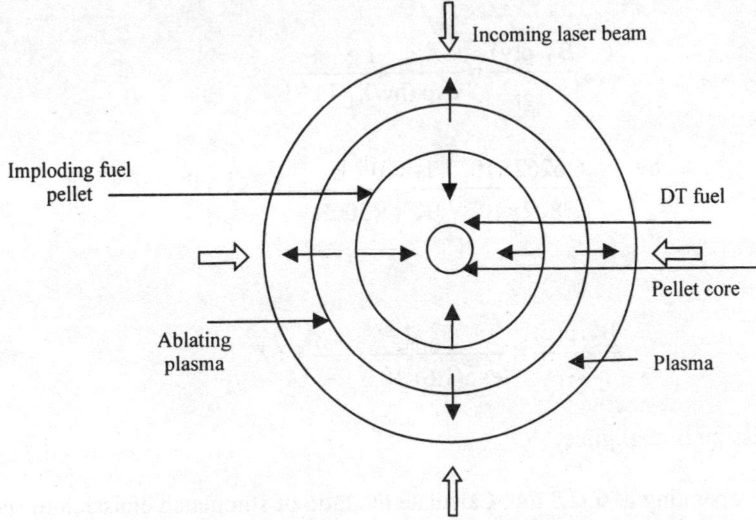

Fig. 16.46 Schematic illustration of the laser fusion process

energy inwards to the solid boundary of the fuel pellet. The rapid heating of the solid boundary ablates material. The ablated material explodes away from the rest of the pellet and expands into vacuum. The equal and opposite reaction force of Newton's third law exert an enormous pressure (~ 10^{12} atmosphere) and compresses the remainder of the fuel pellet inwards. The core of the pellet is compressed as much as 100 times solid density. Such a high compression leads to a fusion reaction and fusion energy is released.

Examples

1. A He-Ne laser emits light of wavelength 632.8 nm and has a power output of 4 mW. How many photons are emitted per second by the laser ?

Energy of the photon

$$E = h\nu = \frac{hc}{\lambda} = \frac{6.6262 \times 10^{-34}\,\text{Js} \times 2.9979 \times 10^{8}\,\text{ms}^{-1}}{632.8 \times 10^{-9}\,\text{m}} = 3.1392 \times 10^{-19}\,\text{J}$$

Let the number of emitted photons per second be N, the energy of these photons is equal to 4 mW, i.e.

$$3.1392 \times 10^{-19}\,\text{J} \times N = 4 \times 10^{-3}\,\text{Js}^{-1}$$

$$N = \frac{4 \times 10^{-3}\,\text{Js}^{-1}}{3.1392 \times 10^{-19}\,\text{J}} = 1.274 \times 10^{16}\,\text{s}^{-1}$$

2. Calculate the relative rate of spontaneous and stimulated process in the microwave region ($v = 10^{10}$ Hz) at 300 K.

From Eq. (16.13)

$$\frac{B_{21}\,\rho(v)}{A_{21}} = \frac{1}{\exp\,(hv/k_B T) - 1}$$

$$\frac{hv}{k_B T} = \frac{6.6262 \times 10^{-34}\,\text{Js} \times 10^{10}\,\text{Hz}}{1.3807 \times 10^{-23}\,\text{JK}^{-1} \times 300\,\text{K}} \approx 1.6 \times 10^{-3}$$

Thus

$$\frac{B_{21}\,\rho(v)}{A_{21}} = \frac{1}{\exp(1.6 \times 10^{-3}) - 1} \approx 625$$

spontaneous emission is negligible.

3. A He-Ne laser is operating at 632.8 nm. Calculate the ratio of stimulated emission to spontaneous emission coefficient.

From Eq. (16.15)

$$\frac{A_{nm}}{B_{nm}} = \frac{8\pi h v^3}{c^3} = \frac{8\pi h}{\lambda^3} = \frac{8 \times 3.14 \times 6.6262 \times 10^{-34}\,\text{Js}}{(632.8 \times 10^{-9}\,\text{m})^3} \approx 6.57 \times 10^{-14}\,\text{Js/m}^3$$

$$\frac{B_{nm}}{A_{nm}} = \frac{1}{6.57 \times 10^{-14}\,\text{Js}/\text{m}^3} = 1.52 \times 10^{13}\,\text{m}^3/\text{Js}$$

4. In He-Cd laser, the laser transition takes place between $^2D_{3/2} \rightarrow {}^2P_{3/2}$ and $^2D_{3/2} \rightarrow {}^2P_{1/2}$ with wavelength 353.6 nm and 325.0 nm, respectively. The radiative probabilities for these transitions are $1.6 \times 10^5\,\text{s}^{-1}$ and $7.8 \times 10^5\,\text{s}^{-1}$, respectively. Compute the lifetime of the upper laser level involved in these transitions.

From Eqs. (16.3) and (16.6)

$$\tau_n = \frac{1}{\sum_m A_{nm}} = \frac{1}{(7.8 + 1.6)10^5\,\text{s}^{-1}} = 1.06 \times 10^{-6}\,\text{s}.$$

5.   In Nd:YAG laser , the lifetime of the upper level of the laser line 1.064 μm is ~ 230 μs . Determine the spontaneous transition probability.

From Eq. (16.6)

$$A_n = \frac{1}{\tau_n} = \frac{1}{230 \times 10^{-6}\,s} = 4.35 \times 10^3\,s^{-1}$$

6.   The quantum yield of the S_1-S_0 transition for Rhodamine 6G dye is 0.87 and the corresponding lifetime is ~ 5.0 ns. Calculate the radiative and non-radiative lifetime of the S_1 level.

Quantum yield φ is

$$\varphi = \frac{\tau}{\tau_{sp}}$$

$$\tau_{sp} = \frac{\tau}{\varphi} = \frac{5 \times 10^{-9}\,s}{0.87} = 5.747 \times 10^{-9}\,s$$

and from Eq. (16.45)

$$\frac{1}{\tau_{nr}} = \frac{1}{\tau} - \frac{1}{\tau_{sp}} = \frac{1-\varphi}{\tau}$$

$$\tau_{nr} = \frac{\tau}{1-\varphi} = \frac{5 \times 10^{-9}\,s}{1-0.87} = 3.846 \times 10^{-8}\,s.$$

7.   For a cavity of volume $10^{-6}\,m^3$, calculate the number of modes which fall within bandwidth of 10.0 nm centred at 600.0 nm.

From Eq. (16.15), the number of modes is

$$\Delta N = \frac{8\pi \nu^2 \Delta\nu}{c^3} V = \frac{8\pi \Delta\lambda}{\lambda^4} V = \frac{8 \times 3.14 \times 10 \times 10^{-9}\,m}{(600 \times 10^{-9}\,m)^4} \times 10^{-6}\,m^3 = 1.94 \times 10^{12} \text{ modes}$$

8.   A laser cavity consists of two mirrors with reflectivity $R_2 = 1$ and $R_1 = 0.5$. Calculate the threshold inversion if the length of the active material is 7.5 cm and the transition cross section is $8.8 \times 10^{-19}\,cm^2$.

From Eq. (16.33)

$$N_n - N_m = \frac{1}{2l\sigma_{se}} \ln \frac{1}{R_1 R_2} = \frac{1}{2 \times 7.5\,cm \times 8.8 \times 10^{-19}\,cm^2} \ln(2) = 5.25 \times 10^{16}\,cm^{-3}$$

9.   Consider GaAs laser diode with length of the active material 500 μm and reflectivity of the mirrors equal to 0.3. Calculate the threshold gain if the cavity losses are 5.0 mm^{-1}

From Eq. (16.32)

$$\gamma_0(v) = \alpha + \frac{1}{2l} \ln \frac{1}{R_1 R_2} = 50\ \text{cm}^{-1} + \frac{10^2}{2 \times 5\ \text{cm}} \ln \frac{1}{0.3 \times 0.3} = (50 + 24.08)\text{cm}^{-1} \approx 74\ \text{cm}^{-1}$$

10. For He-Ne laser line 632.8 nm, stimulated emission cross section is 3.0×10^{-17} m^2 and inversion density is 5.0×10^{15} m^{-3}. Determine the small signal gain coefficient.

From Eqs. (16.32) and (16.33) assuming no cavity losses

$$\gamma_0(v) = \sigma_{se}(N_n - N_m) = 3 \times 10^{-17}\ \text{m}^2 \times 5 \times 10^{15}\ \text{m}^{-3} = 0.15\ \text{m}^{-1}$$

11. A semiconductor hetrojunction is grown with GaAsP active region layer with band gap ~1.85 eV. Assume that the laser transition occurs from the bottom of the conduction band. Determine the laser wavelength of the material

Since the laser transition occur from the bottom of the conduction band to the top of the valence band, the laser wavelength will correspond to the band gap energy. From

$$E = hv = \frac{hc}{\lambda}$$

$$\lambda = \frac{hc}{E} = \frac{6.6262 \times 10^{-34}\ \text{Js} \times 2.9979 \times 10^8\ \text{ms}^{-1}}{1.85 \times 1.6022 \times 10^{-19}\ \text{J}} = 670.2\ \text{nm}$$

12. A Coumarin 6 dye in a solution emits over a spectral bandwidth from 506-558 nm with peak spontaneous emission at 535 nm. Determine the energy width of the laser level S_0 to which transition occur from S_1 assuming emission originate from lowest energy S_1 and is homogeneously broadened.

Since the emission originates from the bottom of energy level S_1, then entire tuning range occurs as a result of transition. We have $E = hv$ and

$$\Delta E = h\,\Delta v = \frac{hc\,\Delta\lambda}{\lambda^2}$$

It is given that

$$\lambda = 535\ \text{nm}$$

$$\Delta\lambda = (558 - 506)\ \text{nm} = 52\ \text{nm}$$

with these values

$$\Delta E = \frac{6.6262 \times 10^{-34}\, Js \times 2.9979 \times 10^8\, ms^{-1} \times 52 \times 10^{-9}\, m}{(535 \times 10^{-9}\, m)^2} = 3.609 \times 10^{-20}\, J = 0.225\, eV$$

13. The lineshape of ruby 694.3 nm line is Lorentzian with full width at half maximum of 330 GHz. The peak transition cross section is 2.5×10^{-24} m^2. The refractive index of ruby is 1.76. Determine the radiative lifetime and if observed lifetime is 3.0 ms, calculate the fluorescence quantum yield.

From Eq. (16.24)

$$\sigma_{se} = \frac{A_{nm} c^2 g(v)}{8\pi v^2}$$

$g(v)$ is Lorentizian and is given by $g(v) = \dfrac{0.637}{\Delta v}$ and $A_{nm} = \dfrac{1}{\tau_{sp}}$. Thus

$$\tau_{sp} = \left(\frac{c}{v}\right)^2 \frac{1}{8\pi \sigma_{se}} \frac{0.637}{\Delta v} = \left(\frac{694.3 \times 10^{-9}\, m}{1.76}\right)^2 \frac{1}{8 \times 3.14 \times 2.5 \times 10^{-24}\, m^2} \frac{0.637}{330 \times 10^9\, Hz} = 4.78\, ms$$

quantum yield φ is

$$\varphi = \frac{\tau}{\tau_{sp}} = \frac{3ms}{4.78ms} = 0.627$$

14. Two mirrors with radii of curvature of 1.8 m are separated by a distance of 2.0m. Determine whether this mirror arrangement leads to stability or not ?

From Eqs. (16.90) and (16.91) we have $g_1 = 1 - \dfrac{l}{R_1}, g_2 = 1 - \dfrac{l}{R_2}$ and from Eq. (16.95) the stability condition is $0 < g_1 g_2 < 1$. It is given that $R_1 = R_2 = 1.8m$ and $l = 2.0m$. From this $g_1 g_2 = 0.123$ and $g_1 = g_2 = -0.1111$. Thus mirror arrangement is stable.

15. One mirror with radius of curvature of 2.0 m and other with radius of curvature 3.0m are separated by a distance of 2.3m. Determine whether this mirror arrangement leads to stability or not ?

From Eqs.(16.90) and (16.91) we obtain $g_1 = -0.15$ and $g_2 = 0.2333$ and $g_1 g_2 = -0.035$. Thus from Eq.(16.95) the mirror arrangement is not stable.

16. A resonator is formed by a convex mirror of radius $R_1 = -1.0m$ and a concave mirror of radius $R_2 = 1.5m$. What is the maximum possible separation if this is to remain stable resonator.

From Eq. (16.95), the stability condition is $0 < g_1 g_2 < 1$ and from Eqs.(16.90) and (16.91) we have

$$g_1 g_2 = \left(1 - \frac{1}{R_1}\right)\left(1 - \frac{1}{R_2}\right) = 1 - l\left(\frac{1}{R_1} + \frac{1}{R_2}\right) + \frac{l^2}{R_1 R_2}$$

for $g_1 g_2 < 1$

$$-l\left(\frac{1}{R_1} + \frac{1}{R_2}\right) + \frac{l^2}{R_1 R_2} = 0$$

$$l = R_1 + R_2 = (-1 + 1.5)m = 0.5m$$

The maximum possible separation is 0.5m.

17. For a confocal geometry of the resonator with mirrors separated by 2.0m, obtain the spot size at resonator centre and at the mirror for wavelength of 10^{-4} cm.

From Eq. (16.88)

$$w_0 = \left(\frac{L\lambda}{2\pi}\right)^{1/2} = \left(\frac{200\,cm \times 10^{-4}\,cm}{2 \times 3.14}\right)^{1/2} = 0.0563\,cm$$

The spot size at the mirrors from Eq.(16.89) is

$$w = 0.0563\,cm\left[1 + \left(\frac{2}{2}\right)^2\right]^{1/2} = 0.0792\,cm$$

18. Consider a hemiconfocal resonator of length 2.0m used for CO_2 laser at wavelength 10.6 µm.Calculate the spot size on both mirrors.

In a hemiconfocal arrangement one of the mirror is plane mirror with radius of curvature infinite and second mirror of radius of curvature R. The plane mirror is denoted by R_1 and concave mirror by R_2.
From Eqs.(16.90) and (16.91) we have $g_1 = 1$ and $g_2 = 0.5$. From Eq. (16.93)

$$w(z_1) = \left(\frac{10.6 \times 10^{-6}\,m \times 2\,m}{3.14}\right)^{1/2}\left(\frac{0.5}{1(1 - 0.5)}\right)^{1/4} = 2.598mm$$

$$w(z_2) = \sqrt{2} \times 2.598\,mm = 3.674\,mm$$

19. What is temporal coherence length of a He-Ne laser operating at a wavelength of 632.8 nm with an emission width of 10^6 Hz ?

We have $v = (c/\lambda)$, $dv = |cd\lambda/\lambda^2|$ and

$$d\lambda = \frac{\lambda^2 d\nu}{c} = \frac{(632.8)^2 \times 10^{-18} \times 10^6}{2.9979 \times 10^8} \, m = 1.3357 \times 10^{-6} \, nm$$

From Eq. (16.38)

$$L_c = \frac{\lambda^2}{d\lambda} = \frac{(632.8)^2 \times 10^{-9}}{1.3357 \times 10^{-6}} \, m = 299.8m$$

20. A sodium lamp has a coherence time ~ 10^{-10} s. Determine the coherent length.

The coherent length $L_c = c\tau_c = 2.9979 \times 10^8 \, ms^{-1} \times 10^{-10} \, s = 2.9979 \, cm$

21. Determine what emission frequency width would be required to have a temporal coherence length of 10.0 m at a source wavelength of 488 nm.

From Eq. (16.38)

$$\Delta\nu = \frac{c}{L_c} = \frac{2.9979 \times 10^8 \, ms^{-1}}{10m} \approx 3 \times 10^7 \, Hz$$

22. A laser rated at 0.10 J can generate radiation in 3.0 ns pulse. What is the power ouput per pulse ?

The power output P is energy per unit time and is expressed in Watts . Therefore,

$$P = \frac{0.10J}{3 \times 10^{-9} s} = 3.3 \times 10^7 \, Js^{-1} = 33MW$$

23. Estimate the threshold pumping power required to start laser oscillation in ruby laser. It is given that N= 1.6 x 10^{19} cm^{-3} , $A_{21} = 333.3$ s^{-1} , $\nu_p = 6.25 \times 10^{14}$ Hz ,the efficiency of pumping source is 25% and only 25% pump light passes through the ruby.

The threshold inversion required is usually small in comparison with N, i.e. $N_2 - N_1 \ll N$. Thus the threshold value of W_p required to start laser oscillation from Eq. (16.59) is $W_p = A_{21}$. The number of atoms being pumped per unit time per unit volume from level 1 to level 3 is $W_p N_1$.

Let the average pump frequency be ν_p corresponding to excitation from level 1 to level 3. The power required per unit volume

$$P = W_p N_1 h\nu_p = A_{21} N_1 h\nu_p$$

Since $N_2 - N_1 \ll N$, thus $N_1 = N_2 = (N/2)$ and

$$P = \frac{NA_{21}h\nu_p}{2} = \frac{1.6 \times 10^{19} \, cm^{-3} \times 333.3s^{-1} \times 6.6262 \times 10^{-34} \, Js \times 6.25 \times 10^{14} \, Hz}{2} = 1104 Wcm^{-3}$$

Since the efficiency of pumping source is 25 % and only 25% of the pump light passes through the ruby rod, therefore the threshold power required is 1104 x 4x 4 W cm^{-3} = 17.7 kW cm^{-3} .

24. Estimate the threshold pumping power required to start laser oscillation in Nd:YAG laser. It is given that $N_2 - N_1 = 4 \times 10^{15}$ cm^{-3} , $T_{21} = 4347.8$ s^{-1} , $v_p = 4 \times 10^{14}$ Hz , the efficiency of pumping source is 25% and only 25% pump light passes through the ruby.

Nd:YAG laser is a four level laser. The level 1 is almost empty and the atoms lifted from level 0 to level 3 go to level 2. Thus the number of atoms lifted from level 0 goes to level 2 via level 3 and hence $N_2 - N_1 \sim N_2 = N$ Thus the threshold value of W_{pt} required to start laser oscillation from Eq. (16.84) is $W_{pt} = T_{21}$. The number of atoms being pumped per unit time per unit volume from level 0 to level 2 is $W_{pt}N_2$.

Let the average pump frequency be v_p corresponding to the excitation from level 0 to level 2 via level 3. The power required per unit volume

$$P = W_{pt} N_2 hv_p = T_{21} N_2 hv_p$$

$$P = N_2 T_{21} hv_p = 4 \times 10^{15} \text{cm}^{-3} \times 4347.8 \text{s}^{-1} \times 6.6262 \times 10^{-34} \text{Js} \times 4 \times 10^{14} \text{Hz} = 4.61 \text{ Wcm}^{-3}$$

Since the efficiency of pumping source is 25 % and only 25% of the pump light passes through the Nd:YAG rod, therefore the threshold power required is 4.61 x 4x 4 Wcm^{-3} = 73.75 W cm^{-3} .

25. Calculate the cavity lifetime for the He-Ne laser of cavity length 40.0 cm. Cavity loss per unit trip due to scattering and other mechanisms is zero. The reflectivity of the mirrors is 0.999 and 0.98.

Cavity lifetime t_c is the time in which energy in the cavity reduces by a factor of 1/e. In the absence of an amplifying medium, the intensity I_0 at a point is reduced by a factor

$$R_1 R_2 \exp(-2\alpha_c l) = \exp[-(2\alpha_c l - \ln R_1 R_2)] = \exp(-k)$$

where R_1 and R_2 are the reflectivity of the mirrors, l is the length of the cavity , α_c is cavity loss per unit trip due to scattering and other mechanisms and k = $2\alpha_c l - \ln R_1 R_2$. The fractional loss per round trip would be

$$x = \frac{I_0 - I_0 \exp(-k)}{I_0} = 1 - \exp(-k)$$

$$k = \ln\left(\frac{1}{1-x}\right)$$

One round trip time corresponds to

$$t = \frac{2l}{(c/n_0)} = \frac{2n_0 l}{c}$$

where n_0 is the refractive index of the amplifying medium. Hence if the intensity reduces as $\exp(-t/t_c)$, then in a time $t = (2n_0l/c)$ the intensity reduces by a factor of $\exp(-2n_0l/c\, t_c)$ i.e.,

$$\exp\left[-(2\alpha_c l - \ln R_1 R_2)\right] = \exp\left[-(2n_0 l / ct_c)\right] = \exp(-k) = \exp\left[-\ln\left(\frac{1}{1-x}\right)\right]$$

$$t_c = \frac{2n_0 l}{c[2\alpha_c l - \ln(R_1 R_2)]} = \frac{2n_0 l}{c\,\ln\left(\dfrac{1}{1-x}\right)} \tag{16.96}$$

Substituting the values of l, $n_0 = 1$, R_1, R_2 and $\alpha_c = 0$ in Eq. (16.96), we obtain

$$t_c = 125 \text{ ns.}$$

26. Calculate the cavity lifetime for a ruby laser of cavity length 10.0 cm where the cavity loss per unit trip due to scattering and other mechanisms and also due to reflectivity of the optical resonator is 10%. The refractive index of ruby is 1.76.

Substituting the values in Eq. (16.96)

$$t_c = \frac{2 \times 10 \text{ cm} \times 1.76}{2.9979 \times 10^{10} \text{ cm s}^{-1} \times 0.1} = 11.74 \text{ ns.}$$

27. Determine the value of cavity lifetime and threshold population inversion for a He-Ne laser using the following parameters:
$\lambda = 6328$ Å, $\tau_{sp} = 10^{-7}$ s, $n_0 = 1$, $l = 20$ cm, $R_1 = R_2 = 0.98$, $\alpha_c = 0$ and $g(v) = 10^{-9}$ s.

From Eqs. (16.23) and (16.32)

$$\gamma_0 = \frac{A_{nm} c^2 g(v)}{8\pi v^2}\left(N_n - \frac{g_n}{g_m} N_m\right) = \alpha - \frac{1}{2l}\ln(R_1 R_2)$$

$$\Delta N = N_n - \frac{g_n}{g_m} N_m = \frac{8\pi v^2 n_0^3}{A_{nm} c^3 g(v)}\frac{c}{2n_0 l}\left(2\alpha l - \ln R_1 R_2\right) \tag{16.97}$$

where c is the velocity of light in vacuum. From Eqs. (16.96) and (16.97)

$$\Delta N = \frac{8\pi v^2 n_0^3}{A_{nm} c^3 g(v)}\frac{1}{t_c} = \frac{8\pi n_0^2}{c\lambda^2 g(v)}\frac{\tau_{sp}}{t_c} \tag{16.98}$$

Substituting the values in Eqs. (16.96) and (16.98) we obtain

$$t_c = 3.30 \times 10^{-8} \text{ s. and } \Delta N = 6.34 \times 10^8 \text{ cm}^{-3}.$$

28. For a Nd:YAG laser, calculate the population inversion densities required to give a gain of $1.0\,\text{m}^{-1}$, where $\lambda = 10600\ \text{Å}$, $\tau_{sp} = 2.3 \times 10^{-4}$ s, $n_0 = 1.83$, $g(\nu) = (2/\pi\Delta\nu)$ and $\Delta\nu = 2 \times 10^{11}\,\text{s}^{-1}$.

From Eq. (16.23)

$$\Delta N = N_n - \frac{g_n}{g_m}\,N_m = \frac{4\pi^2\,\gamma_0\,n_0^2}{A_{nm}\,\lambda^2}\,\Delta\nu = \frac{4\times(3.14)^2\times 1\text{m}^{-1}\times(1.83)^2\times 2\times 10^{11}\,\text{s}^{-1}\times 2.3\times 10^{-4}\,\text{s}}{(1060\times 10^{-9}\,\text{m})^2}$$

$$\Delta N = 5.41 \times 10^{21}\,\text{m}^{-3}$$

Problems

16.1   Calculate the relative rate of spontaneous and stimulated process as in the visible region ($\nu = 10^{15}$ Hz) at 300 K.

16.2   In ruby laser, transition takes place between ^2E and $^4A_2$ levels which gives rise to 694.3 nm line. The lifetime of the upper level is 3 ms. Determine the spontaneous transition probability.

16.3   In Nd:YAG laser , the spontaneous transition probability for laser line 1.064 μm is $\sim 4.3\times 10^3$ /s . Determine the upper level lifetime.

16.4   For a cavity of volume 10 cm^3, calculate the number of modes, which fall within bandwidth of 1 GHz centred at 500.0 nm.

16.5   A Rhodamine 6G dye in a solution emits over a spectral bandwidth from 570-684 nm with peak spontaneous emission at 593 nm. Determine the energy width of the laser level S_0 to which transition occur from S_1 assuming emission originate from lowest energy S_1 and is homogeneously broadened.

16.6   A laser cavity consists of two mirrors with reflectivity $R_1 = 0.5$ and $R_2 = 0.95$. Calculate the threshold inversion if the length of the active material is 7.5 cm and the transition cross section is $8.8\times 10^{-19}\ \text{cm}^2$.

16.7   One mirror with radius of curvature of 5.0 m and other with radius of curvature of 3.0m are separated by a distance of 4.0m. Determine whether this mirror arrangement leads to stability.

16.8   For Argon ion laser line 488 nm, small signal gain coefficient is $0.5\,\text{m}^{-1}$ and stimulated emission cross section is $2.5\times 10^{-16}\ \text{m}^2$. Calculate the inversion density.

16.9   For N_2 laser line 337.1 nm calculate the stimulated emission cross section if small signal gain coefficient is $10.0\ \text{m}^{-1}$ and inversion density is $2.5\times 10^{17}\ \text{m}^{-3}$.

16.10   Consider a resonator consisting of two concave spherical mirrors one with radius of curvature 4m and other with radius of curvature of 1.5 m. The two mirrors are spaced by 1.0 m. Calculate the spot size on the two mirrors when the cavity is oscillating at the wavelength 514.5 nm

16.11   Calculate the degree of non-monochromaticity of a gas laser having a bandwidth of 10^{-8} Å at a wavelength of 6000 Å.

16.12   A laser beam of wavelength 600 nm on earth is focused by a lens of diameter 2.0 m on to a crater on the moon. How big is the spot on the moon if the distance of the moon from the earth is 4.0×10^8 m ?

16.13   What is temporal coherence length of a mercury vapour lamp at a wavelength of 546.1 nm with an emission bandwidth of 6×10^8 Hz ?

16.14   Calculate the temporal coherence length for a white light source that covers the spectrum from 400-700 nm.

16.15   Compare the threshold pumping flux required for the four level Nd:YAG laser with that of ruby laser.

16.16   In the 2P-1S transition in hydrogen atom, the lifetime of the 2P state for spontaneous emission is 1.6×10^{-9} s. Determine the stimulated emission coefficient.

16.17   Calculate the cavity life time for a ruby laser of cavity length 5.0 cm where the cavity loss per unit trip due to scattering and other mechanisms and also due to reflectivity of the optical resonator is 10%. The refractive index of ruby is 1.76.

16.18   Calculate the cavity lifetime for a Nd:YAG laser of cavity length 7 cm where the cavity loss per bounce due to scattering and other mechanisms and also due to reflectivity of the optical resonator is 5%. The refractive index of Nd:YAG is 1.82.

16.19   A semiconductor laser has length of 350 μm, reflectivity of mirrors is 0.50, $\alpha_c = 300$ cm^{-1}. The refractive index is 3.3. Calculate the photon lifetime.

16.20   Determine the value of cavity life time and threshold population inversion for a He-Ne laser using the following parameters:

$$\lambda = 6328 \text{ Å}, \ \tau_{sp} = 10^{-7} \text{ s}, \ n_0 = 1, \ l = 40 \text{ cm}, \ R_1 = 0.98,$$

$$R_2 = 0.999, \ \alpha_c = 0 \text{ and } g(v) = \frac{2}{\Delta v}\left(\frac{\ln 2}{\pi}\right)^{1/2}, \ \Delta v = 1.5 \times 10^{-9} \text{ s}.$$

16.21   Determine the threshold population inversion for a ruby laser using the following parameters:

$$\lambda = 6943 \text{ Å}, \ \tau_{sp} = 3 \times 10^{-3} \text{ s}, \ n_0 = 1.76, \ l = 5 \text{ cm},$$

$$R_1 = 0.9, \ \ R_2 = 0.9, \ \alpha_c = 0, \ g(v) = 6.908 \times 10^{-12} \text{ s}$$

Selected Bibliography

Abragam, A. and Bleaney, B. Electron Paramagnetic Resonance of Transition Ions, Clarendon Press, Oxford (1970).

Akitt, J.W. NMR and Chemistry, 3rd Ed. Chapman & Hall, London (1992).

Arya, A.P. Elementary Modern Physics, Addison -Wesley Publishing Company, London (1974).

Atherton, N.M. Electron Spin Resonance,John Wiley, New York (1973).

Atkins P W, Physical Chemistry, W H Freeman, New York (1994).

Ayscough, P.B. Electron Spin Resonance in Chemistry, Methuen & Co.Ltd. London (1967).

Banwell, C.N. and McCash, E.M. Fundamentals of Molecular Spectroscopy, 4th Ed. Tata McGraw Hill, New Delhi (1995).

Barrow G M, Introduction to Molecular Spectroscopy, McGraw Hill, New York (1962).

Beiser, A. Concepts of Modern Physics, 5th Ed. McGraw Hill Inc. New York (1995).

Bovery, F.A. Nuclear Magnetic Resonance Spectroscopy, Academic Press, New York (1969).

Bransden, B.H. and Joachain, C.J. Physics of Atoms and Molecules, Longman, London (1983).

Cagnac, B. and Pebay-Peyroule, J.C., Modern Atomic Physics: Quantum Theory and its Applications, MacMillan, London (1975).

Carrington , A. and McLachlan, A. D. Introduction to Magnetic Resonance, Harper and Row, New York (1969).

Davis, C.C., Lasers and Electro - Optics, Cambridge Univ. Press (1996).

Eisberg , R. and Resnick, R. Quantum Physics of Atoms, Molecules, Solids, Nuclei and Particles, John Wiley & Sons, New York (1974).

Gordy, W. Theory and Applications of Electron Spin Resonance, John Wiley & Sons, Inc. (1980).

Greenwood, N.N. and Gibb, T.C. Mössbauer Spectroscopy, Chapman & Hall Ltd. London (1971).

Guillory, W.A., Introduction to Molecular Structure and Spectroscopy, Allyn and Bacon Inc. Boston (1977).

Herzberg, G. Atomic Spectra and Atomic Structure, Dover Publications , New York (1945).

Herzberg, G. Spectra of Diatomic Molecules, van Nostrand, New York (1950).

Hollas, J.M. High Resolution Spectroscopy, Butterworths, London (1982).

King, G.W. Spectroscopy and Molecular Structure, Holt, Rinehast and Winston Inc. New York (1964).

Kuhn, H.G. Atomic Spectra, 2nd Ed. ,Longman, London (1969).

Laud, Lasers and Non-Linear Optics, 2nd Ed. Wiley Eastern Ltd. New Delhi (1991).

Linne M A, Spectroscopic Measurement, Academic Press (2002).

Mc Hale, J.L. Molecular Spectroscopy, Ist Ed. Prentice Hall, New Jersey (1999).

Marfunin, A.S. Spectroscopy, Luminescence and Radiation Centres in Minerals, Springer-Verlag, Berlin (1979).

Orton, J.W. Electron Paramagnetic Resonance, ILIFFE Books Ltd. London (1968).

Pake, G.E. and Estle, T.L. The Physical Principle of Electron Paramagnetic Resonance, Benjamin, New York (1973).

Pilbrow, J.R. Transition Ion Electron Paramagnetic Resonance, Clarendon Press, Oxford (1990).

Richtmyer, F.K., Kinnard, E.H. and Cooper, J.N. Introduction to Modern Physics, 6th Ed. Tata McGraw Hill, NewDelhi (1995)

Ruark, A.E. and Urey, H.C. Atom, Molecules and Quanta, Dover Publications Inc. New York (1964).

Schwabl, F. Quantum Mechanics, Narosa Publishing House, New Delhi (1998).

Shore, B.W. and Menzel, D.H. Principles of Atomic Spectra, John Wiley & Sons Inc. New York (1968).

Silfvast, W.T. Laser Fundamentals, Cambridge Univ. Press (2003).

Slichter, C.P., Principle of Magnetic Resonance, 3rd Ed. Springer-Verlag, Berlin (1990).

Svanberg, S. Atomic and Molecular Spectroscopy, Springer-Verlag, Berlin (1992).

Svelto, O. and Hanna, D.C., Principle of Lasers, 2nd Ed. Plenum Press, New York (1986).

Thyagarajan, K. and Ghatak, A.K., Lasers, MacMillan Ind. Ltd. Delhi (1997).

Verdeyan, J.T., Laser Electronics, 2nd Ed. Prentice-Hall of India Pvt.Ltd. New Delhi (1993).

Wertheim, G.K. Mössbauer Effect: Principle and Applications, Academic Press (1964).

White, H.E. Introduction to Atomic Spectra, McGraw Hill, London (1934).

Woodgate, G.K., Elementary Atomic Structure, McGraw Hill, London (1970).

Young, M. Optics and Lasers, 4th Ed. Springer-Verlag , Berlin (1992).

Subject Index